D0154198

Construction Supply Chain Economics

Also available from Taylor & Francis

Construction Supply Chain Management Handbook
W. O'Brien *et al.*   Hb: ISBN 978–1–4200–4745–5

Procurement in the Construction Industry
W. Hughes *et al.*   Hb: ISBN 978–0–415–39560–1

Procurement Systems
D. Walker *et al.*   Hb: ISBN 978–0–415–41605–4
          Pb: ISBN 978–0–415–41606–1

Construction Project Management
P. Fewings   Hb: ISBN 978–0–415–35905–4
          Pb: ISBN 978–0–415–35906–1

Basics of Supply Chain Management
L. Fredendall *et al.*   Hb: ISBN 978–1–57444–120–8

Information and ordering details

For price availability and ordering visit our website **www.tandf.co.uk/builtenvironment**
Alternatively our books are available from all good bookshops.

Construction Supply Chain Economics

Kerry London

Taylor & Francis
Taylor & Francis Group

LONDON AND NEW YORK

First published 2008
by Taylor & Francis
2 Park Square, Milton Park, Abingdon, Oxon OX14 4RN

Simultaneously published in the USA and Canada
by Taylor & Francis
270 Madison Ave, New York, NY 10016

*Taylor & Francis is an imprint of the Taylor & Francis Group,
an informa business*

© 2008 Kerry London

Typeset in Sabon by
Newgen Imaging Systems (P) Ltd, Chennai, India
Printed and bound in Great Britain by
Antony Rowe Ltd, Chippenham, Wiltshire

British Library Cataloguing in Publication Data
A catalogue record for this book is available from the British Library

Library of Congress Cataloging in Publication Data
London, Kerry
    Construction supply chain economics / Kerry London.
        p. cm.
    Includes bibliographical references and index.
        1. Building materials – Purchasing – Procurement. 2. Business
    logistics – Industrial organization economics. 3. Construction
    markets industry structure – management. 4. Construction Firm
    behaviour. I. Title.
    TH437.L66 2007
    690.068'7—dc22   2007012991

ISBN10: 0–415–40971–3 (hbk)
ISBN10: 0–203–96248–6 (ebk)

ISBN13: 978–0–415–40971–1 (hbk)
ISBN13: 978–0–203–96248–0 (ebk)

Contents

1 Introduction to supply chain economics and procurement

What is supply chain economics and how does it relate to procurement?

1.0 Orientation

> **Box 1.1 Chapter orientation**
>
> **WHAT**: Chapter 1 introduces ideas and concepts about the supply chain, supply chain economics and procurement in relation to project-based environments for construction systems.
>
> **HOW**: It describes the rationale for the text, highlights of past research, the key ideas explored and how the text is structured. It takes the reader from some observations about the industry to the research response which is then encapsulated in the research questions and the research strategy. Finally, the chapter provides summaries of each of the subsequent chapters.
>
> **WHO**: The focus is on supply chains in relation to project-based industries and constructed systems. It is particularly relevant to those students, academics and practitioners (including industry participants and government decision makers) who are involved in the property and construction industry. However, the underlying theory and practice is relevant and applicable to a much wider group. Those involved in mining, shipping, aerospace and information technology will find ideas and examples which resonate and are readily applicable.

1.1 Rationale for text

The construction of any built system is a complex problem involving numerous firms who temporarily work together in response to individual projects – or so it seems. This text is focused particularly on the property and construction industry; however, the principles and practices are widely applicable to project-based industries where there is a life cycle process of design, construction and maintenance of an object. Therefore, many of the concepts and case studies that are described throughout this text are intended to provide a trigger for the student, academic or practitioner's thinking in the application to their own situation. The supply chains and the ensuing customer–supplier relationships that are explored in this text are initiated in the project environment, but we soon travel down the chain into a variety of different sectors, and associated procurement scenarios, linking numerous industries.

It is well accepted in the property and construction industry as well as in academia that the practice of continual association and disassociation – that is, forming and reforming for individual projects – is the common mode of economic organization within the property and construction industry. Many of the functional and technical specializations required for project contracts reside within numerous individual firms which are co-ordinated on a project-by-project basis. Firm interdependencies can become particularly critical since construction industry firms rarely act as isolated and independent entities. The degree to which co-ordination is required may vary from project to project and from country to country, as the degree of specialization and vertical integration can differ across locations and projects.

As highlighted, although this text is written based upon case studies from the property and construction industry and in particular case studies of supply chains associated with buildings, the underlying theory and practices which are described are equally applicable to many constructed systems and the sectors associated with them; for example, civil, mining, industrial design, shipping, aerospace, information technology and asset management. The theoretical principles can be extrapolated and applied to many sectors and indeed, because the supply chains extend into numerous sectors, the firms encountered in the case studies do supply to not only more than one sector, but also because 'project thinking' and 'procurement' permeates many of our business systems. The text is concerned with developing theory and providing practical examples to support an area of study I have termed *supply chain economics* and this is of fundamental importance to all firms who engage suppliers in contractual relationships to assist them in their own core business practices. Supply chain economics is a new term – the most common term we have heard of previously is supply chain management.

So what is construction supply chain economics? I suppose the motivation to coin the term has grown out of a very simple idea that I had which was that the supply chain metaphor was a useful way to understand the structural and behavioural characteristics of the construction industry – the economic market structural characteristics and the firm and sector behavioural characteristics. As a past policymaker and now an academic, I found it a useful way to organize my thoughts about this large and diverse industry and its fundamental makeup. As it happens, industrial organization economic theory is a branch of economics that provides some quite useful thinking towards market structure, firm conduct and industry performance. I have used industrial organization economics theory to provide some background theory towards developing a conceptual framework. Of course, I feel nervous and presumptive about defining this term...as soon as one defines a term someone challenges it – but that is largely a major goal of research and scholarship – to be constantly developing new ideas and pushing the boundaries of our current knowledge base.

Box 1.2 Supply chain economics

Supply chain economics is the manner in which the economic market structural characteristics and the firm and sector behavioural characteristics interact to produce attributes which describe types of supplier firms, procurement relationships, supply chain industrial organization and supply chain performance.

Structural and behavioural characteristics produce

- supplier firm classes which rely upon two key related attributes of **commodity significance** and **countervailing power** in the customer–supplier relationship.
- procurement relationship classes which rely upon three key interconnecting attributes of **formation** based upon risk and expenditure, **transaction significance** based upon control requirements and then **negotiation strategies** based upon strategies.
- supply chain classes which are reliant upon four key attributes of **uniqueness, sector type, internationalization** and **fragmentation.**

The act of procurement is a fundamental building block of supply chain economics. Supply chain economics provides the context for supply chain management decision making.

This definition is loaded with specific terms that at this stage have not been explained. It is 'put up front' and the remainder of the text is concerned with exploring and explaining how I arrived at this definition.

I am ultimately interested in the way we can make better procurement and management decisions in the supply chain on a large scale thus improving the performance of a sector and region as a whole. To achieve this I felt that it was important to begin by understanding and describing what and who makes up the chain and how some of those decisions are made and then move into developing a model which represents the negotiation and formation process of links in the chain. The research which I have conducted in this area and which forms the basis of this text does take that next step; however, it does so by exploring another body of literature, namely object oriented modelling and using concepts from that knowledge domain. The logic and argument and discussion towards constructing that model and then taking the empirical findings and refining that model is set aside for now and not presented in this text. That area has been published in my doctoral dissertation and I suspect by the time this book is published will be more accessible in journal papers or books. First, let us dig deeply and explore the links in the chains, how they came to be and what was their context in an economic sense – as this is lacking in both economic policy decision making in practice and in the academic research.

The text focuses upon the simple act of procurement – sourcing of suppliers – how this comes about at each tier and the context of these decisions – and how this is one way we can build an holistic picture of an industry. I would not want to limit what construction supply chain economics can be to what I have discussed in this text – this text represents my thinking and my interpretation and there are various diverse aspects and paths which can and will more than likely be explored – and I discuss some of these in the final chapter.

Procurement by an upstream customer of a downstream supplier, and all the associated negotiation and engagement activities, is always embedded within a structural and behavioural context that is impacted upon by the industrial organization economic system. It is these systems that are explored and described in this text – the ability to understand the structural and behavioural context and the industrial organization economic envi-ronment allows customers and suppliers to apply strategic procurement practices suitable to their own firm. I have typically focused upon 'the project' as the starting point for the supply chains – but it soon becomes evident that, as we move down the chain, there are different economic environments other than the project environment. This mix of short- and long-term business environments is perhaps common to most industries but is rarely considered.

Currently, one of the pervading perceptions in the construction industry is that a lack of cohesion and vertical co-ordination between firms during the project contract is a problem. Indeed, it is often argued that the lack of integration between firms is probably the cause of low industry productivity. There is discourse on construction industry productivity and how it is measured and comparisons against other industries – this is a worthy debate but left for another time.

It is well acknowledged that the acute degree of specialization of the productive functions and the distillation of associated functions into separate firms appears to produce a fragmented, unpredictable and high-risk environment. The construction industry is often associated with adversarial and opportunistic relationships, lack of continuity between projects and firms and low industry productivity. The temporary project organization that characterizes the construction industry is seemingly unstructured, fragmented, chaotic and fractious and is cited as a cause for a poor performing industry. However, it is the mode of organization that is typically fairly common and I have taken the stance that it might be interesting and perhaps more productive to move beyond these arguments and explore a little more what really does go on at each tier between firms during the highly dynamic period of procurement.

Almost pendulum like, in recent times larger industry clients, particularly public sector organizations, have sought out more collaborative relationships for projects as the ideal model to solve these perceived problems. Developing more harmonious business relationships between the key firms in the project team – for example, the client and the contractor – has been a keen consideration in recent years. To this end, various alternative strategies have been sought; for example, the partnering style model of the 1980s and, more recently, the project alliances have been attempts to create environments of trust and longer-term business relationships within a project environment. Even the rhetoric surrounding public and private consolidated financial strategies is 'public–private partnerships', giving the impression at least of the creation of a harmonious business relationship.

The early collaborative nature of the project partnering model has now been extended beyond the individual project to include such concepts as: strategic partnering, alliancing or sequential contracting, which aim to develop long-term relationships between clients and selected firms outside the boundaries of the individual project. These types of alliancing contracts were first borrowed by the construction industry from the mining sectors: oil, gas and mineral. These so-called long-term collaborative relationships between clients and contractors help to create an illusion of more stable long-term relationships within the short-term environment. It helps to create the illusion of stability in the sector at the project level. It is anticipated that with such stability shall come higher productivity.

Examples of long-term relations between firms have been prevalent in manufacturing, retail, mining, electronics and information and communications technology sectors. Some of the most famous are those in the automobile industry originating in Japan. If we look to Japan for such firm-to-firm relationships in the construction sector, we find that such long-term relationships exist outside the boundaries of the individual project and extend beyond the individual projects. The Japanese 'kieretsus' system is a governance structure whereby major contractors often have long-term vertical relationships with key materials suppliers/equipment installers and specialist subcontractors based upon part-ownership or long-standing 'special' relationships (Hasegawa *et al.*, 1978). This so-called stability sits within an economic environment which relies heavily upon the support of major financial institutions having a financial stake in the sector.

However, simplistically appropriating a model of behaviour is fraught with difficulties without understanding the political, cultural, technological and economic context that gives rise to those relationships. It is problematic to attempt to recreate the kieretsus from the Japanese construction environment without this contextualization. Equally, it can be said that the application of the strategic partnering or alliancing concept can be problematic, since successful firm-to-firm relationships is contingent upon understanding the underlying structural and behavioural characteristics of the particular economic environment within which those firm-to-firm relationships are embedded. The project nature of the industry does not particularly appear to lend itself to long-term relationships and this is not a particularly new idea; however, many of these models are suggestive of the need for long-term stable relationships in the construction industry. There are different government policies, demand patterns, business ideologies and market structures operating within various construction environments (Cox and Townsend, 1998) which need to be understood before such ideal types can be imprinted on projects for long-term effect.

There appears very little research which explicitly explores the numerous underlying economic organizational structures of the construction sector and which accepts and perhaps celebrates that firms are procured and products and services are supplied within markets on a project-by-project basis. Firms respond and organize themselves according to this unique economic environment. Perhaps the performance of the sector is being judged through an inappropriate lens – one that has been developed through a tradition of stability and long-term production rather than a deeper understanding of the industrial context of a project-oriented sector. Perhaps we are more productive or more efficient than we credit ourselves – but, as I noted earlier, this is a discussion left for another day.

Industry analysis for performance improvement has traditionally focused on sectors, which includes groups of firms with similar characteristics, engaged in similar production processes, producing similar goods or services and occupying similar positions (AEGIS, 1999). According to Marceau (AEGIS, 1999), the attention is now on chains, clusters and complexes. This represents a shift from the purely mechanistic conceptions of the nature of industrial organization, as a market consisting of a collection of establishments producing homogenous outputs (Scott and Storper, 1986) to a more complex interconnected and interdependent set of markets and firms. This perspective is well suited to the construction sector but has proven difficult to grasp.

In the quest to improve the performance of the construction industry, in the last decade the research community has pursued three paths. First, at a micro level focusing upon transactions at a 'project' level. Second, at the macro level and the relationships between patterns of transactions between whole industries at an 'industry' level. Finally, and more recently, a market or sectorial perspective has found some support; where the industry or market performance is considered as a result of firms and their behaviour at a 'market' level. Each of these approaches has merit and answers many questions about industry performance, yet each misses two critical factors.

Firstly, the importance of the complex and varied connexions between the project, the firms, the industry markets, the project markets, the commodities and secondly that firm–firm procurement relationships are in some manner associated with all of these entities and is the 'glue' which ties these entities together.

Firm–firm procurement relationships on projects arise from the underlying structural and behavioural characteristics associated with each of these entities and it is the project contractual relationship which encapsulates the context between the two firms which gives rise to the legal contractual relationship. The contract is embedded within a context which involves historical relationships, demand, sourcing strategies, supplier responses and current market competitiveness. A great deal of attention has focused on the firm–firm procurement relationships between clients, contractors and consultants; those firms involved in project procurement. On an individual project it is suggested that this may represent less than 5% of all the contractual relationships between firms and so this emphasis on the top level relationships is useful but does not give a good indication of the underlying nature of the project nor the entire industry. The characteristics of the chain of firm–firm procurement relationships is more indicative of the performance of the entire industry. Therefore, understanding the structural and behavioural characteristics of various chains of firms and the associated industrial organization economic context which gives rise to certain types of firm–firm relationships will provide a much-needed detailed understanding of construction industry structure, firm conduct and associated industry performance.

This chain of firm–firm contractual relationships is typically termed the supply chain. Supply chain management for an individual organization is an emerging field of research in the construction management discipline. Supply chain management is the management of upstream and downstream relationships with suppliers and customers to deliver superior customer value at less cost to the supply chain as a whole (Christopher, 1998, p. 15). Less attention has been devoted to investigating the nature of the construction supply chains and their industrial organizational economic environment and, therefore, the full potential of the supply chain metaphor has yet to be realized. The holistic lens of the supply chain metaphor provides a theoretical, intellectual and practical framework.

Understanding the industry at the 'supply chain level' has been on the national agenda for many countries. Such an approach assumes that construction projects are located within an industry whereby transactions are interdependent and the interdependency impacts upon the performance of the industry. A characteristic of the interdependency is that a customer and supplier are both located within particular market sectors, each with their own market 'rules' of behaviour and underlying structure. Because each project has some unique characteristics, firms typically form new firm–firm relationships for each project contract. The business of procuring firms (i.e. suppliers) to assist in fufilling contracts is an activity that occupies a great deal of time, energy and resources on a daily basis throughout the industry and in every construction industry in every country.

I am concerned with modelling of procurement in the supply chain and the rationale for this modelling lies in the argument that firm–firm relationships extending right throughout the chain of contracts underpin the performance of projects and thus ultimately the 'industry' and that the structural characteristics of individual firm markets impact upon these relationships. Therefore, it is of relevance to construction industry policy makers at a regional, national and international level that procurement behaviours are modelled at each level of the chain as this will assist in understanding the nature of performance in the industry upon which their decision making impacts.

1.1.1   Improving supply chain performance: government intervention

There has always been a relationship between macro and microeconomic theory and government policy. However, industrial organization economic theory, a specific branch of economic theory, is explicit in that its primary purpose is for policy consideration and intervention related to firm conduct, market structure and market performance. Government policies affect the state of the construction industry. The relationship between government policies and the construction industry affect the character of the industry (Cox and Townsend, 1998). Policies, either espoused or direct, can impact upon the construction industry in three main ways.

- First, through direct intervention and the development and monitoring of policy and regulations, governments can encourage competition or alternatively restrict practices in certain areas (Warren, 1993).
- Second, governments, as a large client of infrastructure projects, can impact upon the state of the industry by virtue of their demand patterns through the allocation of budgets and capital works programmes.
- Finally, as a large client their purchasing power provides a mechanism to induce certain types of behaviour from firms that they contract with through various procedural systems and practices which they undertake.

The degree of government involvement and policy planning can differ from locality to locality; however, it is one of the underlying elements that assists in shaping the construction industry of a region. Policies to some degree can affect both the dominant business practices and the nature of competition within the industry. Although attempts have been made to use the industrial clusters of firms and firm networks concept to inform construction industry policy, there has been no formal attempt to examine the potential of a supply chain procurement model as an economic tool or instrument.

This discussion is taken up in more detail in Chapter 2, The rationale for the modelling of procurement in the construction supply chain.

1.1.2 Firm behaviour and supplier procurement

The industry is typically presented as a large mass of firms within the one market. Comments about the industry often are reduced to a description that it has a fragmented structure and is made up of a large number of small- to medium-sized enterprises. Descriptions of the industrial organization economic environment do not extend to discussions of any of the following:

- specific chain organization;
- distinct chains of commodities and their markets;
- firm behaviour within markets in relation to vertical governance structures;
- firm behaviour towards procurement of suppliers; and
- supplier responses to customers.

Industry performance discussions do not consider these intimate details of the different supply chains in different sectors. It is not an entirely new concept that the industry is made up of numerous sectors, sub-industries or markets (Hillebrandt, 1982; Groak, 1994). Within each construction project numerous markets for a wide variety of products and services are represented (Hillebrandt, 1982; Murray and Langford, 2003). However, an explicit examination of the industry using the supply chain concept with such a detailed and systematic approach as that described in this text is novel.

The behaviour of firms in markets and the degree of competition is relevant to the supply chain concept. Competition usually relies upon the bargaining power of customers and suppliers, the threat of substitute products or services or the threat of new firms entering the market. These forces reflect the degree of pressure that firms exert upon each other to obtain market share. The bargaining power of suppliers depends upon the degree of product differentiation between firms and the number and relative importance of firms in the market.

Many suppliers in construction would be attempting to convince customers that their product is differentiated from their competitors. It has been assumed that in the construction industry attempts at product differentiation normally will cease when the project reaches the tendering stage. At the tendering stage it is assumed that all products are homogenous and competition takes place purely on price (Hillebrandt, 1982). More and more this assumption is challengeable and it is suspected that the way firms behave does not support this premise. It may be correct if each market is viewed in isolation and not within the context of the chain, but it is suspected that the decision making process during procurement is not like this.

It has been the belief in the past that there is little differentiation between firms in the industry and this belief forms the basis upon which the majority of construction projects are procured. The industry practice of awarding construction project contracts through tendering and the ensuing competitive bid process for cost leadership (Runeson, 1997) assumes a perfect competition market. This practice ripples through the industry and provides the framework upon which other contractual relationships along the construction supply chain are based. It follows then that across the entire industry the competitiveness of firms is based upon cost leadership alone and not differentiation. In this environment firms have no real power and are price takers and as such contractual relationships are based upon the 'arms length' philosophy.

If the vast majority of construction work is procured in this manner it suggests that the industry is composed of only arms length supply relationships. In other industries such arms length supply relationships are normally only suited to non-strategic, low-value and infrequent purchases, where there is a great deal of choice from a market of expert and capable suppliers (Cox and Townsend, 1998).

There is little evidence yet to support this theory that the entire structure and behaviour of the industry is based upon only arms length supply relationships. Although price is a critical decision criterion, it is proposed that the firm–firm procurement relationships may not be as simplistic as previously represented; that there are strategic high value and frequent purchases and that the decision making process in procurement at each tier in the chain is much more dynamic and much more complex than this. The construction environment in some instances appears more complex than assumed and it is suspected that it is composed of firms across a variety of market segments, in which strategic and non-strategic purchases may be both evident. It then may seem somewhat illogical for all relationships to be treated in the same manner.

The arms length approach to procurement impacts dramatically upon our conceptualization of the supply chain and the theoretical position of strategic procurement; which suggests that there is differentiation within the supply chain or that differentiation can be created. Construction clients appear to naively frame their actions towards purchasing a single product without understanding the chain of events they trigger and which lead to their purchase. There is evidence to suggest that some clients have orientated toward thinking that they are purchasing a supply chain rather than a single product or service (Townsend and Cox, 1998). This is certainly the case in other industries (Hines, 1994), but evidence suggests that this is really not so widespread and perhaps the construction industry is not really at that stage of maturity of 'chain procurement'. However, the research reported in this text reveals that regardless of how the client behaves the industry can and does deal with procurement in a strategic manner and as strategically as the upstream and downstream market influences will allow.

It is the understanding of the relationships between the markets along the chain which is critical to improving industry performance as a whole.

A premise for managing the supply chain is that it should be managed for competitive advantage rather than to reduce costs (Hines, 1994). Almost two decades ago it was noted that the interdependencies between customer firms and suppliers is the largest remaining frontier for gaining competitive advantage and that nowhere has such a frontier been more neglected (Drucker, 1992). In the last 20 years the recognition of an altered competition model for many industries has prevailed, whereby supply chains compete rather than single organizations (Christopher, 1998) and yet supply chains as a whole typically don't compete in the construction industry.

To understand how relevant this notion is to the construction industry is a complex issue. There are a wide variety of suppliers on projects operating with so many different levels of technical and managerial expertise that much of the theoretical discussion of construction supply would be more meaningful with descriptions of the types of suppliers, procurement relationships and supply chains that exist. Ultimately, effective supply chain management requires the ability to be able to identify and locate differing levels and types of differentiation across various supply chain options. It is suspected that very few firms have this holistic perspective on supply chains and typically manage only one tier; which is their immediate suppliers.

Upon entering this world of construction supply the bewildering array of different relationships between different firms appears at best fragmented and at worst almost chaotic to the client. To make sense of this fragmented world, various firms play an 'ordering' or gatekeeping role for the client. The most significant gatekeeper roles are the architect, the contractor and the project manager who, in turn, may co-ordinate many of the other firms on the client's behalf and control the messy unstructured design and construction process. The role of some firms to control other firms on projects represents significant positions of authority and responsibility in the construction supply chain since they are effecting a degree of power over much of the critical assets that constitute the supply chain.

The story of the supply chain really begins with clients; as originators of the design and construction process, they are an integral part of all projects and potentially can have the greatest impact on the firms supplying services or goods through both their demand characteristics and then their procurement decisions. Client demand patterns are dependent upon the number of projects and the relative value of projects. Different clients have different demand characteristics and therefore are able to impact upon their suppliers in varying degrees. The degree of impact is concerned with the amount of leverage that the client has within the supply chain. Leverage is the ability to obtain control over particular resources in a supply chain and then to manage those resources in such a way that it becomes possible to appropriate value (generally profits) for oneself, against the interest of customers, suppliers, employees and competitors (Cox and Townsend, 1998).

There are a number of variations to the method of organizing the contractual relationships on individual projects. A range of strategies including traditional, construction management, design and build, turnkey, build-own-transfer, build-own-operate and transfer, strategic alliances, partnering and joint ventures are available. The decision of appropriate project procurement and/or organizational strategy is important and requires a degree of procurement competence to understand the differences. The real difference between the methods lies in different parties assuming authority, responsibilities and risk. Each party is trying to extract maximum reward with minimum financial risk along the supply chain.

There has been a growing awareness of the potential for a range of contractual relationships and the value of strategic procurement practices for the client. However, there is little clarity in the construction industry regarding what supply chain procurement actually entails and the fundamental problems it seeks to solve. For clients to move from project procurement strategies to supply chain procurement strategies on projects would require a detailed understanding of the nature of procurement as it currently occurs in the industry. It is the problem of describing the complex system of procurement along the supply chain within the construction industry which provides the setting for this text.

In recent years it has become accepted wisdom that complex problems such as supply chain procurement cannot be dealt with by some form of prescribed 'routine strategic analysis'. A complex problem such as this will involve the design of a 'strategic thinking process' (Eden and Radford, 1990). If we approach the construction supply chain procurement problem from a similar perspective, it means that the success of using a supply chain procurement methodology is reliant upon designing a broad methodological framework. This will involve investigating, describing and representing how supply chains are organized to enable our future questioning of why they are organized in such a manner and how they can be organized more effectively. It is within this intellectual environment that this text rests.

1.1.3   Current research

The research fields which informed this text primarily fall within two fields of literature: supply chain management and industrial organization economics. These are now briefly outlined and then explored in more detail in Chapter 3, Supply chain theory and models; Chapter 4, Industrial organization economics methodology and supply chain industrial organization approaches; and Chapter 5, Project-oriented industrial organization economics supply chain procurement model.

Supply chain management research emerged from the logistics and management disciplines and has been borrowed by construction researchers. The supply chain management literature reviewed indicated that the research tends to fall within four main streams, these being distribution, production, strategic procurement management and industrial organization economics. Of course, there may be various other ways to organize the literature and with such an explosion of interdisciplinary research in all disciplines it is perhaps problematic to attempt such a broad categorization; however, that is perhaps a worthy debate for another day. For the moment, it is easy to claim based upon the literature review that construction supply chain management literature is lacking in the industrial organization economics field. There is also a lack of empirical research, which is usually needed to test or 'ground' theory within the field. The construction research related to the supply chain concept has to date focused upon management and development of normative ideal types for effective supply chain management; and have used largely a deductive epistemology. Little attention has been paid to supply chain economics and developing positive models which are often inductive and aim to describe what currently exists.

The supply chain research which has taken an industrial organization economics approach has assisted in theoretical development of the field through the development of methods for mapping the structural organization of chains, supplier procurement historical, technical and economic environments, supplier types and procurement relationships (Ellram, 1991; Hines, 1994; Nischiguchi, 1994; Lambert *et al.*, 1998). These studies have developed our knowledge about the underlying structural and behavioural characteristics of firms and the supply chains in those sectors. A select group of construction research has considered topics of interest to industrial organization economics, such as transaction cost economics and project governance (Winch, 2001); sector knowledge flows and innovation (AEGIS, 1999); flexible specialization (Tombesi, 1997), quasi-firm (Eccles, 1981) and vertical integration (Clausen, 1995). However, no previous study to date has comprehensively explored the fundamentals of the theory of industrial organization economics and considered the explicit implications of such an approach either theoretically, methodologically or empirically.

Industrial organization economics theory primarily relies upon the interrelationship between three concepts: firm conduct, market structure and market performance (Martin, 1993). It is concerned with describing structural and behavioural characteristics within industries. Firm conduct in terms of strategies towards firm governance and the identification of firm boundaries is of particular importance to an industrial organization economic perspective of supply chains. Firm decisions about suppliers relies upon an interaction between firm strategy and market structure. In the construction industry, the market is both the broader industrial market and then often a unique project market.

Difficulties were identified in a direct application of the industrial organization economic methodology. Many of the empirical studies in this field are at an industry or market level and therefore do not readily relate to the project focus that is found in the construction industry. One of the major implications is that there is a high degree of multiplicity of firm–firm procurement relationships to be found in a project-based industry; that is, many interactions between firms on many projects; whereas this is not so prevalent in other industries. The industrial organization methodology in the past has accommodated a more stable perspective of firms and markets, whereas the construction industry is more dynamic. To overcome this, the 'industry' is recast and described as a system of supply chains with various significant real world 'objects' that have both structural and behavioural characteristics, which supports the industrial organization modelling philosophy and also accommodates multiplicity between objects that comes with a project-based industry. The objects are described and their relationships to other objects are also described. The underlying premise to the interacting system of objects in the supply chain is that groups of objects form a 'class' and the various class of objects all have similar attributes – which although similar also differ with each individual instance. For example, firms have similar attributes which help us to define characteristics about firms but then they may each differ (or not) with respect to a particular attribute value; for example, firm size. This accommodates project uniqueness and project similarity and the multiple interactions which occur between individual objects. The conceptual framework which is described in detail in Chapter 5, Project-Oriented Industrial organization economics supply chain procurement model, assists in modelling both unique situations and individual instances and common instances in procurement; through its object and class concept. In summary, the object is a unique instance of a class; for example, the firm class is a grouping of all the firm objects that have similar structural and behavioural characteristics. In many ways this description of the system is a way to intellectually and practically develop a set of representation techniques for capturing, specifying, visualizing, understanding and documenting dynamic scenarios which underpin the very fabric of the industry in terms of its structural and behaviour properties. It is the beginnings of a construction chain information procurement (CHIP) model based upon the object oriented methodology.

1.2   Ideas explored in this text

The research which underpins this text was concerned with two ideas: first, the modelling of procurement within construction supply chains between firms in successive tiers and then the development of a methodology to achieve this.

It is suspected that there are a variety of construction supply chains with different characteristics involving different firms that are organized and managed in certain ways. The manner in which they are organized and managed are considered accepted construction industry practices. It is suspected that major shifts to the current practices would require fundamental changes to the underlying structure, yet little is really known about the structural and behavioural characteristics from an industry-wide perspective.

The second central theme of the study was to establish the intellectual basis and practical framework for guiding the description and analysis of procurement related to a project-based industry across many firms, markets and chains. Towards understanding procurement and chain organization the industrial organization economic theory is useful as it suggests structural and behavioural concepts related to firm conduct and market structure. In this theory the interrelationships between market structure and firm behaviour are considered important; all 'things' – that is, entities or objects in systems – have certain structural characteristics and certain behavioural characteristics. It is this interrelation between these two constructs that needs to be encapsulated. Industrial organization economics places a high priority on understanding explicitly the structure and behaviour in a system. However, the economic theory does little to accommodate the high level of different instances of interactions between various objects (for example, firms, projects and market) associated with procurement and the dynamic nature of procurement in the construction supply chain. Multiplicity is largely ignored. The construction industry is underpinned by firms constantly changing relationships and firms forming many relationships. It is this combination of supporting the structural and behavioural construct associated with objects in procurement as well as the many changes that occur which needs to be encapsulated in a methodology.

Strategic supply chain procurement for construction participants may challenge current thinking, practices and accepted norms. Strategic supply chain procurement is concerned with organizing and managing the assets along the supply chain for effective leverage for the client against the interests of other participants. To manage these assets it seems critical to understand what they are and the economic context within which they are placed. This involves an understanding of the firms, their commodities, their markets, the project and the relationship between the firms that make up the chain. The approach taken in this book is to lay out a rather unique model and then refine it with information based upon real world case study projects. The practical application of such a model and methodology would, of course, require some degree of refinement, however; the links between this approach and the world of information management that we currently face are quite close.

1.2.1   *Theory and methodology*

The development of theory for the supply chain concept is a critical component of this text and that is said without apology. There is no need for a justification of theory and the value of theory to the real world of construction. I have conducted too many research projects where I was faced with the wall of hard-nosed construction practitioners who have been somewhat sceptical about 'theory' and academic literature and philosophical thinking who have then changed their position after the project has been completed. To develop a conceptual position with a theoretical framework and then to explore the value of that theory against empirical observations of some form has left me with something of value to say. The balance is to provide the 'content' of the findings against a backdrop of theory – so my intention is not only to 'find out' about supply chain procurement, but also to 'make sense' of what I found out.

Therefore, we embark upon this book with a concerted balance of mixing the theory with the practical. There is a need to develop theoretical and methodological constructs to underpin the supply chain field. Much of the past research is about what should be done in the ideal supply chain and very little attempts to explain the reality of what is done in the vast majority of supply chains. This requires a methodology that attempts to move beyond simple descriptions of case studies that represent an ideal type that the industry should aspire to. It requires identification and analysis of fundamental properties of a number of real world scenarios and the development of a methodology to achieve this.

In summary, the book develops a methodology for describing the economic organizational structural and behavioural characteristics in supply chains and it draws methodologically from industrial organization economic theory. It pursues the line of inquiry that structure and behaviour are encapsulated in 'things'. That these 'things' that are important to the topic of inquiry have relationships and that these relationships help to create structural characteristics; that scenarios in the real world can be representative of occurrences and interactions between 'things' and also give an indication of the behavioural characteristics of the topic of inquiry. The real challenge, however, is to first describe and make sense of the practices in the 'real world' topic of inquiry of procurement in the construction supply chain.

1.2.2　Research questions and objectives

As stated, the purpose of this text is to describe structural and behavioural characteristics associated with procurement in the construction supply chain and to define a methodology to develop a structural and behavioural view of the procurement that takes place across any construction sector related supply chain. There were two research questions posed.

Box 1.3　Research study question

What are the structural and behavioural characteristics of the key objects associated with procurement in construction supply chains?

To what extent can an interdisciplinary study merging industrial organization economics and object-oriented modelling assist construction supply chain procurement representation and analysis?

This text focuses on the first research question only.

The objectives of this work include the following:

> **Box 1.4 Study objectives**
>
> - To investigate construction industry policy in relation to the supply chain concept.
> - To identify the applicability of industrial organization economics theory and methodology to the construction industry and to develop an adaptation of the industrial organization model for an industry that is project-based.
> - To propose an empirical methodology to explore the real world supply chain procurement in light of this model and represent structural and behavioural model views of supply chain procurement across many instances of firm–firm relationships, supply chains, projects, commodities and markets.
> - To refine our notions about the structural and behavioural view of procurement and to further develop a supply chain economics model based upon an empirical study.

In today's economic climate, where a competitive edge is necessary, the effective control of supply chains can make the difference between success and failure on projects. The research reported in this text examines the formation of relationships during the construction process. This field of investigation is of concern to organizations at any level in the supply chain in the construction industry who wish to be more successful than their competitors in terms of understanding their construction supply chains. It is also of concern to any strategic thinkers in the industry working in any type of organization. Construction industry policymakers are always looking for information about the industry and in particular 'what happens over the fence' when we throw projects and various policies out there. The book is concerned with making more transparent the inner workings of the industry, confirming suspicions and challenging accepted myths. The following sections describe in brief the research strategy used to investigate industry practices in relation to procurement.

1.2.3 Research strategy

This book is the result of a research study. The methodology is not described in detail in the text, as I decided the focus should be on the results – however, every now and then a brief summary is given at strategic places in the book so that the claims can be contextualized. The research strategy of inquiry relies upon two main elements:

• the development of a theoretical model and methodological framework
• the representation and validation of the model through real world examples.

In simple terms, this means that the supply chain procurement model is described and represented through a structural model view and a behavioural model view. Object-oriented modelling provides the framework and language for structural and behavioural model views. The structural model view encompasses the static, or structural, aspects which largely relies upon a static description of the object and class model and associations. The behavioural model view encompasses the dynamic, or behavioural, aspects and relies upon descriptions of interactions or collaborations among objects and class elements. The development of the model is an iterative process and also largely interdependent. The structural model is incomplete without methods and operations being described which require input from the behavioural model. Although this may seem confusing, it is explained in more detail in Chapter 5. The objects and classes which are the structural elements of the model include the description of characteristics and the relationships between the following entities:

• project
• firms: both supplier and client
• markets: both industry and project
• commodities and
• firm–firm procurement relationships.

The model described in Chapter 5 was translated into object oriented modelling but this is not presented in this text, as mentioned previously.

1.2.4   Ontology, epistemology and methodology

For those readers interested in the ideas of the strategy for research design, the following discussion is probably of most interest. Increasingly, the boundaries in interpretive frameworks are being blurred. More and more researchers are accepting that interpretive frameworks for research studies are not dichotomous and that the associated ontological, epistemological and methodological assumptions may lie along a continuum – or, at the very least, answers to research questions can be enriched by a variety of approaches (Lincoln and Guba, 1985; Newman and Benz, 1998; Groat and Wang, 2002). Interdisciplinary research is often difficult because ontology, epistemology and methodology are difficult to reconcile across various disciplines-accepted norms.

As an aside, it is probably useful to consider ontology, epistemology and methodology from the perspective of the questions that each seeks to answer.

Box 1.5 Ontology, epistemology and methodology questions

Ontology: What is the form and nature of reality and therefore what is there that can be known about it?

Epistemology: What is the nature of the relationship between the knower or apparent knower and what can be known and what is valid to be known?

Methodology: What is the interpretive framework and associated practical techniques and the set of assumptions which underpin the way in which a study is conducted?

The research reported in this text lies largely within the postpositivism inquiry paradigm and sociological qualitative research techniques are considered appropriate. Postpositivist inquiry is somewhat objective, but less rigid in its claims than, for example, positivist social science. Positivist social science is 'an organized method for combining deductive logic with precise empirical observations of individual behaviour in order to discover and confirm a set of probabilistic causal laws that can be used to predict general patterns of human activity'. Where positivism contends that there is a reality out there to be studied, captured and understood, postpositivists argue that reality can never be fully apprehended, only approximated. It relies on multiple methods as a way of capturing as much of reality as possible. At the same time emphasis is placed on the discovery and verification of theories.

Neuman's views in this regard are set out in Box 1.6.

Box 1.6 Approach to investigation

Positivism assumes that there are incontestable neutral facts on which all rational people agree. Its dualist doctrine says that social facts are like objects. They exist separate from values or theories. The interpretive approach, however, sees the social world as made up of created meaning, with people creating and negotiating meanings. It rejects positivism's dualism, but it substitutes an emphasis on the subject. Evidence is whatever resides in the subjective understandings of those involved. The critical approach tries to bridge the object–subject gap. It says that the facts of material conditions exist independent of subjective perceptions, but that facts are not theory neutral. Instead, facts require an interpretation from within a framework of values, theory and meaning.

Supply chains are complex 'things' (objects) and firms are 'things' (objects). Firms form firm–firm relationships and these procurement relationships on projects are integral to supply chain structural characteristics. Some characteristics of these relationships may require interpretation, even if they appear strictly data-oriented. Even something apparently concrete in nature, such as market size, can become abstract and interpretive. For example, a relationship may be embedded within a market of ten other firms; however, for a particular project the customer reduces the market size to three. The supplier, however, does not know that the project tender market is now three. This interpretive discussion assists in understanding the nature of market competition for that particular commodity for that particular project. Even more open to critique are the characteristics of a relationship that are related to decisions of choice between suppliers; for example, is it price, performance, trust or a combination of these? Is it performance that reduces the market from ten to three and then the final decision is price? Such fine-grained nuances are an important part of the discussions that follow and this is why the research was assisted by being located within a social science postpositivism/interpretive research framework.

One of the most interesting aspects to our industry is that every time we are faced with a project it represents a specialized environment whereby we synthesize selectively a range of various threads of knowledge, skills and capabilities and provide an holistic 'answer' which responds in some way to the demands of the project, client and society. This is an environment that those within the industry are acculturated from the very early days in the University institutions. Our discipline-specific knowledge is formed by both a synthesis of what we do (design, construct or manage) and how we do it. The living and breathing changes that occur on projects and dealing with this constant change in a consciously flexible thinking manner is some-thing that all built environment professionals learn to live with and excel within – it is our *modus operandi*. The level of maturity, responsibility and initiative required to achieve this is unique to our industry and to be valued. To investigate this industry we should not devalue our knowledge from the inside – the storehouse of experiences assist us in our insights into our industry. Of what relevance does this have to the topic in hand? There are two explanations to be considered. First, that when writing about the industry and how people behave it is useful to remember our distinctive features and our context. Second, on a more philosophical level, the approach to studying the makeup of the industry in this book is echoed by professional practices – that is, the research project becomes informed by numerous other disciplines which are selectively examined, critiqued and synthesized – the parallel of our professional practice with the research practice which underpins this text is aligned.

A philosophical dilemma with the idea of presenting material on construction supply chains has been this dichotomy between the individual and the general. In research terms it is an important issue related to ontology and it is concerned with whether the work is nomothetic or ideographic in nature. The previous discussion suggests that an ideographic approach is suitable and that it is the unique elements of the individual phenomenon which are important. The argument for an individually orientated, or 'ideographic', construction supply chain model is strengthened by empirical observation of real world project supply chains. It is contended that a realistic supply chain model has to be based upon individual firm to firm transactions within unique market and project perspectives. However, there are some patterns of structure and behaviour in the real world that were identified in this study but it is beyond the scope of this text to provide the detail in relation to the nomothetic part of the study. Therefore, a focus has been taken to present the individual case study material so that readers can understand the context, background and the key issues and then reflect and relate the case study to their unique circumstances.

The validity of the model is only realized through observations of real world scenarios. The work is grounded in empirical observations. Empiricists hold that all our knowledge must ultimately be derived from our sense experiences. This represents an inductive approach, which begins with detailed observations of the world and moves toward more abstract generalizations and ideas. An inductive approach was largely taken for this research. In practice most researchers are flexible and use both approaches at various points in a study.

Past empirical investigations of the construction supply chain have tended towards studying a small focused group of firms supplying to each other, with the intention of proving that a particular model of inter-firm behaviour yields better performance on projects. That is, boundaries are drawn around only those firms that have certain characteristics to support the development of a particular interpretation. Therefore, the cases are instrumental case studies.

The inductive approach taken here aims at developing the model of supply chain behaviour and structure based upon a wider collection of scenarios. In this text the inductive approach is taken, because the intention is that detailed observations of many every-day occurrences in supply chains in the construction industry will lead towards more abstract generalizations about the supply chains. A number of observations are made about many examples of supply chains and supply chain firm–firm relationships and scenarios. The author does not particularly wish to manoeuvre readers towards a particular way to manage their supply chains – rather to provide examples of how others have and to provide a broader view of all the interacting elements. It is intended to equip our decision makers with a way to see the big picture and perhaps understand the implications of various decisions at many different locations along our supply chains.

The material presented in the case study Chapters 6–10 is interpreted in a qualitative manner, in an attempt to understand structure and behaviour about both the individual objects and the classes. The information was derived from over 47 structured interviews and 44 written questionnaires. The interview material represents information on sourcing strategies from 6 major projects, 4 different client types, 4 different contracting firms, 10 component suppliers and 5 manufacturers. It represents a number of different supply chains with 10 trade supply clusters (concrete, masonry, structural steel, etc.) and consultants. The questionnaire material represents sourcing strategies from 28 subcontractors, 4 component suppliers, 5 manufacturers and 7 consultants and 17 trade-orientated supply clusters and consultants. The case study material primarily draws upon the interview information. The length of interviews varied – ranging from a minimum of 1 to 4 hours.

Clearly, this work is interdisciplinary drawing from a number of research fields: construction supply chain management, industrial organization economics and sociological research methods. It is, however, located within the construction management and economics research tradition. It is allied to supply chain management; however, perhaps the research begins to identify an associated field of supply chain economics – which to date has not been recognized.

1.3   Structure of text

The following diagram in Figure 1.1 indicates the overall structure of how the chapters are organized and how they relate to each other.

1.3.1   Overview

Chapter 2, The rationale for the modelling of procurement in the construction supply chain contains a more particular discussion on the rationale for modelling of procurement in the construction supply chain with special relevance to its relationship to worldwide government investigations, policy and intervention for improvement of the productivity of the sector. Three case studies were conducted using a document analysis technique which mapped selected government policies, investigations and reports. The cases were international, national and regional and were drawn from 1984 to 2003. The charting of policies was based upon a critique of how the documents reflected the way in which the industry was being interpreted as either fragmented or specialized and then as a result of this interpretation whether the underlying models to improve the performance of the industry were calling for normative or positive models.

1.3.2   Model development

Chapter 3, Supply chain theory and models reviews the supply chain management literature and locates this work relevant to others within the supply chain literature. The review includes selected key works in both

OVERVIEW

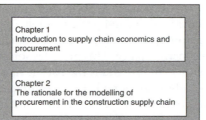

Chapter 1
Introduction to supply chain economics and procurement

Chapter 2
The rationale for the modelling of procurement in the construction supply chain

MODEL

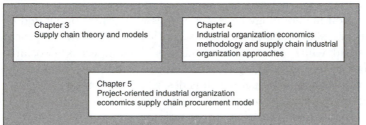

Chapter 3
Supply chain theory and models

Chapter 4
Industrial organization economics methodology and supply chain industrial organization approaches

Chapter 5
Project-oriented industrial organization economics supply chain procurement model

CASE STUDIES

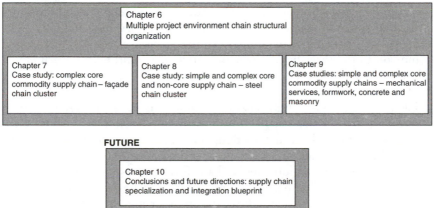

Chapter 6
Multiple project environment chain structural organization

Chapter 7
Case study: complex core commodity supply chain – façade chain cluster

Chapter 8
Case study: simple and complex core and non-core supply chain – steel chain cluster

Chapter 9
Case studies: simple and complex core commodity supply chains – mechanical services, formwork, concrete and masonry

FUTURE

Chapter 10
Conclusions and future directions: supply chain specialization and integration blueprint

Figure 1.1 Overall structure of the organization of the chapters in this book.

the construction management and economics literature and the wider supply chain-related literature in other industries. It provides an overview of the main streams of research. It highlights the lack of broad-based empirical work which is required to develop the field. It also highlights the lack of research in construction literature associated with developing an industrial organization perspective of the chain as opposed to the research in other sectors.

Chapter 4, Industrial organization economics methodology and supply chain industrial organization approaches reviews the industrial organization economics literature and provides a brief overview of the historical development and the strong division of the field into two main schools of thought; namely, the Chicago School and the structure-conduct-performance school. Some fundamental principles relevant for understanding industrial organization concepts are described, including: market structure, firm conduct and market performance. This provides some detail on economic structural and behavioural concepts and is the background for understanding existing supply chain industrial organization approaches and the model developed in this thesis. To develop the model the procurement relationship between two firms is identified as the key concept that ties the industrial organization and supply chain fields. This chapter highlights that those models that have merged the supply chain concept and the industrial organization methodology have not addressed markets orientated towards projects and short-term production scenarios. Much of the empirical work in the field which validates methods and techniques is associated with manufacturing and retailing and not construction. This chapter develops the principles for the model which is described in Chapter 5.

Chapter 5, Project-oriented industrial organization economics supply chain procurement model gives an outline of a new project-oriented industrial organization economic model for procurement in the construction supply chain. It develops the model through the synthesis of the principles within the industrial organization literature and the supply chain literature with the specific characteristics of the project-oriented industry. The principal components of the model include: project attributes; firms: their commodities and their market structure; attributes of firm–firm procurement relationships; structural organization of firms and events in the formation of the chain.

Each construction supply chain is composed of a contractual chain connecting firms which relate to a construction project. The contractual chain links firms who are providing services and/or products along the chain. The product and/or service is termed a commodity for the purposes of this text. A construction supply chain forms in response to a construction project which has particular characteristics; has firms with various attributes that provide commodities that may or may not be homogenous that reside within different types of markets and has firms that are linked through relationships that have certain attributes. The model is described in relation to three key elements: projects, firm–firm procurement relationships and the various associated entities that link these elements. The intrinsic structural and behavioural characteristics are described as the structural elements of the model are assembled throughout the chapter.

1.3.3   Case studies

Chapter 6, Multiple project environment chain structural organization gives an overview of six projects; the characteristics of the firms and the projects.

The projects are building type projects from residential, commercial and public sector and were all being designed and built during the late 1990s to early 2000s in Melbourne. Melbourne is the capital of the state of Victoria in Australia. Victoria is the second largest state, of the seven Australian states, by infrastructure spend. The first project was categorized as a mixed-use arts building with a construction cost of $262mAUS with an expected construction duration of thirty-six months. The second project was categorized as a commercial building sports stadium with a construction cost of $27mAUS with an expected duration of twelve months. The third project was a commercial mixed-use residential high-rise, high-density apartment building of construction cost $32mAUS and a duration of sixteen months. The fourth project was also a commercial sports stadium similar to project two, but with a much higher construction cost of $460mAUS and a construction duration of thirty-six months. The fifth and sixth projects were both residential similar to the third project, however of a much smaller scale. The fifth project was low-rise, medium-density with a construction cost of $20mAUS and an expected construction duration of twenty-four months. The final project was low-rise, medium-density of $0.9mAUS and an expected duration of nine months.

The structural organization maps of the supply chains from the client through to tier one to tier two and then tier three are developed for the six projects. This provides the context for the remaining case study Chapters 7 through to 10. The maps also included the structural organization for various tiers for the case studies. As well as structural organization maps, various matrices were developed which mapped key indicators of suppliers' co-ordinating capability. That is, the following were mapped:

- suppliers were mapped by supplier variety versus number of suppliers
- commodities were mapped for transaction complexity versus commodity type
- firm size versus commodity type
- firm scope by commodity type
- task complexity by commodity type.

The various project procurement strategies are discussed – but it soon becomes clear that downstream linkages are not unduly influenced by the upstream linkage between the client and the contractor and this is discussed in relation to supplier market characteristics and commodity characteristics.

Chapter 7, Case study: complex core commodity supply chain – façade chain cluster provides a detailed description and analysis of the structural and behavioural characteristics of the key suppliers involved in the façade supply chain cluster for project one. This includes a discussion of the façade subcontractor firm and its industrial market environment and then the key suppliers to the façade subcontractor and their markets, including the aluminium extrusions fabricator, the glazing supplier, the glass merchant and the glazing manufacturer. The structural steel fabricator is a complex supply chain in itself and it is left to be discussed in detail in Chapter 8.

The behavioural characteristics of the procurement relationship are described through sequence diagrams. The sequence diagrams describe interactions between the firms and the industrial and project market environments before and during the tendering process and within the context of the various firms' history. The façade subcontractor categorizes suppliers by types and the sourcing behaviour corresponds to this typology. The chapter concludes with structural organization channel maps for aluminium and glazing, which are an aggregated perspective of project supply chains.

Chapter 8, Case study: simple and complex core and non-core supply chain – steel chain cluster provides a detailed description and analysis of the structural and behavioural characteristics of the steel supply chain. It presents the results from the interviews with various project managers, firm executives, production and procurement managers involved in the supply chains for commodities that are clustered around the supply of steel to the construction site. Steel, of course, is involved in numerous products and various supply chains – which were discussed in Chapter 7. A number of structural steel fabricators/subcontractors were interviewed in relation to the projects. After the subcontractor interviews, subsequent interviews with processors, merchants and manufacturers eventually led to a tracing of more general industry chains for the supply of the products. The chapter is organized in a similar manner to Chapter 7 in four main sections, including:

- Firm attributes
- Markets, commodities and competitors, including competitive advantage
- Supplier types
- Procurement relationships, including sourcing strategies.

Chapter 9, Case studies: simple and complex core commodity supply chains – mechanical services, formwork, concrete and masonry, in a similar structure, presents the results from the interviews with various project managers, firm executives, production and procurement managers involved in the supply chains for commodities that are clustered around four key suppliers to contractors. These form four studies, which include the following commodities:

- mechanical services
- formwork
- concrete
- masonry.

Fire products and tiles were also investigated and are discussed in Chapter 9 only to indicate the supply channels. After the clients and/or contractors were interviewed, then specific subcontractors were interviewed and various chains were followed in detail related to a commodity product and they included the following:

Specialist subcontractors supplying complex commodities of products and services design, supply and install

- Mechanical services for project 1, 2 and 3
- Formwork for projects 1 and 3.

Subcontractors supplying simple or moderately complex commodities of product and services supply and install

- Concrete for project 1
- Fire doors and products for projects 1–5
- Bricks for project 6.

After the subcontractor interviews, subsequent interviews with fabricators, processors, merchants and manufacturers eventually led to a tracing of more general industry chains for the supply of the following products with the associated projects:

- Concrete for projects 1 and 3.

1.3.4   Conclusions and future directions

Chapter 10, Conclusions and future directions: supply chain specialization and integration blueprint concludes the text by drawing together some of the key observations from the case studies. The intention of this text was to explore what firms actually do in practice in relation to procurement across a range of supply chains within an economic context. Procurement plays such a large part in what we do in all industries – not just construction. Perhaps those involved in constructed systems are more fixated on procurement as we are in a time of a high degree of specialization – we don't tend to take on more productive functions along the chain unless we are absolutely certain that not only can we create value but, more importantly, that we can be profitable. For this reason – that is, the focus on firm profitability – any notions of the value of 'supply chain management' needs to be intrinsically embedded within the context of the characteristics of the market, the firms involved in that market, the type of commodities which are to be exchanged and thus the firm–firm relationships that evolve. Therefore, the approach has been to explore the economic aspects to the supply chain – thus, supply chain economics rather than supply chain management. A quite wide perspective of a number of firms has been undertaken and served as the supply chain case studies within which to test these ideas in relation to supply chain economics.

This final chapter is organized as follows:

- Aggregated project supply chain organization: supply channels
- Firm–firm procurement relationships
- Chain specialization and integration (CSI) blueprint
- Future research intersections: interdisciplinarity.

The chapter also revisits the model presented earlier and suggests refinements to classifying suppliers, procurement relationships and supply chains and the respective associations between these classes through a comparative analysis of the similarities and differences between the case studies. A more normative approach to supply chain management is presented in the blueprint. The chapter concludes by exploring future interdisciplinary challenges for procurement modelling and the field of supply chain economics, in particular the interdisciplinary context and discourse of this study, the challenges, limitations and also the potential for the development of the fields of supply chain economics and management.

1.4   A final word

Chapter summary

1   The underlying principles in relation to construction supply chain economics and procurement are often equally applicable to many constructed systems with life cycle concepts of design, construction and maintenance and the sectors associated with them; for example, civil, mining, industrial design, shipping, aerospace, information technology and asset management. The theoretical principles can be extrapolated and applied to many sectors and indeed, because the supply chains extend into numerous sectors, the firms encountered in the case studies do supply to more than one sector. Importantly also because 'project thinking' and 'procurement' permeates many of our business systems.

2   Supply chain economics is the manner in which the economic market structural characteristics and the firm and sector behavioural characteristics interact to produce attributes which describe types of supplier firms, procurement relationships, supply chains and supply chain performance.

3   Simplistically appropriating a model of behaviour is fraught with difficulties without understanding the political, cultural, technological and economic context that gives rise to those relationships.

4   Understanding the industry at the 'supply chain level' has been on the national agenda for many countries. The degree of government involvement and policy planning can differ from locality to locality;

however, it is one of the underlying elements that assists in shaping the construction industry of a region. Policies to some degree can affect both the dominant business practices and the nature of competition within the industry. Although attempts have been made to use the industrial clusters of firms and firm networks concept to inform construction industry policy, there has been little formal attempt to examine the potential of a supply chain procurement model as an economic tool or instrument.

5   There appears very little research which explicitly explores the numerous underlying economic organizational structures of the construction sector which accepts and perhaps celebrates that firms are procured and products and services are supplied within markets on a project-by-project basis. Firms appear to respond and organize themselves according to this unique economic environment. Perhaps the performance of the sector is being judged through an inappropriate lens – one that has been developed through a tradition of stability and long-term production rather than a deeper understanding of the industrial context of a project-oriented sector.

6   The story of the supply chain really begins with clients; as originators of the design and construction process, they are an integral part of all projects and potentially can have the greatest impact on the firms supplying services or goods through both their demand characteristics and then their procurement decisions. An understanding of supply chain economics to improve the performance of the supply chain can begin from individual firm–firm procurement relationships; however, many of the ways to improve performance are often out of the range of individual firms.

7   The industry is typically presented as a large mass of firms within the one market. Comments about the industry often are reduced to a description that it has a fragmented structure and is made up of a large number of small- to medium-sized enterprises. Descriptions of the industrial organization economic environment do not extend to discussions of any of the following: specific chain organization; distinct chains of commodities and their markets; firm behaviour within markets in relation to vertical governance structures; firm behaviour towards procurement of suppliers and supplier responses to customers.

8   This book is the result of a research study and the research response is in terms of the research question and the strategy to undertake the investigation; however, the text tends to focus on the results of the study. What are the structural and behavioural characteristics of the key objects associated with procurement in construction supply chains? The subsequent chapters explain in detail the argument behind this research question.

The research strategy of inquiry relies upon two main elements: the development of a theoretical model and methodological framework and the representation and validation of the model through real world examples. In simple terms, this means that the supply chain procurement model is described and represented through a structural model view and a behavioural model view. The structural model view encompasses the static, or structural, aspects of a problem and solution which largely relies upon a static description of the object and class model and associations. The behavioural model view encompasses the dynamic, or behavioural, aspects of a problem and solution and relies upon descriptions of interactions or collaborations among objects and class elements. The development of the model is an iterative process and also largely interdependent. The structural model is incomplete without methods and operations being described which require input from the behavioural model. Although this may seem confusing – it is explained in more detail in Chapter 5. The objects and classes which are the structural elements of the model include the description of characteristics and the relationships between the following: project, firms; both supplier and client markets; both industry and project, commodities and firm–firm procurement relationships.

2 The rationale for the modelling of procurement in the construction supply chain

I wonder what governments think about the supply chain concept?
A construction industry policy analysis

2.0 Orientation

Box 2.1 Chapter orientation

WHY: Chapter 2 develops the argument for the rationale for modelling of procurement in the supply chain through an analysis of selected government-sponsored investigations, policy and interventions which were aimed at sector productivity performance improvement.

HOW: The textual analysis draws from a range of national documents related to project and non-project procurement from five countries to highlight content and illustrate two key themes of fragmentation versus specialization and normative management versus positive economic models. The project procurement-related documents focus on major capital works expenditure programmes typically found in the construction industry and include civil and building projects. We all know that national policy is not the only place where procurement is affected and therefore an analysis on the next tier of government regional or state-based documents was also undertaken for one country as a specific case study. The investigation historically maps key trends in policy thinking.

WHERE: The documents which were examined were from South Africa, United States, United Kingdom, Singapore and Australia. The specific case study was Australia and the states were New South Wales and Victoria.

WHAT: The results indicate a reinforcement of the common simplistic and unhelpful description that the construction sector is a fragmented sector. This is an unproductive approach and fraught with problems at the policy level. More importantly though, the results also illustrate a trend towards the need for positive economic models which underpins the problem with being fixated upon the fragmentation concept rather than a specialization approach. The international analysis and the case study also demonstrate that there is a gap in useful models available for policymakers and therefore demonstrating the need for the modelling of procurement behavioural practices in the construction supply chain at each tier within a particular sector contextualized against the market competitive environment and the structural conditions.

WHEN: It is acknowledged that this only provides a 'snapshot' of the situation. The study is limited through its selection of countries and its selection of studies from 1984 through to 2003. There are, of course, other limitations – a discussion on some of the political context could have been useful. It is not intended though to be a political discourse.

WHO: This chapter is of particular interest to those within public sector organizations and those who deal with public sector organizations. However, policies and thus policy analysis does not need to be limited to government organizations – all large organizations have documents which attempt to explicitly describe their strategic intent in relation to objectives and strategies to achieve those objectives. Most large organizations are on a constant state of change in response to 'doing things better or smarter or more efficient'. Therefore, this chapter is of relevance to anyone seeking to critique and make sense of the underlying perceptions and ensuing strategies made by decision makers in large organizations about what they are expressing as the root cause of the so-called productivity, efficiency and effectiveness ills of their organization and the strategies to redress their problems.

2.1 Introduction

For nearly two decades, from the mid-1980s through to the early part of this century, governments worldwide have made attempts at construction industry policy development related to the concept of supply chains. Understanding industries in terms of chains, clusters and networks had become increasingly important in economies around the world (AEGIS, 1999). It is even now more important than ever before, given the free-flowing trade movement practised by firms through the effects of globalization, the growth of various international markets – particularly construction – increased development of reciprocal agreements between professional Boards and associations and the creation of various international trade agreements. Firms in chains, clusters and networks are linked by contractual relationships. Therefore, underpinning this movement is the assumption that firm-to-firm procurement relationships across chains, clusters and networks are central to economic performance in sectors. All industries are affected by government policies in some manner. All governments have policies for sectors – whether explicit or implicit and whether or not documented and transparent. The investigation of the documents related to the capital works programmes, and typically civil and building construction, can be equally applicable to other sectors. It is critical to analyse government trends and describe an environment which so many work in and yet how few construction supply chain economic tools there really are to support the economic decisions that impact upon so many. This chapter is only really the beginning of analysis on government policy in relation to construction supply chain economics and serves only to describe overall trends in the last decade and to initiate discussions.

The aim of this chapter is to establish the need for a positive economic procurement model for procurement in the construction supply chain through a closer examination of the text of selected government policy documents. The chapter is organized as follows: first, a definition for procurement which is then followed by a brief description of the method used for the textual document analysis. A discussion is then presented on approaches used by selected national governments towards improving the performance of their construction industry. Within this context, the discussion then proceeds by focusing on a particular country – the efforts towards construction policy development within Australia at a national level and then a regional (state) level.

One of the main themes which underpins construction industry policy implicitly is concerned with the question of performance and its relationship to industry structure. The construction industry performance *problem* tends to be interpreted as either one of an industry *fragmentation problem* or a *firm specialization balance*. Those who consider it a fragmentation problem espouse, or are the catalyst for, models of supply chain integration at the project level in the quest to solve productivity and performance problems and seem to ignore the practical realities of economic markets. Those who consider it a firm specialization balance issue tend to accept the industry as it is: a sea of firms of buyers or sellers in a variety of competitive markets exchanging commodities along a supply chain. These policymakers espouse or are suggestive of positive models that accept and describe both scenarios of co-operation and competition at the firm level. This chapter presents the selected investigations and policies reviewed into these two main themes and highlights that the trend in recent years is towards seeking positive models.

The discussion concludes with the rationale for the development of more refined principles and further investigation of a series of economic models of performance that specifically relate to the construction sector. The discussion suggests that the current indicators of performance and the ensuing policies that rely upon these indicators lack an understanding of the reality of the construction sector. In particular, current economic models and indicators of performance do not address supply chain modelling, and yet this is clearly the direction of government policy worldwide.

The chapter concludes with a discussion on the significance of understanding the industry at the 'supply chain level' through procurement, which has clearly been on the agenda for many countries. The supply chain procurement model makes transparent the structural organization of the industry and unveils the interdependent nature of structure and behaviour. The trend towards seeking positive models for procurement is implicit. Currently, there are no explicit models that enable descriptions of procurement in the construction supply chain supported by an industrial organization economic approach.

The chapter is organized as follows:

- Background to policy analysis
- Charting the government quest for improved industry performance: international analysis – national studies and frameworks
- National case study: Australian initiatives
- Government economic models of performance.

2.2   Background to policy analysis

What do we mean by procurement? A wider view of the term is used in this text than is traditionally taken in the property and construction industry. Traditionally, project procurement in the industry has tended to focus upon the procurement method that is designed for the exchange between the client and the key firms for design and construction contracts. Largely, this involves first-tier firms and generally includes architects, project managers and main contractors and perhaps major specialist subcontractors and suppliers. Project procurement strategies are then often categorized as either design and construct, traditional, build-own-operate, construction management, project management, etc. The classifications for project procurement identified and in general use around the world are useful; however, they have little meaning as we move away from the client–supplier relationship and downstream in the supply chain.

Rowlinson and McDermott (1999) attempted to define procurement and began with a general definition from the *Oxford English Dictionary*, 'the act of obtaining by care or effort, acquiring or bringing about'. Other definitions were suggested and his discussion ended in the citing of the CIB W92 working definition of procurement: 'the framework within which construction is brought about, acquired or obtained'.

For the purposes of this chapter this definition is limiting as it is aimed at project procurement strategies. Procurement is the activities required to execute the exchange of a product and/or service between any firms at any point in a project and between any firms. In this chapter, a wider view of the firm–firm procurement relationships between various firms involved in the chain is taken. This is a slightly different interpretation to project procurement or construction procurement than has generally been accepted by the construction community. This more fundamental interpretation is not unlike Huston's (1996) definition: 'the procurement process includes all of the activities that are required to obtain the goods and services required for a project.' However, procurement in Huston's interpretation was viewed only from the main contractor's perspective. It is noted that Chapter 3, as it explores supply chain theory, identifies some further definitions of supply chain management and considers fields such as logistics as well as strategic procurement.

2.2.1 Method for document critique

The method taken in this particular chapter to explore the various policy documents is document analysis and it is not that common in the disciplines of construction management and economics. Document analysis is primarily a data collection and analysis strategy common to sociology.

The government policy documents are considered because governments can play a significant role in impacting the performance of an industry. The texts are examined in relation to the assumed underlying perceptions and strategies by policy makers about what they are expressing as the root cause of the so-called ills of the industry and the strategies to redress this *problematic* industry.

Analysis of the actual development of public policy is a large undertaking in itself; however, this has not been reported on in this chapter. This would form another complete study in itself and is certainly worthy of attention.

The documents were selected based upon the following criteria:

- Public access
- Policy, government investigation or research studies commissioned by government (or direct involvement)
- Representative of construction industry policy at the national level
- Selection of geographical regions – AsiaPacific, Americas, Africa, Europe
- Evidence of change.

A key part to this analysis is that the author is also an instrument – which is both a limitation due to bias and then an attribute which assists credibility and validity. To explain further, I have been employed in an organization which developed construction industry policy – although at a state level. I was, however, also part of various committees/working parties of the peak body which included representation of the various governments (state and national level) which was responsible for the co-ordination of policy. The body is known as the Australian Procurement Construction Council (APCC). I have 'lived' through parts of the development of some of the documents analysed in this paper. However, a limitation is that I was not part of the development of the documents in other countries. For this reason, the first part of the analysis of international documents is to provide a context and the second part focuses on the Australian scene. The method for analysis involved reading the documents and identifying and coding through the text examples of the themes as in Table 2.1.

The documents were then mapped in a diagram on a spectrum of normative fragmentation versus positive specialization models. The next step was to trace through the relationships between various documents and establish connections and identify shifts in thinking. A combination of a content and discourse analysis was conducted to then determine how to represent the degrees of how 'normative' or 'positive' the approach was – this relied upon such things as the number of times the elements in Table 2.1 were reinforced in the document and then the general cohesiveness of the argument and the consistency of the language in the document – that is, sometimes strategies are an array of strategies/tasks without little evidence of an underpinning coherent economic policy or theory. If the approach was direct it made its way well into the Fragmentation or Specialization circle and the more direct then the further to the right or left as the case may be. There are times when there is evidence of both approaches. This is, of course, an interpretation and it has its limitations – it is perhaps worthwhile to consider the positive versus normative categorization as a spectrum.

2.3 Charting the government quest for improved industry performance: international analysis – national studies and frameworks

Various governments have instituted national investigations and examinations on the perceived performance of this sector. The underlying assumption of the state of the sector is that it is poorly performing. In the quest for improved industry performance, the notion of industry fragmentation has often been discussed as a cause of this poor performance. This notion of

Table 2.1 Themes to categorize approaches

Themes	Normative fragmentation model	Positive specialization model
Industry descriptors	Small- to medium-sized enterprises, fragmented and adversarial as a negative attribute	Small- to medium-sized enterprises, fragmented and adversarial as an attribute; not necessarily negative
Strategies	Strategy to change Focused integration at a project level	Strategy to describe and accept Diffuse and varied strategies to enable and support business environments for co-operation or competition
Perspective	Project	Firm

fragmentation has often been discussed in terms of a characteristic of the structure of the industry; however, as a characteristic indicative of a poorly performing industry with low productivity.

Governments have commissioned studies into the construction industry for decades in the quest to identify problems and initiate new policies. For example, since 1944 successive UK governments have commissioned seven studies into the industry, which is almost one every 7 years. According to Rogan (1999), calls for improvement are not new to the UK construction industry and he cited and compared findings from the following: Simon (1944), Emmerson (1962), Banwell (1964), Wood (1975), NEDO report (1988), Latham (1994) and the most recent Egan report (1998). More recently, Murray and Langford (2002) have published an extensive critique of the UK government reports from 1944 to 1998 which clearly identified that the interest in reforming the construction industry performance is not entirely new.

Although these were all government commissioned studies produced at different times, it is surprising that there are a number of common themes. Likewise, in Australia, South Africa, Japan, Singapore, Sweden, Denmark and the United States, governments have attempted to analyse their own construction industries at various times in the quest to improve performance. Selected investigations across these countries that demonstrate general approaches worldwide are now considered in the light of the importance of modelling of procurement in the supply chain for analysis of industry performance.

In summary, the issues central to the findings of the investigations, or the ensuing policies, focus on firm specialization, fragmented industry structure, short- versus long-term relationships and the role of strategic procurement in the supply chain. The Figure 2.1 summarizes and charts the selected investigations and policies from 1984 till 2003 against the two dimensions of industry interpretation: *fragmentation* and *specialization* for the United Kingdom, United States, South Africa, Australia and Singapore. The constructs underpinning fragmentation is the view that project integration needs to be achieved to solve the problems and it is normative in its approach. The constructs underpinning specialization is that co-operation and competition needs to be both considered and it is positive in its approach. Note in this context the term 'positive' refers to an approach which primarily focusses on describing accurately characteristics of a situation versus 'normative' which refers to an approach which focusses primarily on the need to change a situation assuming situation characteristics without deep investigation. Table 2.2 summarizes the documents that were explicitly analysed and are reported on in this chapter.

The first interpretation of the industry is that it has a fragmented structure with many small- to medium-sized firms and that ultimately this fragmentation is the industry's cause of poor performance (FAR, 1994; Atkins, 1993; CCF, 1994; Latham, 1994; DITR, 1997b; CIDB, 1999; Tan, 1999; CRC, 2001; CRC, 2003); refer Figure 2.1. The response was then to impose normative

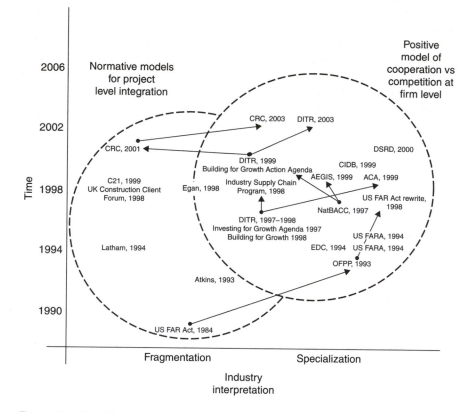

Figure 2.1 Specialization versus fragmentation: selected international government construction industry analyses mid-1980s–2003.

models for project supply chain integration with the assumption that this will ensure industry development (ISCP, 1998; AEGIS, 1999; Tan, 1999; CRC, 2001; CRC, 2003).

The second interpretation of the industry that is evident in various documents is that the firms in the industry have become more and more specialized and that growth in the SME sector is a healthy indicator of a strong economy (Atkins, 1993; OFPP, 1993; US FARA, 1994; DITR, 1997; Egan, 1998; ACA, 1999; AEGIS, 1999; CIDB, 1999; DITR, 1999). The response is then to impose positive models for industry development that support various scenarios of firm co-operation and competition down the supply chain (OFPP, 1993; NatBACC, 1997; CIDB, 1999; DSRD, 2000; DITR, 2003).

Figure 2.1 also indicates various connections between investigations and policies. In some cases the same investigation results in diverse interpretations of the industry. In some cases an early strategic direction is modified and moves towards supporting firm specialization in the industry. The following

Table 2.2 Summary of documents analysed

Country	Document	Year	Acronym
US	Federal Acquisition Regulation Act	1984	FARA
UK	Atkins Report	1993	Atkins
Australia	Construction Industry Development Association	1993	CIDA
UK	Latham Report	1994	Latham
US	Office of Federal Procurement Policy	1993	OFPP
UK	Construction Client Forum	1998	CCF
Australia	Department Industry Trade & Resources Action Agendas	1997	DITR
Australia	National Building and Construction Council	1997	NatBACC
Australia	Industry Supply Chain Programme	1998	ISCP
Australia	Australian Construction Industry Forum	1998	ACIF
Australian	Australian Expert Group Industry Studies	1999	AEGIS
South Africa	Construction Industry Development Board	1999	CIDB
Australia	Australian Consulting Association	1999	ACA
Singapore	Strategic Economic Plan for Singapore, Construction 21	1999	C21
Australia	Department of Industry, Science & Resources Action Agendas	2000	DISR
Australia	Cooperative Research Centre for Construction Innovation	2001, 2003	CRC

Section 4, National Studies and Frameworks, discusses some of the key points from these national reports and policy documents in detail in terms of the fragmentation versus specialization perspectives. This selection represents examples of both types of industry interpretations. The Australian federal government initiatives are discussed in detail in Section 2.5 but are noted on Figure 2.1 as bolded text (London, 2005). As noted previously, the next level of state or regional policies was also undertaken for the Australian case study and is also discussed in Section 2.5. In recent years in Australia there has been a general trend towards more positive policy frameworks and models; however, these are still unsupported by methods to describe the industry state prior to policy intervention and then benchmark performance post intervention.

2.3.1 *Industry fragmentation and normative models of integration*

The theme of fragmentation is interesting to consider from the series of UK studies. In Rogan's (1999) comparison across each study conducted since 1944, one of the most interesting features is the suggestion of a theme of *fragmentation*. This problem of fragmentation is not really new. As early as 1944, Simon highlighted that 'change in the industry now means that 30–40 subcontractors are now needed for the site works' and that 'the industry was made up from large national contractors to small jobbing builders.' Again, in 1975, Wood indicated that the construction industry was of a fragmented nature. He drew from a survey of 50 building and civil engineering firms in that study and a survey of 2000 firms from an earlier research report.

Atkins (1993) also concluded that the construction industry was very fragmented and this was caused by a diversity of technology, customers, projects and market sectors. He claimed that fragmentation of markets has prevented the development of industry standards and fragmentation of the professional level inhibits the interchange of information.

However, he also noted that, as projects become more complex, a deeper specialization is needed by individual professionals along with a greater need to work in multidisciplinary teams. Atkins (1993) implicitly identified what Egan (1998) explicitly concluded some 5 years later; namely, that the high degree of fragmentation caused by specialization is indeed a positive attribute of the industry. Specialization, however, requires appropriate management.

In the Atkins report, it was suggested that the diffusion of technology in construction is handicapped by small fragmented markets, uncertain demand, complex legislation and that, although specialization is needed, it should also be managed. Therefore, it was suggested collaboration between separate sides of the industry was needed and closer links should be established. Atkins (1993) then cited the automotive sector as an exemplary model. This was the first report to suggest that the automotive industry model of the large assembler with an integrated supplier network was an appropriate model for the construction industry to emulate – whether or not this is the case is debatable. Implicit in this recommendation is that assumption that the large automotive assembler is analogous to the contractor and that the first and second tier component suppliers in the automotive sector are analogous to the small subcontracting firms on construction projects.

This approach to controlling and stabilizing the industry's fragmentation problems is also reflected by a general trend by the academic research community towards considering the *construction process* as analagous to the *manufacturing process* and that equally *manufacturing production*-oriented theories were useful to *construction production* (Koskela, 1992; Aouad, 1999). However, it probably wasn't until after the release of the Egan (1999) report in the United Kingdom that this approach was used as a means for performance improvement in the construction sector, when the lean construction and supply chain management movement gained a great deal of momentum.

The oft-cited Egan Report (Egan, 1998) also stated that fragmentation inhibits performance. The UK industry at that time had 163,000 companies, of which most were employing less than eight people. However, it was the first time that it was suggested explicitly that fragmentation be regarded as both a strength and a weakness. One of the positive attributes of fragmentation is that it provides flexibility to deal with varying workloads. The negative attribute is the difficulty of continuity of teams from project to project and the inherent inefficiencies of this type of system.

The recommendation was for integration of the process and team 'around the product' (Egan, 1998). There was a significant emphasis by Egan (1998) on understanding the process of construction and that the integration of the process was important to improving the performance of the industry. Project process was viewed as separate sequential operations undertaken by individual designers, constructors and suppliers who had no stake in the long-term success of the product. Changing this culture was fundamental to increasing efficiency. The rationale behind the *integrated process model* was that efficiency was constrained by the separation of the process of design and construction and that these processes reflect the fragmented structure.

It is noted that the worldwide movement in the concept of buildability and constructability some 10 years earlier had already identified this and sought to address this issue of fragmentation of process (CIDA, 1995; NSW PWD, 1993; McGeorge and Palmer, 2002) and to a large degree had made little impact (McGeorge and Palmer, 2002).

In 1998, the UK Construction Client Forum reported that there were too many contractual interfaces in the projects and that poor communication down the supply chain often led to costs spiralling out of control. There was a call for finding ways of solving interface problems in supply-side and encouraging firms to innovate and share the resulting benefits. There was also the recommendation that the industry must develop supply chains that focus on client values with continuous improvement.

Both the Construction Client Forum (1998) and the Egan Report (1998) discussed improved management of the supply chain as a way of addressing process and industry fragmentation. Adversarial relationships, fragmentation and lack of integration were destroying the construction industry. The assumption was that future improvements in productivity will arise from restructuring the supply chain from raw materials to site erection. It was also suggested that early involvement with specialist contractors was necessary for this to occur and that improved relationships would follow. At this stage, *lean production* principles from the automotive and steel industry were explicitly recognized as an important concept to adopt. Much of this may seem naive with hindsight.

In the late 1990s, improved subcontractor relationships were being proposed and yet we can trace back a further 30 years where the same findings were being reported and suggested. Banwell (1964) suggested that new relationships are essential if the kind of advice needed is to be made readily available and that specialist trade contractors should be brought in early to enable the development of close relationships with other partners.

The UK government is not dissimilar to many other governments in their concern for the perceived poor performance of their construction industry, the understanding of the national importance of the industry and the attempt to investigate, plan and improve performance. Likewise, other national governments have been concerned by the apparent lack of performance caused by a fragmented and adversarial construction industry and have developed policies or strategies or commissioned investigations in attempts to address this.

2.3.2   *Procurement strategies and industry fragmentation*

Client project procurement has been seen as a way to address the poor performance of the industry. In particular, procurement strategies that integrate the major first tier suppliers to clients have been advocated.

The Australian government used a project alliance contract for a major capital works building project, the National Museum in the national capital. This is discussed in more detail in Section 2.5 of this chapter. At a similar time, in 1999, the Strategic Economic Plan for Singapore, Construction 21 (C21), was released. This is a strategic blueprint spelling out the vision and strategies to re-invent Singapore's construction industry. The C21 is underpinned by the concept of integration as a means for improving productivity (Tan, 1999).

The blueprint aims to change that country's construction industry from one that is plagued by negative productivity growth and a heavy reliance on unskilled foreign workers, to one that is professional, productive and progressive, and whose workforce is able to exploit knowledge for competitive advantage. The vision was to transform Singapore's construction industry into a world-class player in a knowledge age. The Construction 21 Blueprint is the construction industry's response to Singapore's economic vision of becoming a globally competitive knowledge economy.

Hence, in early 1998, the Minister for Manpower initiated the Construction Manpower 21 Study. The study sought to address two problems:

- the low level of productivity and negative productivity growth in the construction industry
- an over-reliance on unskilled foreign workers.

The Construction Manpower 21 Study was later merged with the Ministry of National Development's Committee on 'Practices in the Construction Industry', and expanded to become the Construction 21 (C21) Study. This joint plan aims to address the current inefficiencies in the industry and transform it into a knowledge industry. Given that upstream decisions have an impact on downstream construction processes, the C21 Study determined that it needed to address issues across the construction value chain, from design to construction and to maintenance (Tan, 1999).

Similar to the UK studies, one of the main concerns with the industry was that of low productivity. According to the Construction 21 Report

'One of the main causes of low productivity in the industry is the lack of integration of activities across the construction value chain where design is segregated from construction or other downstream processes. Closer integration among the industry players in carrying out a project would facilitate the adoption of good practices as many of such practices (buildability, safety and maintainability) have to be considered or specified at the design stage. This will bring about higher efficiency and productivity' (Tan, 1999).

To achieve their vision, six strategic thrusts were developed and Strategic Thrust 4 is aimed at procurement to achieve productivity.

Box 2.2 Strategic Thrust 4: Integrated Approach, Singapore

Strategic Thrust 4: Integrated Approach advocated the adoption of progressive procurement methods that can integrate the activities of industry players to achieve synergy and attain productivity breakthrough (Tan, 1999).

In Strategic Thrust 4, the government urged the promotion of Design and Build. The approach taken in the Singapore investigation was to recommend an active promotion of Design and Build (D&B) methods to foster closer integration.

To solve the productivity problems of the Singapore construction industry, the assumption is that it is related to procurement methods and the integration of the chain of industry players involved on a project by project basis. The solution was a policy directed towards a design and build procurement method. This is a contractual arrangement between the contractor and the client. This may or may not impact upon any other procurement relationships down the chain; therefore, it is suspected it may have little effect on the subsequent levels in the chain. Likewise, it is suspected that the project alliance contract used by the Australian government is unlikely to have any significant long-term effect on players downstream in the supply chain; albeit with the best of intentions.

2.3.3 *Specialization and positive models of competition/ co-operation*

The assumption underpinning normative models of project integration has been that fragmented industry structure leads to fragmented project activities which ultimately are the cause of poor productivity. The resultant models and, thus, policies, are concerned with developing blueprints for the whole industry based upon the large firm client and large firm contractor and the individual project relationship. Specifically, for example, this means that concepts and strategies such as single-source solutions through project alliances and design and build; integrated process maps or lean supply solutions based upon the large firm automotive assembler model will provide the answers to the poor performing industry. These solutions are suggested by governments who assume that through the first tier the numerous contractual interfaces between firms in the supply chain will be managed better and productivity of the industry will improve. The naivety of this assumption is astounding. Whether we like it or not, unless there is a raft of explicit incentives, rewards and/or punitive measures developed within the contractual relationship between the client and the contractor, it is suspected that short-term project integration, let alone any long-term industry integration, will not be achieved.

An alternative view of the fragmented structure of the industry is offered by those who see the industry past this first tier of suppliers and the reality of the wide variety of firm to firm procurement relationships that underpin the industry. This results in positive models which describe the industry and accept a wide variety of co-operative and competitive behaviour between firms in their procurement relationships. Allied to this behaviour is an industry structure of firm specialization coupled with various firms' sizes and market characteristics. Typically, the ensuing policies use the procurement concept as an instrument of government intervention that reaches beyond the first tier of the supply chain. In these situations there is also an attempt to monitor the impact of the intervention. These situations are rare indeed!

The United States's Federal Procurement Reform from 1993 to 1998, the South African Construction Industry Development Board (CIDB) White Paper of 1999 (CIDB, 1999) and the directions of the Australian Department of Industry, Training and Resources (DITR, 1997) are examples of this approach. It is noted that this chapter only attempts to describe the approach and does not analyse the success of these approaches – although in Section 2.5 the lack of success in the Australian situation is described. The US and South African examples are discussed in this section and the directions of the Australian government are discussed in detail in Section 2.5.

2.3.4   *Procurement as a government instrument*

In the last decade, significant changes have taken place in the US federal government's procurement system. The impetus for such changes has been the realization that, rather than a secondary administrative function, procurement or contract management is actually a core business function of the majority of agencies. For example, Department of Defense, Department of Energy and the National Aeronautics and Space Administration in the United States spend 46, 94 and 78% respectively of their annual budgets on contracted products and/or services. In the twenty-first century, contract management is not regarded as a subsidiary function and procurement is regarded as a major government instrument used by agencies to implement a wide range of policies. This explicit approach to strategic procurement is not dissimilar to the South African CIDB.

US procurement reform began in 1993 within the Office of Federal Procurement Policy (OFPP) and resulted in changes to the US Federal Acquisition Streamlining Act (1994) and the Federal Acquisition Reform Act (1995). Both made fundamental changes to the rules-based, highly codified 1,900 page 1984 US Federal Acquisition Regulation. The major thrust of the changes was related to business strategy and source selection.

The argument for the philosophy of rules-based procurement policy till 1993 was that 'rules for procurement retained knowledge and allowed reuse of solutions to problems' (Kelman, 2003). Ultimately, the question underpinning the rules-based approach was why did procurement managers have to rediscover the theory of economics and competition when they were procuring products and/or services? For a more detailed discussion on the US procurement reform and the transformation from the traditional rule-based objectified system to a discretionary empowerment system, refer to Kelman's 'Remaking Federal Procurement' – working paper No 3 from the John F Kennedy Harvard School of Government. Kelman was the Director of the OFPP from 1993 to 1998 and was responsible for creating and implementing procurement reform during this time.

An example of a rule that impeded strategic procurement is that of *full and open competition*, which translates to contract managers being fearful of giving advantages to suppliers through, for example, their performance in supply chain management as evidenced by past performance. Ultimately, it was so difficult to *reward* good past performance by suppliers and equally it was difficult to *punish* poor performers as questions of probity and fairness would be raised.

In the US situation this led to a system of safe and mediocre procurement practices. Contractor or consultant performance assessment reports are common; however, they can be problematic in reality. Government employees are ill equipped or ill supported in providing objective assessment of past performance. Quantitative measures are one mechanism to support an assessment. More qualitative assessments are difficult to document and leave employees fearful of competition policies, limitation of trade practices and equity and probity issues. A cultural change and an education within agencies were required. My personal experience supports this statement in the Australian context; unless a government agency who is responsible for awarding and managing contracts has a well developed, equitable, transparent and up-to-date information management system, then it is extremely difficult to 'prove' any assertions about past performance. Internal to the agency there has to also be a common language amongst employees and clear categorizations and definitions of performance so that one employee can rely upon another's assessment. There also needs to be an element of trust.

To enact change in the United States, two major strategies included the concepts of *best value* and *government-industry co-operation*. The important point is that procurement was taken on a case-by-case basis rather than one-rule-fits-all scenarios. Not all rules were discarded; however, it was recognized that following inflexible rules was not enough to achieve best value for government procurement. Procurement reform and the changed agency culture and environment allowed such past problems (among others) as *buy the lowest bid* and *don't consider past supplier performance* to be addressed. In many situations the rules were not changed but the informal rules or business practices changed as there was a change in attitude supported by the strategic direction of the agency leaders.

Although this seems somewhat familiar to the previous discussion on models based on integration, it differs in that project integration may or may not be the solution that provides best value – it goes much further; the changes are systemic. A project-strategic alliance or any other form of project procurement contract may only provide best value to those involved in the contract; it does little for the numerous other contracts related to the construction supply chain. This procurement policy has limited impact upon the industry beyond the individual project and beyond those firms involved in the project in contracts closest to the client.

In all cases of procurement reform in the United States, the driving motivation was that, in the past, rules had inhibited responding to procurement situations in a manner that would ultimately provide best value to government. Reform was underpinned by the idea that an endorsement of the wide variety of collaboration and/or competition scenarios as offered by the marketplace would produce better results.

A much less radical approach was outlined in the White Paper of 1999 by the South African government. The strategy as detailed by the Construction Industry Development Board (CIDB) also reflects an underlying interpretation of policy development responding to the nature of the industry rather than imposing normative rules. The most significant aspect of the South African approach is the explicit intervention at the subcontractor level through policy instruments. This again reflects a much more sophisticated attempt at addressing the co-operative and competitive nature of the industry at successive tiers in the supply chain. It is also underpinned by an understanding of the range of firm-to-firm procurement relationships available. The approach to policy development has largely been through the development of government discussion papers and ensuring industry consultation rather than any national commission or investigation.

In 1994, South Africa embarked upon a range of initiatives aimed at developing a 'comprehensive construction industry development policy as part of its contribution to the national project of reconstruction, growth and development' (CIDB, 1999, p. 5). From 1994 till 1999 a range of industry consultations, interdepartmental government initiatives and discussion documents have been developed which led to the formation of the Construction Industry Development Board, which includes private and public sector members on a range of specialist Focus Groups. Then, in 1999, a White Paper consolidated the CIDB's consultation process and defined the vision and strategy. The CIDB is now mandated through an Act of Parliament (SA DWP, 1999). The driving force for the policy development was that the construction industry was considered a national asset and as such required a deliberate and managed process to optimize its contribution to the economy.

Similar to the Singapore C21 Plan, the development of the construction industry policy was aided by the establishment of 10 Focus Groups, each with their individual themes. Public sector procurement reform was high on the agenda. Three important points arise from an analysis of the Focus Groups' aims, objectives and strategies, and they include:

- government's approach to procurement as a policy instrument
- government's analysis of the industry and
- targetted procurement strategies.

First and foremost, the approach to procurement and how it is considered an instrument of government intervention for long-term productivity performance echoes an approach similar to the US's policy for Federal procurement reform. The interpretation of how the industry operates and how policy should be developed is an explicit recognition of the co-operative and competitive nature of the industry. This interpretation is reflected in the aims and strategies of their procurement policy in particular.

Second, the South African government attempted to analyse the structural conditions of the sector and, like so many other governments, realized that the construction sector contributes significantly to the Gross Domestic Product (GDP). For example, the sector contributes 35% of the GDP, and of that 35% public sector procurement expenditure accounts for just over 22%. Realizing this, the government recognized that they had a significant potential leverage in their role as a client and so developed the Affirmative Procurement Policy (APP). This policy uses procurement as an instrument to achieve socio-economic objectives during a time of significant change for the country.

Box 2.3 South African Government Policy White Paper

'In defining Government policy for the construction industry, the White Paper proceeds from an understanding of the industry and the environment in which it operates. Analysis of this environment captures the specific trends and structure of the industry, as well as the current opportunities and constraints which enable and impede its development. Informed by extensive consultation and by practical experience, the White Paper locates the requirements of industry development within the context of Government's mandate and the regional and global context' (SA DPW, 1999).

It is interesting to note that in all the various papers, documents and reports there is then no explicit description of how procurement takes place in the industry. However, regardless of this, many attributes of the APP suggest that there was an implicit understanding. It is suspected that the wide industry consultation process in the development of the policy assisted in some understanding of the structural and behavioural characteristics of the industry, although given its strategic significance it would have been worthwhile in a more considered and explicit investigation and description.

The South African policy borrowed from the Singapore and the Australian state of New South Wales models in developing contractor registration systems based upon prequalification and tender preference policy and design and build procurement methods. However, the South African approach went somewhat further than both of these policies and developed the Affirmative Procurement Policy, which extended to the subcontract agreements which included subconsultants and subcontractor levels in the supply chain.

The APP enabled *targetted* procurement and this was then monitored and directly related to a change in the underlying structural characteristics of the industry.

Box 2.4 South African targetted procurement

Targetted procurement is a system of procurement which provides business and employment opportunities for marginalized individuals and communities, enables procurement to be used as an instrument of social policy in a fair, equitable, competitive, transparent and cost-effective manner and permits social objectives to be quantified, measured, verified and audited (CIDB, 1999, p. 9).

As a result of targetted procurement, which related projects to economic and social clusters and regions, the market share of SMEs increased from 0.5% in 1993 to 32.5% in 1998. This begins to measure the changes in the structural characteristics of the industry; however, it does not address the behavioural characteristics.

However, the failing of the policy is the lack of a cohesive plan for outcome-based indicators that would monitor and evaluate the procurement reform as an instrument of policy. For example, recommendations were made for modification to industry standard subcontract agreements to align them with the APP and yet there was no discussion on how this policy might change the underlying structural or behavioural characteristics or how to describe this change in a cohesive and rigorous manner. The targetted procurement measurement is a small attempt to measure and monitor change; however, it only describes that SMEs have more projects and raises other questions. For example, what impact does this have on other construction supply chain participants in terms of structural and behavioural characteristics and what impact does this policy really have on productivity, profitability and innovation?

Therefore, although procurement reform is clearly on the agenda and in many cases pointing towards a positive model of procurement, there is no overall economic model that allows for description of the industry throughout the construction supply chain. There is no method to develop descriptions of procurement throughout the supply chain which would include attributes of firms, their markets and firm–firm relationships.

2.3.5   Summary

There has been a trend of construction industry policy development directed towards the supply chain concept. Further to this is the explicit emergence of the significance of procurement as an explicit instrument for industry intervention and in many instances the link between the supply chain concept and procurement is made. The activity of procuring products, materials and/or services consumes the industry – it is speculated that we spend half the time bidding and putting in place suppliers and the other half actually doing the job. Although normative models of project supply chain integration have typically been supported, there is a shift towards more positive-based models, particularly in the United States and Australia.

The discussion until now has focused on the various governmental investigations and policy directions in terms of strategies for improvement of the performance of the construction industry. However, with unique interpretations and strategies, many are united in the assumption that the industry has a fragmented *structure* and fragmented *process*. Fragmentation is either the key determinant of the poor performance of the construction industry, or an attribute of the industry that indicates firm specialization but requires careful management.

These organizations typically then perpetuate the same approach to understanding the industry structure through the same methods. For example, their view of the structure of the industry is based upon simple descriptions matching the number of firms in the industry to the number of employees in the firm to the annual turnover. The same descriptions are often repeated and resurface in many reports. The underlying structure of the industry is not progressed any further than this, yet claims are made regarding the need to understand the underlying causes through understanding the underlying structure. The descriptions simply rely upon grouping all firms in the industry and do not consider any differentiation between any groups on any basis and any consideration that there are different supply chains with different characteristics.

This chapter has highlighted that there seems to be a cycle of investigations followed by a surge of strategies of how to solve the problems identified.

The strategies are supported with quite clichéd understandings or at least clichéd descriptions of the industry as fragmented with numerous small- to medium-sized enterprises (SMEs) and as low productivity and poor performing without ever progressing forward on the analysis of the structure and behaviour of the industry and subsequent ways of measuring the change in performance if new strategies are put in place. There seems little creative or critical thinking about what models of economic performance we should be developing for this particular industry which takes into account quite disparate sectors with their own market structure, conduct and performance characteristics which all converge on an industry which is embedded in the 'project'. Perhaps it is too difficult a task to be tackled solely at a national level for many countries and there needs to be a harmonized system of policies within a country. For this reason it is worthwhile to consider a specific country and the national and state/regional policies. The following case study is a critique of the various Australian government initiatives.

2.4   National case study: Australian initiatives

In the mid-1990s in Australia, attention was focused on the development of the construction industry by the federal government. The ensuing years have seen a considerable amount of activity and attention on policy development.

Currently, two peak bodies have emerged with clearly identified responsibilities for developing construction industry policy in Australia. The first is the Australian Procurement and Construction Council (APCC) and the second is the Department of Industry, Tourism and Resources (DITR). The difference lies in that DITR is a federal (national) government body and the APCC is at the state (regional) level, and is a peak body for the state government agencies. More recently, the APCC membership has grown to include representation from the national level of government from both Australia and New Zealand in a further attempt at policy harmonization.

This section explores the federal approach through DITR and the state approaches through APCC and the two states with traditionally the largest capital works programmes, New South Wales and Victoria.

2.4.1   National level approach

The DITR is the agency at the national level of government which is responsible for policy development for the construction industry. The *industry* portfolio mandate implies that they provide national leadership across many industries or sectors, of which the construction industry is just one. Since the late 1990s this agency has developed some 29 industry Action Agendas (DITR, 2004), including one for the Building and Construction industry. The selected approach to industry analysis through a variety of investigations produced a mix of industry interpretations and a blend of normative and positive models. However, in recent times the shift has been towards more positive models of the industry.

The DITR's drive from the late 1990s finds its roots in the 1997 strategy 'Investing for Growth' – a Commonwealth Government strategy designed for the development of all Australian industry. The mechanism through which this is implemented is the Action Agenda process and it is the primary method by which individual industries are targetted. Action Agendas are a key part of the Commonwealth Government's long-term strategy to develop Australian industry. The Action Agendas are in varying states of implementation and the detail of the Building and Construction Action Agenda which was outlined in Building for Growth (DITR, 1998) is now discussed. It is important to caveat the discussion in this chapter with the fact that status of the 1997/1998 Agenda is that 'all initiatives have now been implemented' (DITR, 2004).

There were a number of strategies developed, but there was no method developed to monitor the achievement or success of these goals; or at least no publication of the evaluation process. The assumption underpinning these strategies was the belief that improved competitiveness and/or efficiency would be the result. At no time was there a method to monitor the changing structural or behavioural characteristics that the strategies purport to achieve.

2.4.2 *Government and industry policy development*

In 1997, DITR established the National Building and Construction Committee (NatBACC), which comprised of national leaders in government and industry, including representatives from state government agencies, professional groups, major clients, government contractors, consultants and subcontractors. Their role was to alert industry and government to the future trends impacting on the industry and to identify appropriate actions required to improve the performance of the industry (NatBACC, 1998).

The Building for Growth policy for the building and construction industry produced in 1998 by DITR was developed from both the Investing for Growth Agenda 1997 (DITR, 1997b) and the advice from the NatBACC (NatBACC, 1998). The strategy recognizes that boosting growth in particular sectors will require the joint efforts of industry and government. DITR considered that 'through these combined efforts Australia will be in a strong position to build on its assets and develop the industries of the future'. Therefore, lengthy consultation processes took place, perhaps not unlike the consultation processes undertaken in Singapore and South Africa. (Note: for the purposes of historical record, at the time of the development of the agenda the agency was known as the Department of Industry, Science and Tourism – DIST – and then shortly afterwards the Department of Industry, Science and Resources. In 2003 it gained its current title of Department of Industry, Tourism and Resources – DITR. The agency's documents are referenced as DITR.)

The strategy for the building and construction industry, described in Building for Growth 1998, was 'aimed at creating a framework which allows the building and construction industry to achieve its full potential and maximize its contribution to the growth of an internationally competitive and outward looking Australian economy' (DITR, 1998). The supply chain concept was introduced explicitly in the strategy for the first time in Australia and prior to the Egan Report.

Box 2.5 Australian policy aim

The underlying supply chains and business systems will need to be redefined and move away from the current short-term project-to-project culture to one which is more strategic, long term and enduring. This will include an understanding of the increasingly complex financial arrangements for projects and the links with manufacturers and construction service providers (DITR, 1998).

The Building for Growth policy identified six key issues, including:

- innovation
- information technology
- procurement and project delivery
- workplace issues
- the environment
- regulatory reform and exports.

Throughout the document there was either explicit or implicit discussion revolving around the need to address the fragmented industry structure. It was claimed that fragmentation is caused by so many firms and the short-term project-by-project approach by firms when dealing with each other in the supply chain. For example, even the rhetoric regarding the information technology revolution draws connections between electronic tendering and procurement and the creation of virtual teams. This was underpinned by the logic of the argument to capitalize upon using the information technology revolution as an enabler to *re-engineering the supply chains* (DITR, 1998). Not a particularly revolutionary concept – but indeed in hindsight is proving difficult to achieve (London, 2006).

2.4.3   *National government sponsored cross industry supply chain programme*

The Supply Chain Programme was an early initiative by DITR to facilitate improved performance across all industries. This programme was designed for organizations in all industry sectors, not specifically for construction. Developed in early 1998, it was designed to provide financial support to companies that wished to be involved in improving supply chain partnerships for a specific project. The aim of the programme was to improve competitiveness in Australian business through forming closer and more effective customer–supplier relationships, thereby providing improved competitiveness against the threat of the 'emergence of fewer and larger firms operating in the global market' (DITR, 1997).

Twelve case studies were conducted involving firms across various industries and their suppliers. The results of the programme were reported at a national industry conference in 1999. Unfortunately, there is no publication from that conference nor from the programme.

From my personal observations made at the conference and discussions with programme participants, the successful case studies focused upon an individual company identifying the health of their own organization, mapping the existing suppliers, making explicit sourcing strategies and mapping sourcing strategies down the supply chain. It appeared that the greater the understanding of the individual focal firm, the firms participating along the supply chain, their competitive markets, structural industrial organization and the power relationships between firms, the greater success there was for the project.

Two case studies were deemed unsuccessful in this programme. One of these unsuccessful supply chain projects was the construction industry project and the other was from the tourism sector. A DITR employee and the director of the firm involved in the construction industry case study presented the results of the supply chain project based upon their experiences and personal views. The presenters explained the extent of the success of the programme through their individual projects. It is noted that both the construction and tourism sector projects did not report anything about the firms, their capabilities, the markets or the type of relationships that existed in the particular supply chain; whereas, without exception, these issues were reported by the other ten successful projects. Suppliers and sourcing strategies were not mapped in the construction and tourism projects.

The reason claimed by the DITR Director for lack of success was that those projects were located in a single-project industry versus a continuous-process industry and therefore long-term relationships were difficult to establish. The explanation for lack of success in supply chain management was the short-term adversarial relationships brought about by project-based industries and that the supply chain concept may have little relevance to project-based industries.

Rather than saying 'it is all too hard', this may actually suggest that it is not as simple and straightforward as other sectors and that some deeper thinking is required for the development of project-based supply chain-related industry policy rather than that the supply chain concept has little relevance. Ultimately, the construction chain moves out of the project environment and into sectors where there are longer-term relationships – at the moment we have very little knowledge about any of the supply chain. The supply chain programme was not a part of the Building for Growth policy for the building and construction industry – a distinct lack of co-ordination within a government agency!

2.4.4   Government studies related to procurement in the supply chain

It appeared from the supply chain programme by DITR and the initial consultation process with NatBACC that the supply chain concept was on the agenda for the construction industry. However, that focused understanding and investigation of specific industries may be required to progress the understanding of the particulars of supply chains in project-based industries. Further to improving construction industry performance and to progress Building for Growth (DITR, 1998) towards developing Action Agendas, more detailed investigations of the industry were sought. The result of these discussions reaffirmed that this multibillion dollar industry needed to adapt rapidly and embrace change to fulfil its potential and take advantage of emerging opportunities (NatBACC, 1998).

To assist NatBACC in responding to the draft Building for Growth strategy, DITR commissioned a series of research studies to examine and analyse specific issues in greater detail; four of these studies are related to procurement in the supply chain. Two of these studies were to assist NatBACC in responding to the strategy and included:

*   procurement and project delivery in the building and construction industries (Australian Pacific Projects Corporation)
*   mapping the building and construction product system (University of Western Sydney, Macarthur)

The other two studies included:

*   alliances and networks – national Australian Museum case study by QUT/ CSIRO
*   productivity and subcontracting for ACA 1999.

The following Table 2.3 summarizes and discusses the key themes identified in the four studies.

The Building and Construction Product System research assumed the construction sector was highly fragmented with many small to medium enterprises. Similar to Atkins, Egan and Kelman, this was considered to be an attribute of an industry with a high degree of firm specialization and was to be managed well for innovation. However, rather than developing a deeper insight into the industry structure and firm behaviour, they tended to develop prescriptive norms for a well-functioning innovative building and construction industry based upon an industrial complex model (Gann and Slater, 1998). Conceptually, this research is also building upon OECD work on studies of national systems of innovation. Recommendations to policy makers were made on the basis of a desktop analysis with a distinct lack of empirical evidence to support claims. Similar to the previous study conducted by APPC, the results are of limited value because of the lack of rigour in the method. The supply chain was only described in a broad manner and with little understanding of the detail of the structure and industrial organization of the firms in real or actual project supply chains. The discussion addresses a market view of some key markets and the major players in the key markets, but does not seem to address the firm or project level of supply chains.

In summary, the recasting of the industry in terms of industry group linkages indicated potential for the supply chain concept and procurement; however, in reality, it did not give any description of the *real* supply chains that exist nor the nature of procurement and the *real* linkages that exist.

2.4.5 *Commentary on building and construction product system map*

It is worthwhile taking some time to discuss the study by AEGIS (1999) as it is significant because it illuminated a much wider scope of the industry than had previously been imagined in Australia. It did this by describing conceptually all the supply chain actors and the non-supply chain actors who are influential in the industry. Since innovation, and by implication within the assumptions of the study knowledge flows, was the aim of the research there was an attempt to include all the key actors in the *building and construction product system map*, which introduced regulators and technical education and training infrastructure. It then attempted to rationalize the existing body of statistics of the industry with the new view. When describing the actors, the focus was largely on the leaders in the contracting sector.

With seemingly little description of the empirical reality of the contractual linkages and interdependencies between firms throughout all the supply chains, recommendations were then made to policy makers. The recommendations consistently described policies to assist R&D, training and information dissemination. Typically, the recommendations were broad and with little reference or understanding of the market drivers that impact upon economic performance, productivity, innovation and competitiveness. Much of what was purported as the ideal environment for innovation relied upon the industry participants organizing their own supply chain. The policy recommendations were silent on intervention. They were also silent on theory or methodology relating projects, firm competition, procurement and firm–firm relationships in the supply chain.

Although there was considerable mention of the project-based nature of the industry, in reality the recommendations did not really reflect the fundamental way in which projects drive industry governance structures

Table 2.3 Summary of key themes in Australian national government-driven studies aimed at policy development

Study	Document analysis
Procurement and project delivery	'Procurement and Project Delivery' study was commissioned by NatBACC and conducted by Australian Pacific Projects Corporation (APPC). It was designed to 'examine the current delivery and procurement methods and identify approaches and models to enhance industry efficiency both locally and offshore' (APPC, 1998). The report contains very little rigour in method for examining the current delivery and procurement methods and is shallow in its methodology and discussion. The lack of rigour limits the validity and general applicability of the results of the study. There are primarily two criticisms to be made of this investigation: first, only the procurement methods between the client and the contractor were considered. This is considered to be too shallow to fully understand procurement in the industry and does not really reflect the aims and objectives for the industry as espoused by NatBACC or DITR; second, the study was limited in its consideration of the issues related to procurement as it only identified the types of contractual relationships in use, such as lump sum, construction management, design and construct, build own and operate, etc. The authors did not consider the methods leading to the formation of these contracts nor the procurement methods used by other parties involved in construction projects. So much of the rhetoric in the DITR strategies has been about the supply chain and yet the supply chain was only defined in the report as the client and the contractor, that is, the first level suppliers. They missed the opportunity to examine the procurement between all chain participants; for example, contractors and their various suppliers and the suppliers to the contractor's suppliers.
Mapping the building and construction	The second study of significance that was commissioned by DITR through NatBACC was conducted by a research institution and included the work by the Australian Expert Group for Industry Studies (AEGIS), a research team at University of Western Sydney. This study, Mapping the Building and Construction Product System, has more far-reaching consequences than the previous study and is described in detail. This study aimed at a wider understanding of the construction *chain of production*. The AEGIS model for the Building and Construction Industry Cluster contributes to the development wider of the industrial organization perspective. The approach taken in the research focused upon the linkages and interdependence between actors in a network of production. The linkages were not necessarily contractual; they were assumed linkages made by the researchers based upon a limited number of interviews. The underlying premise is that improved economic and

innovation performance can be 'partly explained by differences in the density and coherence of innovation linkages within industrial chains, clusters and complexes' (AEGIS, 1999). They suggest that exploring industries through chains, clusters and complexes offers a new way of looking at the economy as opposed to more traditional industry mapping such as those provided by input–output data. It is more in line with the changing character of market-based capitalism – from being based on hierarchy to being based on alliance. This approach also attempted to reflect the increasing complexity of modern production which involves more complex product-systems that combine component suppliers, manufacturers and service suppliers. The Building and Construction Product System research assumed the construction sector was highly fragmented with many small to medium enterprises. Similar to Atkins, Egan and Kelman, this was considered to be an attribute of an industry with a high degree of firm specialization and was to be managed well for innovation. However, rather than developing a deeper insight into the industry structure and firm behaviour, they tended to develop prescriptive norms for a well-functioning innovative building and construction industry based upon an industrial complex model (Gann and Slater, 1998). Conceptually, this research is also building upon OECD work on studies of national systems of innovation. The assumption to the prescriptive norms was that improving performance relied upon knowledge and information management. The industry as a chain of production was subsequently mapped in detail as a 2×2 matrix with a vertical product/service dimension and a horizontal knowledge-intensity dimension. This map was considered 'a stepping stone between traditional industry statistics based on hierarchical systems of classification and broader conceptual views of industry activity as depicted in clusters, chains and complexes' (AEGIS, 1999). Recommendations to policy makers were made on the basis of a desktop analysis. Similar to the previous study conducted by APPC, the results are of limited value because of the lack of rigour in the method. The mapping of firms into the matrix was based upon the experiences and knowledge of the researchers and no empirical data was used. Various 2×2 matrices were developed by deduction with little reference to primary data sources and cursory reference to secondary data. The industry was recast in terms of the four main groups of users, producers, training institutions and regulators. Diagrams were developed which indicated the intensity of the knowledge flows between each of these groups. These maps were developed from different public company databases, published documents and limited personal interviews. The major firms were organized and listed according to groups: suppliers, project firms and project clients. There were no criteria as to why particular firms were chosen and discussed. The discussion focused upon

(*Table 2.3 continued*)

Table 2.3 Continued

Study	Document analysis
	general information about size and turnover. There was also no method described as to how the knowledge flows were defined or by what criteria strength of flow was defined. The supply chain was only described in a broad manner and with little understanding of the detail of the structure and industrial organization of the firms in *real* or actual project supply chains. The discussion addresses a market view of some key markets and the major players in the key markets but does not seem to address the firm or project level of supply chains. In summary, the recasting of the industry in terms of industry group linkages indicated potential for the supply chain concept and procurement; however, in reality, it did not give any description of the *real* supply chains that exist nor the nature of procurement and the *real* linkages that exist.
Networks and alliances	Another strategy instigated by DITR was the showcasing to the industry of the Networks and Alliances project. This study was not commissioned by NatBACC. This involved a case study of a project that would 'aim to prove that networks of firms provide greater integration, financial strength and stability, allowing firms to respond to emerging opportunities'. CSIRO and Queensland University of Technology have been involved in empirical work on the National Museum of Australia. The research conducted on the National Museum of Australia construction project (Walker *et al.*, 2002) appears to be aligned with the DITR approach that improving the market leader's performance would improve the construction industry. This is a normative-based model of project integration. The delivery method for the National Museum of Australia was a project alliance contract based upon risk and reward between seven organizations. The Department of Industry, Science and Resources awarded an internationally significant research grant to case study the construction of the National Museum of Australia and the Australian Institute of Aboriginal and Torres Strait Islander Studies on the Acton Peninsula in Canberra. The \$155 million Acton Peninsula Development in Canberra is unique in Australia because it is the first major building project to use an alliance contract. An alliance contract is an agreement between parties to work co-operatively to achieve agreed outcomes on the basis of sharing risks and rewards. Alliance contracts have the potential to deliver substantial cost and quality benefits without the adversarial relationships common in more traditional contracts.

The research will review the construction project delivery method, and the use of advanced information technology in the design, construction and project management of the Acton Peninsula Development. The construction alliance consists of: Bovis Lend Lease, Tyco International, Honeywell, Ashton Raggatt McDougall Architects in association with Robert Peck von Hartel Trethowan Architects, Anway Exhibition Design and the Commonwealth Government (Walker et al., 2002).

Productivity and subcontracting

The investigation was completed by two commercial research firms and the findings indicated that Australia in comparison with other countries had a higher productivity because of the subcontracting system. Subcontracting is portrayed as a positive attribute that is contributing to an efficient Australian industry. The industry requires a huge range of specialist skills. In Australia, these skills are applied under intense and continual competition through the subcontracting system. Rather than developing and retaining intermittently used specialist skills in each competing construction company, the industry shares the skills of a number of subcontractors. In this way efficient usage of skills is possible and a 'cross-industry' development and refinement of those skills is facilitated (ACA, 1999). Subcontracting is, then, presented as a useful method that has encouraged firm specialization rather than the service remaining in the contractor firm and being of a more generalist nature.

The subcontract system employed in the building industry has encouraged specialization. The benefits achieved in applying new techniques and products are transferred from one project to another across company boundaries. The learning curve on many trades has effectively been eliminated. The subcontract system has not been applied so successfully in the heavy industry and civil works side of the industry to the same degree. This is mainly due to the risks involved and the magnitude of the sections of the work inherent in the contracts. To some degree this is due to the form of contracts now being used by government agencies whereby all risk is passed on to the contractor (ACA, 1999). The subcontracting system enables projects to obtain suitable expertise and resources at the time they are required on site. The risk is allocated to the party best able to deal with it – the specialist subcontractor. Some disadvantages exist; for example, in ensuring standards are maintained in areas such as training, safety and environmental management. Recognizing this, the company has established systems and processes to include all site employees in these processes (ACA, 1999).

Further to this, *high productivity and output are achievable where the head contractor can assist subcontractors to manage and organize their resources (ACA, 1999).*

(AEGIS, 1999). The basic premise is on linkages between firms and yet the research did not discuss procurement or firm-to-firm relationships in the chain of production. The relationships that were discussed were at the highest level between client and contractor. There was no reference to any of the major literature in the construction research community on procurement or to supply chains.

Improved economic performance, productivity, competitiveness and innovation appear to be driven by lead contracting or consultant firms in this model. In this manner, the model of *clusters* relies upon smaller firms clustering around large firms. This is one type of industrial organization; however, it is suspected that there are a range of other systems. There is perhaps even a much more subtle and richer understanding of the industrial structure that goes somewhat deeper than this notion of smaller firms clustering around large firms – something that extends our understanding beyond the one size of a firm as the basis to improve industry performance. The model did, however, support the industry as specialized. However, it did not address the co-operative and competitive nature of the market and seemed to lack any comprehension of the way firms worked in the industry.

Similarly, the Networks and Alliances study was a normative-based model of project integration and was typically naive in its understanding of the economic drivers for change across an entire industry. This research concentrated upon the first tier of suppliers and little or no thought was given to the tiers beyond. The premise is that, to achieve performance, first tier suppliers will need to manage their supply chains; that is, the contractual relationships of parties at successive tiers.

Initially, the study was aimed at the use of project-based websites and observations of electronic communication within a project that was delivered with an innovative procurement method, namely a project strategic alliance. After the study had been completed, it was recast as an exemplar of Best Practice Supply Chain Management theory and practice. The main thrust of the report is a focus on information flow concepts derived from supply chain theory. The report was prefaced by a caveat that information flow was only one aspect of supply chain management. This shift indicates the wider acceptance of supply chain concept that has evolved in recent years.

In a contrasting approach to the AEGIS and Alliances study, in late 1999 the Australian Constructors' Association (ACA), an industry association of various construction firms, with the support of the DITR, commissioned a study to benchmark productivity in Australia compared to other nations. Subcontracting, and the ensuing firm specialization, was found to be a necessary and important attribute of the industry.

The study is problematic in some respects. For example, other countries use subcontracting, so the claim that this is different in Australia is challengeable. Although the association commissioning the report is representative of the contractor, the recasting of the subcontractor system as a mechanism for cross-industry development has potential. Even though it is claimed that the specialist subcontractor is the best party able to deal with risk allocated to them, many who are not specialist subcontractors will not have the capacity to deal with the ensuing risk that is allocated to them. Two interesting questions arise here:

- first, to what extent are the claims regarding the subcontractor assistance through management and organization of resources true? What is the nature of the contractor and subcontractor procurement relationship?
- second, to what extent are the claims regarding cross-industry development true? How tightly woven is the actual market and pool of subcontractors that the head contractor is drawing from?

2.4.6 Current national strategy

DITR refined their strategy and developed the Building for Growth Action Agenda 1999–2002 based upon reports produced by both NatBACC and ACA.

From the studies, NatBACC produced a report for government outlining 34 recommendations in the following areas:

- interaction with government
- industry statistics and analysis
- co-ordination of policy
- encouraging innovation
- information technology
- regulations and standards
- project delivery and business improvement
- environment
- workplace relations
- training skills and development
- trade facilitation and export.

These were largely translated into the seven strategies as outlined below.

Box 2.6 Australian industry policy: seven strategies

The Building and Construction (B&C) Industry has developed a Building for Growth Action Agenda in partnership with the Government over 3 years, from May 1999 to June 2002. The Action Agenda focused on seven key issues the industry needs to address to ensure its long-term performance and international competitiveness (DITR, 2001).

- utilizing advances in technology
- innovation, research and development
- environmental issues
- innovative procurement and project delivery mechanisms (benchmarking, alliancing)
- managing the supply chain
- regulation
- globalization.

The supply chain has now become an explicit strategy! Two of the strategies are concerned with the supply chain concept and procurement and are suggestive of a 'best practice' normative model for project integration.

The government committed $3.6 million to a range of initiatives to the implementation of the strategies. The outcomes were identified in 2002 and are described in Table 2.4, along with other outcomes from the Action Agenda which were developed subsequent to this in 2003.

Table 2.4 Summary of key national government initiatives arising from industry studies and consultation 2002–2003

Outcome	Detail of initiatives
1	Establishing the Australian Construction Industry Forum (ACIF) as a peak body to represent the interests of the industry (construction, property developers, investors, trade services – engineers, architects and supply networks sectors) – this was derived from the ACA.
2	An International Benchmarking study with an on-line benchmarking and analysis tool specific to the building and construction industries. GlobalConstruct is a fast, easy way to compare your performance with that of other industry players – by industry segment, country and strategic focus. GlobalConstruct also identifies key industry success indicators and drivers of leading edge performance.
3	An Innovation Report by Price Waterhouse Coopers on the results of the first innovation survey of the B&C industry. The controversial report findings will provide the basis for robust debate by the industry on addressing their perceived and actual innovative practices.
4	Construction Forecasting by Econtech and Andersen will provide efficient, quality, basic market intelligence on non-residential property industry trends and activity cycles free of charge via an easily accessible internet site.
5	A report on Best Practice Supply Chain Management theory and practice with case studies to demonstrate how companies can improve their performance and competitiveness.
6	Research on the successful Project Alliancing contract for the construction of the National Museum of Australia,

Table 2.4 Continued

Outcome	Detail of initiatives
	Acton Peninsula. Alliancing is suited to projects with a budget greater than $20 million, high risk, time and budget constraints and with client support to achieve a quality project. A project alliance delivery strategy requires commitment, a flexible approach, trust and a no-blame-no-disputation culture.
7	Globalization to encourage building and construction firms to establish networks and target overseas markets. Case studies on each grant and the experiences of the participating companies in their target country markets are also available.
8	Government and industry undertook extensive research on the status of, and issues affecting, the industry through a National Building and Construction Committee (NatBACC).
1–4 subsequent initiatives	1 Making the B&C industry more aware of the range of Government Programmes to encourage industry to adopt new technology. 2 Establishing the Co-operative Research Centre (CRC) for Construction Innovation, which received Commonwealth Government funding of $14 million over 7 years for basic and strategic research in five related programmes: (1) Virtual environments for lifecycle design and construction; (2) construction project delivery strategies; (3) environmental sustainability; (4) integrated design and construction support systems; and (5) management, adaptability and future of build assets. 3 Continuing support for the Australian Building Codes Board (ABCB), which is responsible for: developing and managing a nationally uniform approach to technical building requirements, currently embodied in the Building Code of Australia (BCA); developing a simpler and more efficient building regulatory system; and 4 Enabling the building industry to adopt new and innovative construction technology and practices (DITR, 2003).

The table indicates the subsequent initiatives of which one was the establishment of the national Co-operative Research Centre for Construction Innovation (CRC_CI). The initial thrust of the CRC_CI was research programmes related to project integration normative models. A shift occurred in 2003 and the programmes were redesigned into three programmes. Programme A is now Business and Industry Development, which is aimed less at specific projects and project integration and more at industry-wide understanding of the behaviour of firms in the industry. Therefore, it is an indication of further support for the evolution of thinking towards the development of positive models to describe the industry.

There has been a trend of construction industry policy development directed towards the supply chain concept. Further to this is the confirmation of the significance of procurement as an explicit instrument for industry intervention and in many instances the link between the supply chain concept and procurement is made. Although normative models of project supply chain integration have typically been supported, there seems to be a shift towards more positive based models. The approach taken by Australian state-based policymakers is now explored in detail as the states often do not operate in the same manner as the federal government in relation to the capital works programmes and the property and construction industry. The reasons for disharmony are the subject of another nationally funded study which is underway at the writing of this text (London and Chen, 2006).

2.4.7 *Regional case study: state level investigations and construction industry policy*

In Australia, the seven state- and territory-based government agencies have a key role in policy development related to the construction industry and this trend towards positive-based models is even more apparent. Figure 2.2

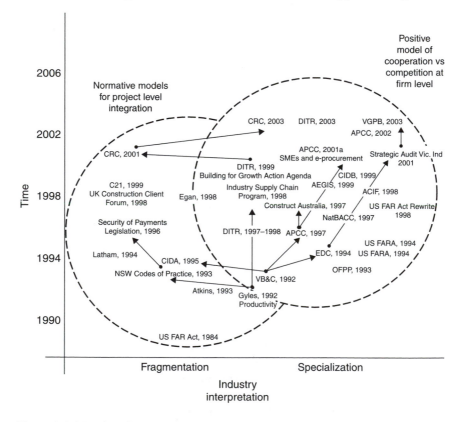

Figure 2.2 Trends of selected Australian state-based investigations and policy development related to the construction industry 1990–2003.

overlays the previous national approaches and the Australian state-based investigations and policies. Beginning with the Gyles Report in 1992 which initially led to normative models to change the industry (NSW Codes of Practice, 1993; CIDA, 1995; CRC, 2001), the historical trend of policy development has moved towards needing more positive-based models that would reflect the industry's structural and behavioural characteristics (APCC, 1997, 1999, 2002a; VGPB, 2003). Refer to Figure 2.2, which maps the Australian material onto the previous diagram. There has been a gradual recognition of the role of the supply chain concept and the role of procurement in policy development, and in particular the role of procurement throughout the industry supply chains (APCC, 1999) however at times it appears that opportunities are not capitalized upon as much as one would expect.

More recently, one of the most significant trends by the New South Wales and Victorian state governments has been the adoption of more and more public–private partnerships (PPPs) or privately financed initiatives (PFIs). These project procurement strategies are aimed at shifting the burden of financing infrastructure to the private sector – they are an internal government strategy and are not aimed at making performance improvements in the industry. In 1998, I was the project manager for a PPP in the Northern Territory. The Victorian state government completed their first prison capital works projects under a PPP in the mid-1990s. This is a topic for a book in itself and, of course, deserves much more attention – however, PPPs are not used by governments in Australia as a strategy to 'manage' or influence the supply chain.

In recent years, the most widely known event with regard to examining the Australian construction industry was the 1992 NSW Gyles Royal Commission. It remains a significant event in the history of the construction industry of Australia, as it sought to unearth many of the 'normal' practices that firms in the industry took for granted as commonplace and were considered part of the culture of the industry. Many of these practices were deemed illegal and legislation has sought to redress some of the more serious activities. One of the most critical of these was the security of payments for firms in the supply chain, particularly firms subcontracting to the main contractor. In 1996, legislation was introduced to attempt to solve these problems, particularly the *paid when paid* clauses found in many subcontracts.

Apart from serious illegal practices and unethical activities, which the codes sought to address, the NSW state government's other main concern was in modifying industry behaviour. The Royal Commission was most significant in that it resulted in reform agendas being developed in the state of NSW. The Construction Industry Policy Steering Committee, led by the NSW Public Works Department, developed Codes of Practice and Codes of Tendering as a means to address the findings of the Commission. The most significant concern was the adversarial relationships that pervaded the industry. On capital works projects the partnering and value management concepts soon found favour as a means to improve those projects. It was assumed that the overall construction industry culture would be improved if partnering charters were developed between clients and contractors and value management workshops were conducted between the design, construction and client teams. The focus was on improving client and contractor relationships. However, it was considered that these measures would have a flow-on effect to the other firms in the contractual chain on the project and trickle down to the rest of the industry, ultimately addressing many of the adversarial relationships and solving the poor performance of the industry. Unfortunately, it has had limited success in altering the performance of the industry, according to the NSW Productivity Commission findings.

2.4.8   Policy development alignment: Australian Procurement Construction Council

The New South Wales state government is just one of the state and territories responsible for construction industry policy development. The state government agencies are aligned through the APCC. The council members comprise the chief executive officers of the state and the territory infrastructure agencies and also the representative from the Department for Economic Development in New Zealand. The council reports to the Australian Procurement and Construction Ministerial Council, which comprises those ministers with portfolio responsibility for procurement and construction policy. Although the DITR is a federal agency, it has on occasions been represented on the APCC council; the council largely represents the 'state level' bodies. The council aims to develop nationally consistent approaches to broader procurement policies, processes and practices.

Box 2.7   Australian Procurement and Construction Council

The Australian Procurement and Construction Council Inc. (APCC) is the peak council of departments responsible for procurement and construction policy for the Commonwealth of Australia, State and Territory governments, and the New Zealand Government (APCC, 2002b).

APCC had its origins in 1967 in the National Public Works Council (NPWC) and was primarily made up of each of the six states or two territory government agencies. In the past, many of these agencies were directly responsible for providing construction services related to capital works and repairs and maintenance expenditure programmes. However, in the 1980s across the country the various agencies have moved to outsourcing much of their design, construction, project and asset management operational management activities. Many agencies that hold budgets for their facilities now engage this 'public works' construction agency on strategic advice related to capital works and asset management and then operational advice on the management of the procurement of consultants and contractors.

The single most significant issue that this holds is that a generally higher focus is now placed upon managing at a strategic level for the industry. They have been relieved of many of their procedural duties and thereby are more focused upon industry policy development and implementation. They possess a dual role in that not only do they hold constitutional responsibilities but by virtue of their purchasing power they have significant influence in the enactment of policy as a 'client' in the industry. The state governments have specifically agreed to use this dual leadership and participant role to support industry development (APCC, 1997).

In 1997, the APCC convened a Meeting of Procurement and Construction Ministers. As a result of this meeting and wide industry consultation at the National Construction Industry Forum, a strategy document, *Construct Australia*, was produced which has subsequently acted as a blueprint for the development and implementation of policy across the state organizations. Once again, the current profile was discussed in terms of an adversarial culture, short-term focus and a fragmented industry.

Construct Australia then outlined what was expected of the industry in terms of what the future might hold for the industry. A response to this future scenario was outlined by describing desirable attributes of the industry. Industry propositions were then explored under each of these attributes and from this 10 National Strategies were developed. The following Table 2.5 summarizes and discusses *Construct Australia* in terms of the following theme: attributes of an ideal construction industry SMEs in supply chains. It also includes a commentary on the APCC Government Framework for National Co-operation and Electronic Procurement (APCC, 2002a). In summary, the role of the supply chain and procurement was central to the policies. The *Construct Australia* document still relied upon integration concepts and based discussion on the difficulties of integrating fragmented SMEs, whereas the Electronic Procurement framework tended to accept fragmentation as a characteristic of an industry which had become highly specialized.

2.4.9 Victorian government approach

The Victorian state government is a member of the APCC and alongside the New South Wales state government has had a great impact upon the construction industry in Australia. The Victorian state government approach to policy development has supported the supply chain concept and the role of procurement in industry performance and development. In particular, the industry interpretation and ensuing policy development supported an understanding of the competitive nature of markets and firm specialization in the industry. They are more market-oriented than the APCC and New South Wales and for this reason, coupled with the fact that the industry case studies which are described in Chapters 6 through to 8

Table 2.5 Summary of key themes APCC policy

Concept	Document analysis
Attributes of an ideal construction industry	Similar to many other investigations across the world, fragmentation was an issue at the policy level. It was discussed in relation to process and industry fragmentation.
	Work is organized into small isolated packages and the outcome is a fragmented approach both in terms of design, where separate small design consultants are used project-by-project and in terms of construction where multiple levels of small specialist subcontractors and suppliers are used (APCC, 1997).
	Further to this fragmentation is seen to be a negative characteristic of the industry with comments such as:
	This fragmentation together with the divisions between design and construction, limits opportunity for efficiency gains and encourages the pursuit of singular interests (APCC, 1997).
	There was much speculation in the document about what the future held for the industry participants. For example, it was argued that clients will favour a response from industries that rely upon integration and single-source solutions whereby the construction services on a project will extend beyond the familiar range of services associated with a single project.
	This might be achieved by a packaged solution that involves re-engineering systems, sourcing, supply chain management, negotiation of cost effective purchasing arrangements and the maintenance of minimum inventories (APCC, 1997).
	The impact on procurement and relationships was highlighted and the virtual enterprise was suggested in terms of both formal contractual linkages and organization and informal relationship management. The impact of developments in information technology was suggested as a key driver for change. Globalization was also suggested as a major influencing factor for the Australian construction industry.
	The APCC identified four attributes of an ideal construction industry: seamlessness, efficiency and profitability, innovation and being environmentally responsible. It is significant that each attribute focused upon the firms or, rather, the enterprises and what the enterprises will be delivering in the future. For example, a seamless industry will involve 'client and focused enterprises delivering seamless services through integrated alliances that provide packaged, high value single sources solutions for clients offering long term contracts' (APCC, 1997). The answer to the improved performance of the industry is seen by improving performance of the participants in the industry.
	As overall performance relies on the performance of all participants, there will be a need for more effective management of supply chain relationships and alignment of all participants' objectives (APCC, 1997).
	Clients would deal with a single entity and not with numerous individual supplies and this may result in horizontally integrated arrangements. In reality, the current clients (including government departments) often typically deal with a main contractor and a primary consultant, albeit on a project-by-project case. There is a preoccupation with integration which suggests that even this arrangement poses difficulties because of the lack of integration that flows down the chain. The response from industry may well be design and construction-type

SMEs in supply chains

procurement methods or equally similar integration-type project procurement delivery methods. As discussed previously, it is doubtful that the single source solution will solve or even impact upon downstream procurement relationships. Single-source solutions would require mechanisms on the part of the client to monitor the nature of the management of the supply relationships that occur down or across the chain. The framework that described the methods that states would use to implement this approach reflected an understanding of the supply chain concept.

Construct Australia has clear intentions with regard to improving the performance of the industry through supply chain management. However, it remains unclear as to the actual detail of the mechanisms that APCC will implement to do so. There is scope in the National Strategies 2 and 9. National Strategy 2 is concerned with establishing national key performance indicators for each of the attributes which would seek to monitor the 'health and structure of the industry' in response to key government initiatives. To reform industry, National Strategy 9 is aimed at encouraging improvement and change in individual organizations, which ultimately requires investment on behalf of the individual organizations. APCC agreed to encourage this investment by rewarding better performers through greater business opportunities to gain government business.

To ensure small to medium enterprises (SMEs) industry participants benefit from this approach all mechanisms will need to incorporate effective supply chain management procedures (APCC 1997).

The SME is important to the construction industry supply chain, particularly with respect to the significant proportion of consultant firms and subcontracting firms that play a critical role in the actual 'design' and 'construction' functions. Therefore, alongside the *Construct Australia* agenda, a National Action on Small to Medium Enterprises was developed since growth and stability with SMEs is a healthy indicator of a strong economy (APCC, 2001a). The 'APCC National Management Framework for Delivering Opportunities for SME in government procurement and contracting' (APCC, 2000b) indicates this:

In 1996–1997 Small and Medium Enterprise (SME) represented 97% of non-agricultural private sector business, according to the figures released by the Australian Bureau of Statistics. SMEs employ 3.2 million people across Australia, with an average increase of 3.2% per year since 1983. Growth in SMEs is a healthy indicator of a strong economy. SMEs contribute significantly to Australia's economic well being (APCC, 2001b).

The APCC then agreed on four national principles and then various national actions designed to assist the growth of the SME sector by improving access to government business.

There seem to be two mindsets by the APCC when approaching the development of SME policy. The first approach is towards enhancing the capacity of SMEs to compete domestically, nationally and internationally. This would be achieved by the following; first, by either reducing the size of the contracts so that they can engage in the contractual chain directly; or, alternatively, encouraging the formation of joint ventures, consortia or partnerships – that might enable SMEs to join the contract chain at the prime or subcontracting levels through the greater

(Table 2.5 continued)

Table 2.5 Continued

Concept	Document analysis
	resource capacity and strength provided within a cluster or network of other SMEs firms. The second mindset involves the concept of single-sourced solutions and packaging of contracts, which in turn ultimately forces SMEs out of a direct contract with the client. The rationale behind this is that SMEs will cluster around larger national and multinational companies and ultimately develop improved relationships as second and third tier suppliers. This also assumes that rationalization of suppliers is the way for the client to manage the supply chain.
	The role of the larger contracting firms as managers of the construction supply chain, and as firms who can be empowered to improve industry performance, is to a large extent the philosophy of the argument in the government discussion paper on Rewards and Incentives (Kenley, London and Watson, 1999). However, the various strategies that were discussed were aimed at the client – in this instance the government – being able to reward efforts of the primary contractor (whether that be a consultant, subcontractor, etc.) for demonstrable efforts in supply chain management. Currently, the understanding of the nature of chain relationships beyond the first tier is largely hidden and yet second, third and even fourth tier supplier markets and relationships are clearly key to much of the discussion worldwide, in improving economic performance. A reward and incentives programme would provide transparency of the behaviour of firms within their markets and therefore behaviour of firms between tiers down the chain.
Electronic procurement and changing supply chain structure	Further to the National Strategy on Small and Medium Enterprise in government procurement and contracting is the 'Government Framework for National Co-operation on Electronic Procurement' (APCC, 2001b). This framework is designed in the first instance to provide a national trading community that is consistent electronically. Implicitly underpinning this approach is the view that governments have a role in preparing firms in Australia to compete globally in international online trading arenas. One of the major aims is to support the SME's ability to be able to easily access the new electronic marketplace and take part in the electronic business-to-business transactions. If the full potential of the electronic marketplace is realized, it will increase the size of markets on a national basis. This explosion of the size of the market would result in more competition. The current view held by policymakers is that a more competitive marketplace means greater efficiencies and higher productivity.
	The electronic marketplace can potentially have major impact on construction procurement and wider choice of suppliers is only one aspect. The power of seamless electronic transactions for the construction supply chain is that clients are empowered to transact with any business along the chain, thus altering the traditional structure of the supply chain. The rhetoric assumes that this type of procurement environment could take place in the future. The ability to describe the product and/or service to the point that it is possible to engage in transactions electronically means that clients can engage in procurement at any point in the chain. It is difficult to predict the degree of penetration of the electronic trading environment concept to the construction industry. The industry is slower in its uptake of information technology in conducting business practices than other industries (Betts, 1999; Finch, 2000), but that does not mean that it is not emerging as a new business practice.

Regardless of this, the electronic procurement strategy acknowledges the importance of differences between supplier offerings and that

Suppliers should, therefore, be able to highlight differences in their product, service and support through electronic catalogues. Details may include service response times, logistic information, maintenance agreements, technical support and other product and service information (APCC, 2001b).

Further to this

Australian governments have identified procurement as a strategic activity which can facilitate the achievement of government, community and industry outcomes in a non-discriminatory framework which focuses on buying outcomes. Australian governments have underpinned these principles by encouraging sound supply chain management, industry development programmes, performance monitoring and robust supplier selection processes including criteria and evaluation methodologies (APCC, 2002).

The advent of electronic procurement highlights that change within markets and between markets along and across the supply chain is imminent. In particular, underlying structural and interfirm behavioural changes to the industry are likely. In many ways, the move towards utilizing the power of IT in procurement accepts that this is a highly specialized industry with a high degree of fragmentation because with specialization and fragmentation comes a high volume of transactions of all types across countless supply chains. It is able to process and manage this more efficiently and effectively. This is not confined to APCC members, however; the nationally co-ordinated e-procurement policy developed by the APCC (APCC, 2001b) was one of the first such government policies in the world to seriously address such issues.

At the OECD 3rd Global Forum for 'Fostering democracy and development through e-government' in Naples in March 2001, a number of speakers stressed the importance of these potential developments for the strengthening of governance, and the framework within which governments bind themselves to act in the interests of their citizens and to promote their well-being. The principles of accountability and transparency lie at the core of good governance. The key finding of the forum was that nothing is more powerful in combating corruption than conducting transactions openly and with public knowledge of the rules and criteria to be applied. Electronic technology can be a powerful tool for good governance through its ability to spread accurate and comprehensive information, to automate processes and to provide a record of each transaction (OECD, 2001).

The APCC likewise stated that:...the emerging use of electronic commerce in government procurement places more demand on the need for the buyer/supplier relationship to be more open, reinforcing Australian government's underpinning principles of value for money, accountability, transparency, equity and fair dealing, and open and effective competition (APCC, 2001b).

Buyer/supplier relationships in the first and second tier would tend to be more open in the *new* electronic procurement environment; however, this really does not address buyer/supplier relationships in subsequent tiers. It is suggested that modelling of procurement in the supply chain is the mechanism to make transparent buyer/supplier relationships throughout the supply chain and it seems apparent that this is how the thinking of the APCC has evolved.

were undertaken in this state, it is worthwhile to explore in detail the policy and procurement environment in Victoria.

Although there is often discussion about the 1992 New South Wales Royal Commission into Productivity, the state of Victoria commissioned an inquiry into the Victorian Building and Construction Industry in 1992 as well. The Economic Development Committee made public the results of the inquiry in 1994 and published their findings and recommendations. The terms of reference were broad and far-reaching (EDC, 1994) and included:

- reviewing tendering procedures for government works and recommending codes of practice
- reviewing and recommending changes to the building and planning process in Victoria
- investigating improvements in industry productivity and identifying and recommending changes to improve productivity.

There were a number of findings that are relevant to our understanding of policy development of procurement in the construction supply chain. First, towards the development of a Code of Tendering for the state, the findings of the Economic Development Committee are set in Box 2.8.

Box 2.8 Victorian Building Construction Industry Policy

[G]iven its importance to the growth of the Victorian economy the Victorian Building Construction Industry (VBCI) should be classified as a key industry sector in terms of the State Government's Industry Policy and be provided with industry and facilitation assistance available to other key industry sectors identified in the Industry Policy (EDC, 1994).

Coupled with this, the recommendations of the Committee are set in Box 2.9.

Box 2.9 Sector strategy

[T]hat the Department of Business and Employment develop a detailed sector strategy for the VBCI industry sector with the aim of improving the industry's efficiency and international competitiveness (EDC, 1994).

To this end, the committee also recommended that market competition should be maximized under the proposed open market system and the tendering system be opened up to both domestic and international firms. This is quite different to the domestic nature of the NSW Royal Commission findings.

Many of the findings are similar to the Royal Commission in NSW in terms of collusive practices during the tendering period and unhealthy and restrictive practices related to trade unions. It was recommended that a code for tendering be developed in the state and that the NSW Construction Industry Development Association and Western Australian Codes of Tendering were useful models to adopt.

Probably most concerning was the finding that *purchasing* is not recognized as a career stream in the public service and this can be detrimental to the efficiency and probity of the tendering process. The committee also recommended that 'strategic management principles should be applied to enhance the development and implementation of uniform tendering policies and principles for all state government works and services.' This is quite similar to the discussions in the US federal procurement reform strategy (Kelman, 2003).

There was very little discussion about the strategies to change the behaviour of firms in the industry. The focus was on recommendations for strategies to change the behaviour of the unions as the trade union's role in creating unrealistic demands for their members is detrimental to performance of the industry in terms of productivity measures, particularly in time lost. There were recommendations to change client's behaviour as the instigator of the tendering process, particularly in developing uniform tendering policies and procedures to improve probity, equity and efficiency. To support this, not only were policies recommended but for efficiency and probity it was recommended that 'purchasing' should be recognized as a career stream.

2.4.10 Victorian government state procurement and industry development

Now, over a decade later, we see the recognition of procurement in terms of its role in developing industry.

Box 2.10 Procurement state policy

The Victorian Government recognizes that government procurement policies and practices can be used to enhance its efforts in promoting and developing competitive Victorian industry (VGPB, 2003).

Further to this, four strategies are identified as the best way to achieve this:

Box 2.11 Victorian state strategies

–   encouraging Victorian businesses to grow, invest, innovate, export and exercise best practice
–   maximizing opportunities for Victorian suppliers to compete (especially small business) for government business on the basis of best value for money over the life of the goods or service
–   minimizing the costs of doing business with government
–   encouraging Victorian industry to develop innovative goods and services to meet emerging government needs

(VGPB, 2003)

In the early 2000s, the Department of Treasury and Finance acquired the role of advice on procurement.

Box 2.12 Procurement branch

[T]he Procurement Branch is responsible for the provision of advice on procurement and tendering policies to the Minister for Finance, the Victorian Government Purchasing Board (VGPB) and State departments and agencies. This includes developing and facilitating contracting opportunities within the public sector, new systems and skills to promote improved procurement and commercial practice, plus managing a range of government contracts and service arrangements (VGPB, 2003).

One of the key elements of the Procurement Branch's role is the advice to the VGPB, which was established by Part 7A of the Financial Management (Amendment) Act 1994 and replaced the State Tender Board from 1 February 1995. Its principal role is the achievement of the government's purchasing reform programme with specific focus on strategic procurement and achievement of purchasing principles and better commercial practices. Within this role, the board is responsible for establishing supply management policies and guidelines; assisting agencies in the acquisition of goods and services; accrediting agencies to undertake their own purchasing and tendering; co-ordinating a limited number of central contracts, and facilitating the establishment of improved skills, systems and business practices.

The philosophy of the Board is to play a guiding and facilitating role in developing a co-ordinated government procurement strategy rather than focusing on providing overall central purchasing and tendering services. However, there is no monitoring of, or evaluation strategy at an industry level on, the manner in which procurement is currently taking place or the manner in which it might change after various government policies have begun to take effect.

However, one of the first initiatives of the Board was to establish an accreditation framework to enable the devolution of procurement functions to agencies. With respect to the construction industry, the Office of Major Projects (OMP) has full responsibility for procurement of the nominated major capital works projects for the Victorian state government. However, the OMP are focused on projects and neither have a role nor a responsibility towards construction industry policy development nor implementation nor improving the sector as a whole.

2.4.11 *Major procurer of state projects*

The OMP manages the nominated major projects within the framework of the Victorian state government's Project Development and Construction Management Act 1994. To procure the project, the OMP has power to contract directly with consultants, contractors and subcontractors – this is not a common practice in Australia; it is unusual for a specific agency to legally contract directly with firms. The majority of major project contracts to date have been awarded based upon competitive tender, usually arranged by selective invitation rather than open tender. Where other state agencies are discussing various strategic alliances and risk and reward methods for projects, the OMP is steadfastly resistant to establishing strategic alliances, primarily due to the government policy of competitive tendering (Roenfeldt, 1999).

> **Box 2.13 Government as client: inward project-oriented perspective**
>
> Each project is viewed in isolation and a successful relationship on any contract only serves to provide the confidence that such an arrangement might be replicated on future contracts, thereby ensuring the supplier's inclusion on a selected tender list (Roenfeldt, 1999).

The majority of OMP contracts are with the head contractor and the primary consultant. It seems that there is little direct engagement with the construction industry and the management of the supply chain is devolved to the first tier of suppliers, namely the primary consultant and the primary contractor. It is also suspected that the nature of the procurement relationships in the chain is not well understood, monitored or evaluated beyond these first tier relationships.

Therefore, it seems that the procurement policy supports supply chain management but the implementation is less advanced. Part of the reason for this could be the project approach of the particular division as opposed to a more strategic approach to procurement. It has always been recognized that the OMP has a key role to play in the state's economy and the construction industry has been relocated to the Department of State and Regional Development (VSG, 2003). This closer relationship may well have an impact upon the manner in which major projects are procured. The policies of this agency are explored in more detail now.

2.4.12   *Victorian state industry development*

The Department of State and Regional Development conducted strategic audits of the Victorian industry, as per the aim set in Box 2.14.

> **Box 2.14   Government as client: external industry development-oriented perspective**
>
> [I]dentifying both current business needs and the long-term strategies needed to realize the growth potential of Victorian industry. The findings and recommendations put forward by the Strategic Audit will provide the basis for the development of industry plans for each industry examined. These plans will reflect the needs of industry and the competitive market environment in which Victorian industry operates (DSRD, 2000).

The strategic audit was conducted in two parts:

- Strategic overview: this part focuses on the structure, performance and the major long-term trends and issues for the Victorian economy and for industry in general, particularly in a global context.
- Industry audit: this includes in-depth analysis of selected key industries. It will include analysis of their competitiveness and the opportunities and challenges of particular sectors (DSRD, 2000).

The approach has been to analyse constraints and opportunities within selected industries. Industries chosen include: automotive, environmental management and renewable energy industries, metal fabrication; precision engineering; professional and technical services; textile, clothing, footwear and leather; and transport, logistics and distribution.

Interestingly, although it was identified as a key sector for the Victorian state in the 1994 inquiry, there is no analysis nor industry plan being prepared for the construction industry. However, other sectors that input into the construction industry are being analysed, namely the metal fabrication industry, precision engineering and professional and technical services. The logistics sector also is a significant supplier in the chain of contracts to the construction industry.

The automotive and building and construction markets are the two most important markets for the metal fabrication industry in Victoria (IBIS, 2000) with construction and building representing 48% of the market share and automotive only 14%. These markets create demand for products such as cast and forged metal components and aluminium and steel building products. Therefore, the health of the metal fabrication industry is reliant to a large degree on the health of the building and construction industry.

There are three major considerations that impact upon downstream demand for the metal fabrication industry. The first is the obvious, and oft-cited, cyclical nature of the building industry. The second is the increase of globalization. The metal fabrication industry was under severe threat in Victoria due to alarming increases in imports from 18% in 1992–1993 to 41% in 1997–1998. Finally, the threat of technological change through the use of new materials and new technology will cause continual pressure.

The examination of this industry resulted in various strategies being suggested that were aimed at:

- increasing metal fabricated exports
- increasing domestic share market; and
- exploitation of specialized niche markets (especially in light metals).

Globalization was seen as both a threat and an opportunity.

In terms of strategies that are directly related to procurement in the supply chain, two suggested strategies were discussed. These included the idea that organizations either become consolidators (i.e. organizations are formed by mergers and acquisitions to create synergies and eliminate investment duplication) or specialists (i.e. those who specialize in products, technical processes or functional capabilities in order to survive long-term) (McKinsey, 1999, cited in SAVI, 2001). According to McKinsey's model (1999), the scale of operations in Victoria and Australia indicates that firms would become specialists. Alternatively, the popular strategy of sections of the industry forming 'clusters' to compete was proposed.

2.4.13 Summary

The discussion until now has typically focused on the various governmental investigations and policy directions in terms of strategies for improvement of the performance of the construction industry. Although there may be unique interpretations and strategies, many are united in the assumption that the industry has a fragmented *structure* and fragmented *process*. Fragmentation is either the key determinant of the poor performance of the construction industry, or a positive attribute of the industry that indicates firm specialization but requires careful management.

The various agencies typically then perpetuate the same approach to understanding the industry structure through the same methods. For example, their view of the structure of the industry is based upon simple descriptions matching the number of firms in the industry to the number of employees in the firm to the annual turnover. The same descriptions are often repeated and resurface in many reports. The underlying structure of the industry is not progressed any further than this, yet claims are made regarding the need to understand the underlying causes through understanding the underlying structure. The descriptions simply rely upon grouping all firms in the industry and do not consider any differentiation between any groups on any basis. The Victorian state government with the development of various industry plans has initiated some thinking towards the sectors as firms within markets – thus trying to develop an industrial organization economic perspective.

However, the Victorian state government did not attempt to develop a plan for the construction industry – it begs the question:

Why not? Perhaps it is too complex, too diverse and just too many different supply chains – where would one start? How would one evaluate the success of the plan?

The following section discusses in more detail the methods and associated difficulties for describing and evaluating the performance of the construction industry currently used by governments.

2.5   Government economic models of performance

It is well documented that the construction sector plays a significant role in the economy. Governments also understand that the construction sector plays a significant role in the economy and they are more and more understanding implicitly why they need to intervene in this particular sector of the economy. One of the greatest difficulties faced by governments is understanding the impact of their policies.

Box 2.15 Commentary

Who does construction industry policy development? It is speculated that in many governments in many countries in the world, construction industry policy is developed by construction professionals who are employed within a government agency and have knowledge and experiences of the industry – they may be an architect, an engineer, a quantity surveyor or a construction manager – they may also have done some additional study – either postgraduate studies in a built environment discipline or from another discipline – either business, management, law or public sector management – perhaps economics. OR – they may not have completed any additional studies at all. Typically, their heart lies within the art and science of design and construction; it does not lie within policy making…this is perhaps both a blessing and a curse.

Two measures that are indicators of the importance of a sector are its contributions to gross domestic product (GDP) and to employment figures. The construction sector makes one of the most significant contributions to a nation's economy in terms of GDP and employment of any of the service industry sectors.

There are three ways of measuring GDP: expenditure approach, production approach and income approach (Scollary and John, 2000). Each measure may give rise to different values of the GDP. The common method is the expenditure approach. The construction sector in Australia, for example, consisting of firms mainly engaged in the construction of residential and non-residential buildings and engineering structures and in related trade services, accounted for 5.5% of GDP for 1999–2000 and employed almost 8% of the national work force (ABS, 2001).

However, the real contribution of the construction sector and the firms directly associated with construction projects has been a concern with governments and organizations such as the OECD alike for a number of years. Although many national accounting systems often cite the GDP for construction in their respective countries as ranging between 2 and 6%, a 1985 United Nations member country study suggested that the GDP is in reality approximately 6–17% (OECD, 1985).

A comparison between the countries using the real value of final expenditure at international prices produces a resultant league table for a selection of countries with the United Kingdom at (6%), United States (10%), Australia (11%) and New Zealand (11%). According to Kenley (2002) these may differ from those obtained from national accounts at any given time, but they serve to illustrate that construction is a major contributor to the national effort and that the role of the sector to that national economy has been undervalued in the past.

2.5.1 Macroeconomic performance measurements

How does GDP really help us to measure industry performance? A number of policies by governments discussed so far have raised the issue of the need to monitor and evaluate the impact of various policies. There is an implicit understanding that the underlying structure of the construction sector influences the way that it performs as a whole, and that the conduct and performance of firms within the sector is also influenced by the structure. However, there are to date few examples which even discuss these inter-relationships of structure, conduct and performance in relation to the supply chain. There is a well developed branch of economics, industrial organization economics, which has various measures relating to a particular sector which considers structure, conduct and performance as a conceptual framework but draws the boundaries around the market. This is problematic as the supply chain has multiple markets – more on this is discussed in Chapter 4, Industrial organization economics methodology and supply chain industrial organization. Sector performance is often considered quite separately and based upon indicators that use aggregated data which bears little relationship to unique markets and specific supply chains.

Instead, governments tend to rely upon information about the performance of the sector which is generated from their national statistical agencies. In many ways, decisions are made based upon information that only reflects part of the reality of the construction sector and more importantly doesn't consider the role of firms and the nature of markets in the sector. In short, the measurement and monitoring typically takes a macroeconomic approach to the sector this is useful. However, it can be problematic as these types of indicators aggregate data and give a false understanding of individual issues in markets related to specific commodities and firms. The supply chain concept does not sit easily with the macroeconomic methods of industry performance. Although governments are moving towards policies that can change the organizational structure of the chains, the market structure within those chains and the behaviour of the chains, markets and the firms, this approach to describing the sector and the various chains is largely unexplored.

In Australia, the method for classifying a construction sector firm (and employer/employee) is typically strictly limited to construction site trade services and excludes all (business) units mainly engaged in providing architectural supervision or consultant engineering services; such units are included in Property and Business Sectors (ABS, Cat 1292.0 ANZSIC Construction Industry Classification). Most countries adhere to an industry classification system developed by the OECD known as the International Statistical Industry Classification (ISIC) system. Firms that supply to subcontractors may or may not be included in this sector. In many situations they would be accounted for in manufacturing and wholesale. Coupled with this, a proportion of the mining and quarrying; agriculture; hunting and forestry; manufacturing; real estate, renting and business; wholesale and retail and transport, storage and communication sectors contribute significantly to construction projects.

The impacts of the construction sector go well beyond the direct contribution of onsite construction activities. There is little denying that the sector's effects are widespread. Whether or not various other sectors are included or excluded in the national accounts, there is little denying that the sector has important linkages with other sectors. These linkages give an indication of the chain of production associated with the construction sector. This is often discussed as inputs and outputs to the sector and whole sectors are aggregated and considered in terms of inputs and outputs. This is now considered in more detail.

2.5.2   *Macroeconomic input–output analysis: aggregated construction supply chain measure*

The econometric input–output analysis is the primary technique for understanding the aggregated relationship between the construction sector and the various sectors that are linked with it either upstream or downstream in the productive chain.

A well-known method for understanding these types of sector inter-dependencies has been developed and is known as the macroeconomic input–output analysis technique. This technique describes how the construction sector interacts with other sectors and seeks to understand the role of the construction sector in the national economy of a country. The input–output technique is based on Leontief's (1936) insight that commodities are needed in the current production of other commodities. The output of a good or service in an economy is either used in the production of goods and services (including itself) or it goes into final consumption (e.g. households, export, government). Each output in an economy can be represented by an equation, with output equal to its final consumption plus the sum of its inputs used in all production activity throughout the economy. It is possible to trace the effects of a change in final demand, or change in output, of one good or service throughout its inter-sector-linked relationships, so that the knock-on effects on other sectors may be measured. Mathematically, an economy can be described as an integrated system of flows and transfers from each activity of production, consumption or distribution to each other activity. Each sector absorbs the outputs from other sectors and itself produces commodities or services which are in turn used up by other sectors, either for further processing or for final consumption. All these flows or transfers are set out in a rectangular table – an input–output matrix.

The construction sector is an important contributor to an economy both as a sector in itself and its relationship to those other industries in the productive chain. There is an overwhelming volume of transactions occurring annually which clearly indicates that procurement in the supply chain is of national importance. This section has established both the significance of the construction sector, the significance of the supply chain and associated with that the significance of the contractual links procurement in the construction supply chain to the national economy. It has also indicated that, although firms and their relationships and the nature of markets is on the agenda for governments, the economic models to describe this are not in use.

To understand the supply chain interdependencies at an industry level the descriptions of the performance of the sectors has traditionally focused on macroeconometric techniques that aggregate numerous and quite diverse subsectors into a construction sector. In this manner, all the construction sector input and output is mapped against all the inputs and outputs of another linked sector – for example, the manufacturing industry. Output and input is usually measured through economic indicators such as share in gross national product and national income.

The importance of a more refined view of the interdependencies between firms and their markets in the supply chain and the inclusion of the manufacturing sector has been recognized by governments and construction researchers alike. The input–output methodology has, in the past 10 years, gained acceptance as a technique to understand interdependencies between various sectors that supply to the construction sector. In this context, the most significant proponent of the input–output analysis technique has been by Professor Ranko Bon (1986, 1988, 1990). The thrust of his work has been to analyse the role of the construction sector and its interaction with other sectors of the economy and then to compare these results across countries. This economic analysis can give an indication of the effect on the construction industry that a change in demand, or output for a good or service in an interlinked sector, will have on the industry. The types of results in this work suggest that the construction sector follows the economic destiny of the manufacturing sector, its primary partner in economic growth and development (Bon, 1990). Sectors that are commonly used in the analysis include agriculture, mining, construction, manufacturing, trade and transportation, utilities, services and government.

In such analysis there is a heavy reliance upon nationally collated statistics and the related definitions for sectors. There has been considerable discussion of this technique and its value to the construction industry, particularly related to the severe problems associated with using input–output tables to measure *productivity* of the construction sector. Does it really measure productivity? The argument tends to suggest that the reliance upon the data that is currently available and collected is inappropriate and unreliable. It is interesting to note that this debate began some 20 years ago. Even proponents of the technique acknowledge the serious deficiencies of the data they are using.

This methodology gives us a global perspective on the chain of production and is quite useful. However, it tells us little about the nature of the individual markets of the firms that are supplying more specific products and services to the construction industry – nor does it intend to. No delineation is made between aluminium supply chains and brick supply chains or air conditioning supply chains – the real 'stuff' that we work with in our daily activities. As noted previously, for descriptions about the specific commodity markets we need to turn to another branch of economics; namely, industrial organization economics.

2.5.3 *Industrial organization economics and market analysis*

The industrial organization economic model as a methodology deals with the performance of business enterprises and the effects of market structures on market conduct (pricing policy, restrictive practices, innovation) and how firms are organized, owned and managed. The emphasis in this field is primarily on the structure, conduct and performance of firms that compete with each other within a particular market and then the associated policy considerations. The most important elements of market structure in these models refer to the nature of the demand (buyer concentration), existing distribution of power among rival firms (seller concentration), entry/exit barriers, government intervention and physical structuring of relationships (horizontal and vertical integration). The role of the industrial organizational model is to give substance to the traditional neoclassical abstract concepts of market structures.

The Australian Expert Group Industry Studies (AEGIS) model discussed earlier in the chapter is an attempt to explore further industrial organization economics and certainly reveals more about the workings of the industry in terms of the firms and the participants of the industry than the high-level atomistic input–output concept of production chains. The connection between supply chain body of knowledge and the industrial organization approach is discussed in further detail in the following Chapter 3, Supply chain theory and models.

2.6 A final word

Box 2.16 Further reading

Government agencies change direction – perhaps a cynic might say in direct response to new leaders. However, the following are early publications on a study conducted on 'Supply Chain Sustainability', which takes the ideas from the results of this text and investigates supply chains in the pre-cast concrete sector and the construction waste sector using the blueprint advocated in the final chapter of this text.

Construction Supply Chain Industry Policy Analysis

London, K. and Chen, J. (2006) Construction Supply Chain Economic Policy Implementation for Sectoral Change: Moving Beyond the Rhetoric, The Construction Research Conference of the Royal Institute for Chartered Surveyors, London, 6–8 September.

Chapter summary

1 Governments worldwide are seeking to improve the construction industry. National plans are being adopted that work towards this and which consider the supply chain concept somehow an important part of their plan. Understanding the industry at the supply chain level is clearly on the agenda for many countries.

2 This chapter provided a selected review of policy documents that described government approaches towards improving the performance of their construction industry. The question of performance tends to be interpreted as either one of an industry fragmentation problem or a firm specialization industry attribute.

3 Those who consider it a fragmentation problem espouse normative models of integration at the project level to solve the productivity and performance problems. Those who consider it a firm specialization industry attribute tend to accept the industry as it is and espouse positive models of co-operation and competition at the firm level.

4 The trend in recent years has been towards more positive models whereby government policies intervene with an appreciation of firm profitability as the way to achieve industry productivity. In many of these models the constraints of the industry are viewed in a different light; for example, the growth of SMEs is viewed as a healthy indicator of economic growth within a sector and one that should be managed, rather than as a problem that should be solved.

5 Governments have three major roles to play in the industry: as controller of the regulatory framework, as a major client, and as policy maker. Industry plans often suggest various strategies that aim to change the behaviour of firms operating in the industry without knowing how to evaluate, monitor and measure that change.

6 The last part of this chapter described the economic models for description of performance and measurement used by governments. It is assumed that the performance of business enterprises will improve the industry, and yet, too few approaches address the real world view of the structural and behavioural characteristics of the supply chain which typically manifest themselves in decision making involved with procurement. There is a gap in the understanding of construction contractual relationships, between the highly aggregated and industry level, and that which focuses upon the boundaries of individual projects.

7 Understanding the nature of only one market sector is naïve when developing construction policies – as the property and construction sector, like the economy at large, is so interdependent and has numerous supply chains. The industry is rarely understood through the characteristics of the individual markets and the interrelationships between markets. Very little analysis relies upon developing a methodology for understanding the industrial structure of the sector. Descriptions of the sector assume that it is homogenous and is fragmented. Too few strategies for industry development and improvement make transparent how the industry is organized, how markets are organized, how firms are organized and why firms behave in the manner that they do.

8 In construction policy, procurement is often only seen within the framework of project procurement and rarely understood as practices that occur at each level in the supply chain and between countless firms. If this is done, performance, efficiencies, productivity, innovation and competitiveness become a part of the procurement model of the supply

chain. It is in the interests of all to understand the underlying structural composition and behavioural practices in the industry throughout the supply chain. A supply chain procurement model would make transparent the structural organization of the industry and unveil the interdependent nature of structure and behaviour. Although the trend is towards seeking positive models for procurement, it is implicit and as yet there are no explicit models that describe procurement in the construction supply chain.

9 Procurement is a major part of the *rules* of behaviour or *rules of the game* in the industry. Modelling of procurement is just one way to comprehend the economics of the construction supply chains. It will allow strategies to be developed which will target growth for particular regions or markets.

10 There is a growing body of literature in the emerging construction supply chain field and it is important to understand where the work in this particular text is located in the supply chain body of knowledge. Therefore, the following chapter charts some of the major movements in supply chain theory. In particular, the chapter reviews trends in the research to determine the level of research involved in supply chain modelling and, in particular, supply chain procurement modelling in relation to industrial organization economics.

3 Supply chain theory and models

Are there any supply chain economic approaches? Where is this research located within the existing body of knowledge?

3.0 Orientation

Box 3.1 Chapter orientation

WHY: Chapter 3 describes the background to the supply chain concept and locates this text within the current body of knowledge of supply chain literature.

HOW: Chapter 3 critically reviews the supply chain management literature in terms of major themes. The review is not intended to provide detail – the purpose is to provide a context and an orientation to the major approaches taken in the field. By providing this broad overview you are then equipped to seek out more specialist material which is particular to your own needs in two ways. First, you are given a framework or process in which to evaluate other material and, second, you are given content areas which will guide you to other writers around the world in both the construction-specific and the various other business and management communities. The review includes selected key works in both the construction management and economics literature and the wider supply chain-related literature in other industries. It identifies four main streams of research, including: distribution, production, strategic procurement management, industrial organization economics.

WHERE: The review takes in material from academics located all over the world, including the United Kingdom, United States, Australia, Hong Kong, South Africa, New Zealand, South America, Japan, Sweden and Italy.

WHAT: This chapter highlights the lack of broad-based empirical work which is required to develop the field. It also highlights the lack of research in construction literature associated with developing an industrial organization perspective of the chain as opposed to the empirical and theoretical research work evident in other sectors.

WHEN: The review chronologically maps selected material from the early 1980s till 2005.

WHO: Without a doubt the supply chain concept is relevant and can be applied to any type of enterprise in modern life – indeed, whether in business, non-profit or government activities. In its simplest form in our various situations we receive 'work' from someone – a client or a customer – and we 'engage' suppliers to complete that 'work'. The supply chain concept has a role to play in all industries. There are many examples which we can draw upon, including: retail, construction, ceramics, auto manufacturing, textiles, information technology, concrete and electrical engineering – to name just a few. The principles discussed in this chapter can be modified, adapted and developed and applied to many environments. The particular focus in the chapter is the construction sector – but the breadth and scope of the literature reviewed provides a context to think about the supply chain for any constructed systems.

3.1   Introduction

This chapter serves to provide a broad overview of the current literature on the supply chain concept and organize it according to key themes related to the perspective taken by the researchers. The review indicated that there has been a lack of theoretical and empirical research within the construction community that considers the fundamental structural, economic and organizational nature of the industry's supply chains, but has instead focused upon performance and benchmarking research – in particular, supply chain management. This type of research tends to support a normative model of project integration. An industrial organizational economics perspective, which is developed through Chapters 4 and 5, will provide a framework which would both challenge and complement the current performance and benchmarking of supply chain research. Of course, we always want tools and techniques and 'quick fixes' – this text is unfortunately not about quick fixes. It is more constructive for strategic thinkers – and, let's face it, if you are reading this text you are a strategic thinker – to be provided with a framework, and then some examples and scenarios to match that framework against – most organizations need to work out their own tools and techniques. It would be presumptuous of me to tell anyone how to manage their own firm – this text is about providing a framework and then some insights into how some organizations behave. You can make your own judgements on what I present. This chapter provides some insights on the development of that framework.

Supply chain management (SCM) for an individual organization emerged in the late 1990s as a distinct field of research in the construction management discipline, but less attention has been devoted to investigating the nature of the construction supply chains and their industrial organizational economic environment. A review of key construction and mainstream management supply chain literature is organized around four themes: distribution, production, strategic procurement management and industrial organization economics. The review highlighted that there was a distinct gap in theory and practical examples in supply chain research related to knowledge about procurement practices and the industrial structure within which these are embedded. The merging of the supply chain concept with the industrial organization economics model, as a methodology for understanding firm conduct and industry structure and performance, is an important contribution to both construction supply chain and construction economic theory. It provides a framework to understand procurement practices and behaviour by firms involved in construction supply chains.

Much of the industrial organization supply chain literature has tended to focus upon manufacturing industries where firms are typically permanent organizations. This raises issues as to the differences between industries founded upon temporary organizations compared with permanent organizations. There is potential for the development of an industrial organization methodology applicable to the project-based industry and this is explored in further detail in Chapter 4. Ultimately, industrial organization research seeks to have direct implications for industry performance and policies.

3.2   Supply chain terminology

One of the contentious issues in the supply chain literature is the difficulty in defining the boundaries of the supply chain concept since there are so many different interpretations. It has been suggested (Day, 1998; Hines *et al.*, 1998) that supply chain research is fragmented and that the field suffers from a lack of validity related to this definitional problem.

Nevertheless, there are four major themes which have been identified in the interpretations of the supply chain which revolve around the following: distribution, production, strategic procurement and industrial organization economics. Rather than attempting to negotiate rights and claims over definitions, Section 3.2 provides key definitions related to each approach. Further to this, Section 3.3 provides a broad overview as it charts these major trends in the supply chain movement. The definitional differences typically parallel the major trends in the supply chain movement. It has often been argued that there is a logistics versus supply chain dichotomy that has seemingly fragmented and plagued the concept. It appears that there is not only a logistics and supply chain debate but that there has been a wide divergence in how the supply chain concept is thought about. Yet, it is not quite as fragmented and chaotic as some would suggest, because there are both implicit and explicit connections between the areas. Rather than a negative view of the inter-'disciplinary' discourses taking place, it is this rich weaving of ideas that can ultimately assist in the development of a theory. Interpretations typically draw upon the following orientations as set in Box 3.2:

Box 3.2 Summary of various supply chain definitions

Distribution

- a combination of distribution, logistics and marketing perspectives combining upstream and downstream supplier management of materials and information flow, as best exemplified by various definitions, including those by Baker (1990), Christopher (1998); Bowersox and Closs (1996); Copacino (1997) and Lambert *et al.* (1998)
- and in construction, the materials management literature: Clausen (1995) and Agapiou *et al.* (1998).

Production

- lean production and materials and work flow in organizations; as best exemplified by Krafcik (1988) and Womack *et al.* (1990) models
- and in construction literature, OBrien's production approach (1995) and the lean construction definitions (Howell, 1993; Seymour, 1996).

Strategic procurement management

- an organization's strategic management of their suppliers for competitive advantage; as best exemplified by Ross's strategic versus tactical approach (1997), Lamming's lean supply concept (1992) and Porter's value chain concept (1985)
- and in construction management literature, Cox and Townsend's relational and asset competence model (1998) and Saad and Jones' (1998) key suppliers (specialist subcontractor management) approach.

Industrial organization economics

- a wide industry or market perspective on the research problem; as best exemplified by Harland's (1996) supplier networks, Nischiguchi's Alps (1987) structure of suppliers and Lambert *et al.*'s network of business relationships model (1998)
- and in construction, an early attempt by London *et al.* (1998) on supply chain procurement modelling.

Broad overview

- And then an overall categorization of supply chain definitions into tight versus loose and soft organizational versus hard logistics by Day (1998), who originated from the ceramics industry in the United Kingdom, and Hines' (1998) typology of five types of supply chains, including intra-functional, inter-functional, inter-organizational, network and regional clustering. This is discussed in more detail in Chapter 4 – particularly with reference to network and regional clustering.

The following are a selection of the supply chain definitions and related concepts, including: supply chains, distribution channels, supply chain management, logistics and lean production.

3.2.1 Distribution: mainstream management

The early work in distribution, commenced in industrial dynamics in the 1960s by Forrester (1961), provides a background to much of the analytical modelling distribution-related literature that became important in the 1990s. He simulated channel interrelationships in a production-distribution system for retail chains in order to demonstrate how the total channel behaves and responds in terms of supply and demand. The result of this work became known as the *bull whip effect*, whereby incorrect forecasting can have increasing amplification as information moves down the chain.

The relationship between information and product flow, from raw material through the entire chain, became the central concern of materials management which in turn became known as the logistics movement. In the early 1990s, logistics and supply chain management were being defined as the one concept; however, it soon became apparent that logistics was only a part of the wider concept known as supply chain management. There was quite serious debate and concern in the logistics community internationally about this for a number of years.

For example, Christopher (1998), one of the early recognized logisticians, noted that, whilst the concept of supply chain management is relatively new, it is in fact no more than an extension of the logic of logistics and suggested the following:

Box 3.3 Logistics definition

Logistics is essentially a planning orientation and framework that seeks to create a single plan for the flow of product and information through a business and that supply chain management builds upon this framework and seeks to achieve linkage and coordination between processes of other entities in the pipeline, that is, suppliers and customers and the organization itself (Christopher, 1998, p. 15).

He defines supply chain management as follows:

Box 3.4 Supply chain management definition

Supply Chain Management is the management of upstream and downstream relationships with suppliers and customers to deliver superior customer value at less cost to the supply chain as a whole (Christopher, 1992, p. 15).

To distinguish between supply chain management and the supply chain, the following definition was provided:

Box 3.5 Supply chain definition

The supply chain is the network of organizations that are involved, through upstream and downstream linkages, in the different processes and activities that produce value in the form of products and services in the hands of the ultimate customer (Christopher, 1998, p. 15).

According to Lambert, Cooper and Pugh (1998), the Council of Logistics Management announced a modified definition of logistics based on the emerging distinction between supply chain management and logistics in 1998. The modified definition explicitly declares the council's position that logistics management is only a part of supply chain management. The revised definition was as follows:

> ### Box 3.6 International Council of Logistics Management definition
>
> Logistics is that part of the supply chain process that plans, implements and controls the efficient, effective flow and storage of goods, services and related information from the point-of-origin to the point-of-consumption in order to meet customers' requirements (Lambert *et al.*, 1998, p. 3).

Lambert, Cooper and Pugh then define alternative supply chain management as follows:

> ### Box 3.7 Alternative supply chain management definition
>
> Supply chain management is the **integration of key business processes** from end user through original suppliers that provide products, services and information that add value for customers and other stakeholders (Lambert *et al.*, 1998, p. 1).

Although the two definitions of supply chain management (Boxes 3.4 and 3.7) seem very similar, it is perhaps worthwhile to think of Lambert's call to integrating key business processes throughout the supply chain – it seems somewhat more specific than Christopher's "management," but is probably driving us to achieve the same end. Once again we see a fixation on integration. If we really consider this for a moment we can see that this may be problematic. Think about your own supply chain. How difficult is it to integrate even your suppliers, let alone all the players in the chain from end user to original supplier? If this is the aim – how achievable is it? The case studies in Chapters 6 through to 8 provide a range of examples of a spectrum of 'integration' indicating a realistic attempt at integration. In summary we should understand what position we play in each of our markets and then position ourselves so that we can firstly manage first tier suppliers in our strategic chains. There are examples in the case studies where an organization is able to extend its reach beyond one tier and begin to integrate key business processes.

These definitions arrive at a time when the logistics field was evolving into a much wider concept than the function that the materials management or transport department undertook within a large manufacturing company. This evolution is well discussed by others (Coyle *et al.*, 1996), where it has been suggested that there was a period of much fragmentation during the 1960s, through to evolving integration in the 1980s. If you are interested in logistics then it is worthwhile seeking out Coyle *et al.*'s work, as they provide a succinct historical chronology of the field.

The logistics and supply chain management terms have often been used interchangeably and, *throwing caution to the wind*, Copacino's (1997) definition reflects this:

Box 3.8 Logistics and supply chain management combined definition

Logistics and supply chain management refer to the **art** of managing the flow of materials and products from source to user. The logistics **system** includes the total flow of materials, from the acquisition of raw materials to delivery of finished products to the ultimate users (Copacino, 1997, p. 7).

Poirier and Reiter (1996), similar to Copacino (1997), in their definition, offered the term system to describe supply chains and thus proposed a concept of interacting components with interdependencies. It is also good to see an acknowledgement that it is an 'art'.

Box 3.9 Supply chain

A supply chain is a system through which organizations deliver their products and services (Poirier and Reiter, 1996, p. 6).

Poirier and Reiter's (1996) model of supply chains was then developed based upon the chain as a network of interlinked organizations, or constituencies, that have as a common purpose the best possible means of affecting that delivery. For a practical approach and simply written, this is a worthwhile reference to seek out. Their model is reflected in the following diagram (Figure 3.1):

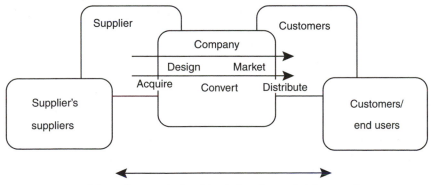

Information, product and funds flow (forward and reverse)

Figure 3.1 Supply chains.

Source: Poirier and Reiter, 1996, p. 6.

Bowersox and Closs (1996), steadfastly entrenched in the distribution field, used the channel concept as defined by the American Marketing Association in their development of the concept of an integrated supply chain process based upon logistics.

Box 3.10 Distribution channel definition

A distribution channel is the structure of intra-company organization units and extra-company agents and dealers, wholesale and retail, through which a commodity, product, or service is marketed (Baker, 1990, p. 47).

Bowersox and Closs (1996) suggest that the 'supply chain perspective shifts the channel arrangement from a loosely linked group of independent businesses to a coordinated effort focused on efficiency improvement and competitiveness'. They suggested that information and inventory (product) flow would be most efficient and effective if there was supply chain integration; such marketing and logistical requirements would rely for successful distribution upon channel-wide co-operation. According to these authors, success or failure is related to relationship management, which is 'the development and management of supply chain arrangements' (Bowersox and Closs, 1996).

3.2.2   *Distribution: construction materials management*

In interpreting the supply chain concept, the construction materials management researchers or logisticians have applied the philosophy in the same literal and operational manner to materials supplier management as the early logisticians. The management of materials to site and on site is an important area of work that, in the evolutionary days of the construction supply chain concept, was central to the supply chain concept.

Clausen (1995) defines construction logistics as follows:

Box 3.11 Construction logistics definition

[L]ogistics comprises planning, organization, coordination and control of the materials flow from the extraction of raw materials to the incorporation into the finished building (Clausen, 1995).

Agapiou *et al.* (1998) have also investigated material flow to construction sites and adopted Clausen's definition. Their research was based upon the logistics concept and considered the flow of materials to site rather than workflow. They also identified the importance of flow of information and an emerging role for the materials supplier as suppliers of information in the early design phase.

3.2.3 Production: mainstream management

I could not in all good conscience go too far into a discussion on supply chains without mentioning the word 'lean'. My colleagues are probably already 'up in arms' that we are into Chapter 3 and have not provided a lengthy discussion on this concept. The supply chain concept has indeed been heavily influenced by production theory and, in particular, *lean* production. I certainly do not want to detract from the enormous contribution that lean production has made to the supply chain movement – however, there are many more texts which have dealt with this subject-matter in a much more comprehensive manner than I intend on doing.

Briefly, the term *lean* was actually first coined to describe this system by Krafcik (1988) to explain the system of production developed by Toyota, the Japanese auto manufacturing company.

Box 3.12 Lean construction

The lean approach attempts customization of high volume production to provide customers with exactly what they want at the time they want it. It is characterized by improving flexibility, reducing waste and improving flow along the supply chain. The flow is improved through management and control of each actor along the supply chain. The concept relies upon some form of integration from raw material supplier to various subcontractors who supply materials or components to the manufacturer.

3.2.4 Production: construction management

In an attempt to address problems of poor productivity in the construction industry, researchers sought inspiration from the successful lean production concept. Koskela (1992) was the first to consider the relevance of production theories and in particular the lean production theory for construction. Perceiving poor workflow as a problem in construction, *lean* researchers attempted to apply *lean* production theory.

Howell, a keen lean protagonist, in 1993 defined it as follows:

Box 3.13 Lean construction

A new way to manage the production in the architectural, engineering and construction industry with implications for commercial relationships and project delivery processes. Lean construction planning and control techniques reduce waste by improving work flow reliability.

Later, in 1996, Seymour *et al.* added to the thinking:

Box 3.14 Lean thinking

[L]ean construction is a philosophy, a radically new way of thinking, talking about and re-forming the processes and organization of construction (Seymour, 1996).

The lean construction movement has been primarily concerned with identifying waste in the construction process and then systematically eliminating that waste. Lean also explicitly recognizes the role of the supply chain in creating waste and this is further explored in the next section on strategic procurement.

It is also a recent concept in the construction research community and therefore the actual boundaries of lean construction and in fact supply chain management are still being negotiated. It would be remiss of me not to mention the early work of Bill O'Brien as one of the keenest supporters of supply chain management. Although not explicitly writing about lean construction, but certainly using theory from production, he produced one of the earliest (if not the first) explicit construction supply chain studies. He analysed resource management by construction firms that supply to construction sites and suggested simply as follows:

Box 3.15 Construction supply chain management

Supply chain management is a set of concepts and methods that have been developed in the manufacturing industry in recent years (O'Brien, 1995).

This definition serves to highlight the borrowing of the concept and also perhaps the lack of theory developed within the construction management and economics research disciplines. But time has marched on and, although still a fledgling field, it is slowly emerging as a robust area with more and more practical examples of case studies associated with supply chain management.

3.2.5 Strategic procurement: mainstream management

Underlying many definitions is a *strategic* perspective of the network of organizations that compose the entire supply chain and associated with this is the management of this network.

At the heart of the lean production concept is the supporting system of inter-firm relationships (Lamming and Cox, 1995). Lamming's empirical research, where he defined the concept of lean supply, identified the importance of the strategic arrangement of the firms that would support such a system. Lamming's (1993) concept of lean supply is important to the understanding of strategic procurement in the supply chain.

Box 3.16 Lean supply

The definition of lean supply goes beyond partnership as it is practically manifested to a balanced value chain in which complementarity of assets is assured through joint analysis of competencies and investments. The lean supply chain is designed to compete with other supply chains, to win business from the end consumer of the product or service. It is a complex matter because the suppliers in the chain are simultaneously in several other chains. Lean supply is the name given to the supply system which is necessary to support lean production (Lamming, 1993).

This strategic versus tactical approach to the supply chain concept was explicitly highlighted by Ross (1997). In his text, *Competing through Supply Chain Management*, he acknowledged that businesses from a wide variety of industries, not simply the automobile manufacturing firms, had become increasingly interested in exploring the opportunities for competitive advantage that can be gained by leveraging the core competencies and innovative capabilities to be found in the networks of business partners within the supply chain.

Although Ross (1997) identified two levels with which to conceptualize supply chain management (SCM), namely the strategic and tactical (Ross, 1997), his text concentrated on the emerging strategic capabilities of the SCM concept.

Box 3.17 Competing with supply chains: strategic perspective

Supply chain management is a continuously evolving management philosophy that seeks to unify the collective productive competencies and resources of the business functions found both within the enterprise and outside in the firm's allied business partners located along intersecting supply channels into a highly competitive, customer-enriching supply system focused on developing innovative solutions and synchronizing the flow of marketplace products, services, and information to create unique, individualized sources of customer value (Ross, 1997, p. 9).

3.2.6   *Strategic procurement: construction management*

If we now turn to construction research, one of the most comprehensive studies that examines specifically construction supply chains was by Cox and Townsend (1998) in their text *Strategic Procurement in Construction*. Their work involved describing the practices employed by six large organizations in the strategic management of their supply chains. These are discussed in greater detail later in this chapter in Section 3.3.8. One of the contributions of this text was expanding the idea of supply chains from the literal construction materials supplier to a broader perspective. It is important to note that the model proposed by Cox and Townsend (1998) does not negate the importance of logistics; however, it was not its central focus.

Box 3.18 Strategic construction procurement

Strategic procurement management is the development of an external sourcing and supply strategy designed to maintain a sustainable position for that organization in the total value chain (Cox and Townsend, 1998).

Saad and Jones (1998) explored conceptually improving the performance of the specialist contractors through a more effective management of their supply chain, within the context of the supply chain management concept, rather than in the logistics concept. The premise was that this would improve the performance of the construction industry as a whole.

Saad and Jones (1998) described supply chain management as the following:

> **Box 3.19 Supply chain management**
>
> Supply Chain Management (SCM) is increasingly being seen as a progression on internal programmes aimed at improving effectiveness. The focus is now not only limited to increasing the internal efficiency of organizations but has been broadened to include methods of reducing waste and adding value across the entire supply chain. The holistic approach associated with SCM is essentially motivated by the benefits to be derived from a more effective management of the interfaces between all organizations involved (Saad and Jones, 1998, p. 453).

The supply chain management research typically orientates itself towards normative and integrative models, whether they be hard systems engineering and analytical models or soft relationship management models. However, selected researchers have approached the supply chain concept using a positive modelling framework whereby they research what is as opposed to what ought to be. Typically, these approaches have been attempting to describe the industrial structure and context. Some of these have been categorized as industrial organization models and are now considered.

3.2.7 Industrial organization economics

One of the significant contributions of the industrial economics approach to supply chains is the attempt to describe and analyse the system of supply chains. New (1997) noted that the development of the idea of the supply chain owes much to the emergence from the 1950s onwards of systems theory and the associated notion of holistic systems. There are many variations to systems theory, but at the core is the observation that a complex system cannot be understood completely by the segregated analysis of its constituent parts (Boulding, 1985). Systems theory and the supply chain concept sit well alongside each other – and indeed systems theory and an industrial organization economic approach are particularly complementary. Although perhaps many have moved on from systems thinking and there have been detractors, it still finds many supporters worldwide and has a resurgence every now and then. I find it particularly interesting and often find it quite pleasing to see new contributions and examples of how systems thinking is applied. Nevertheless, New (1997) provides an interesting discussion regarding the scope for research of the supply chain and explains particularly the twin dichotomy between research and thinking on the supply chain versus supply chain management.

Selected supply chain researchers in mainstream management literature have studied the structure of the production supply chain. There are, as a result, important models that merge the field of industrial organization and supply chain theory. This work has been largely pursued through a group of researchers who are concerned with analyzing issues related to the organizational structure and underlying systems of supply chains. The following are a selection of their interpretations of the supply chain concept.

Harland (1996) widened the perspective of the supply chain by suggesting the term, *the supply network*, as a means for capturing the full complexity of the firms involved through a more holistic view of the process. A supply network can be defined as:

Box 3.20 Supply network

[A] number of entities, inter-connected for the primary purpose of supply of goods and services required by end customers (Harland, 1996).

These 'entities' may be engaged in long-term relationships, but the boundary of the network is ultimately ambiguous. In reality, a spectrum of supply relationships exists, ranging from tight long-term to loose short-term relationships. The term supply network seems to be gaining increased acceptance in the literature (Slack, 1991). The supply network suggests that there are different structures of chains, alternative reconfigurations, potentially numerous firms involved, and a high degree of complexity.

3.2.8 Industrial organization economics: construction

With this in mind, and influenced by Cox and Townsend's (1998) case study work on supply chain management, London *et al.* (1998a) proposed the following working definition specifically for supply chain procurement for the construction industry:

Box 3.21 Supply chain procurement

Supply chain procurement is the strategic identification, creation and management of critical project supply chains and the key resources, within the contextual fabric of the construction supply and demand system, to achieve value for clients.

Although a definition strictly related to strategic procurement and normative in its intent, it identifies that the contextual fabric of the industry is important to procurement modelling. Further to this, the following Figure 3.2 illustrates a model of the supply chain more representative of the major groupings of construction industry actors and is an adaptation of Poirier and Reiter's earlier model. I personally often feel more comfortable when I start to identify the players in my own work environment – the idea becomes more concrete and perhaps less abstract. Another important feature in the construction industry is that there are generic construction supply chains such as the following diagram indicates; however, each project can have unique procurement relationships that alter the chain dramatically. This is particularly relevant with contractors, clients and consultants and design and construct (design build), project alliance and build-own-operate and transfer contracts. Alternative procurement methods, that have included actors which have traditionally been located further down the chain – for example, specialist subcontractors or manufacturers – can begin to play a particularly critical role in supply chain management.

London and Kenley (2000b) extended the development of a wider perspective of the construction supply chain concept with exploratory empirical investigations. Winch (2001) also suggested that there was a wider perspective of the supply chain when he delineated between the *project supply chain* and a *construction industry supply chain* and this is discussed in more detail in Section 3.3.8 and also Chapter 4.

Although a somewhat exhaustive introduction to the various interpretations of concepts associated with the supply chain, it is anticipated that this provides a good fundamental platform for the next Section 3.3 – which charts the supply chain movement. Section 3.3 is organized in a similar manner to this section.

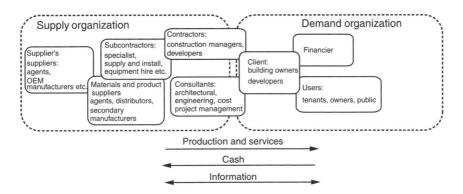

Figure 3.2 Generic construction industry supply chains.
Source: London, 2002.

3.3   Charting the supply chain movement

It is apparent that the supply chain concept is part of an eclectic and developing hybridized interdisciplinary field. It became an explicit area of research in the mid-1980s and originated largely from the two separate management streams of distribution and production, which merged into the field of logistics. The first use of the term was by Oliver and Webber in 1981, who used it as a metaphor to describe the integration of business functions so that production can be geared toward the needs of the customer. The intervening two and a half decades have witnessed an overwhelming amount of analysis and discussion, particularly in the general management literature.

As the previous section indicated, there have been wide interpretations of the supply chain concept. It is now generally accepted that logistics and supply chain management are indeed two separate concepts where the logistics area is subsumed by the broader field of supply chain management.

Since it became an identifiable area of research, the supply chain concept has been widened through the influence of other research frameworks and for this literature review these may be broadly grouped by the same four themes that the definitions were grouped by:

- distribution
- production
- strategic procurement management
- industrial organization economics.

Construction research involving the supply chain concept is a relatively new field, having explicitly emerged in the mid-1990s. Similar to the mainstream management literature, it is evolving with corresponding influences from the theory of production, distribution and strategic procurement.

Significantly, there has been little if any construction industry research merging the supply chain and industrial organization fields, as found in other research communities by Ellram (1991), Hines (1994), Nischiguchi (1994), Harland (1996) and Lambert *et al.* (1998).

Figure 3.3 charts some of the more significant supply chain events, models and definitions against these four influences for the period between 1980 and 2005. It is important to understand some of the major pieces in the puzzle, since the streams are interwoven. The chart highlights key interpretations of the construction supply chain concept. The circled portion of the diagram represents the theoretical origins of the approach to the supply chain concept advocated and described in this text.

3.3.1   Distribution: mainstream management

Supply chain management has long been associated with the management of the physical distribution of products from raw material through manufacturing processes to 'point of sale' for the end product.

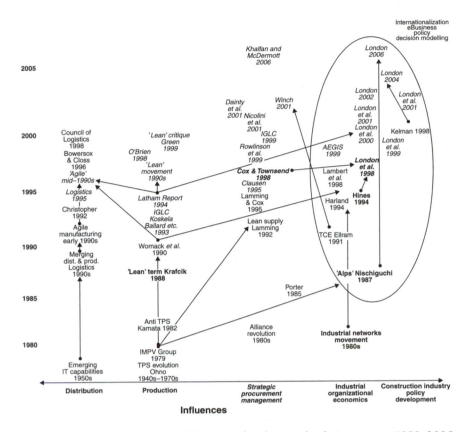

Figure 3.3 Charting the major influences for the supply chain concept 1980–2005 (adapted from London, 2005).

Note
Italicized text is construction management and economics literature.

As noted earlier, Christopher (1998) has been considered one of the pioneers of the logistics and supply chain movement. Borrowing from Porter's value chain concept (1985), he moved the perspective of materials management from a tactical low-level task in the organization to a strategic management concept that supports customer focus and creates competitive advantage.

This emphasized the development of integrated supply chain processes to support planning and co-ordination of complex supply chain systems for efficient and timely movement and storage of products and/or materials. Modelling of these systems has involved mapping of time and cost resources and considering such concepts as *time-compression* and *just-in-time* (Hall, 1983; Wantuck, 1989), relative location of stock and warehouse management (Gold, 1991), transportation analysis and optimization models to improve logistics performance. Many of these models and associated tools aim to improve the flow of information and the flow of goods through analytical and mathematical programming techniques.

This approach has often relied upon hard systems methodologies to model, forecast and predict the product, information and funds flow. Min and Zhou (2002) developed a taxonomy of the mathematical supply chain models that classified models as either deterministic, stochastic, hybrid or IT driven. In their review of these types of models, they concluded that 'reinventing traditional analytical tools will not be the answer for many managerial issues involving real-world supply chain problems.' They suggested that many of these managerial issues are soft (ill-structured, strategic and behavioural) and are not hard (structured, operational and technical) issues, which cannot commonly be addressed by analytical tools afforded by current mathematical programming techniques.

These types of models rely upon the quality of the data and such stand-alone rigid mathematical models are losing ground in both the supply chain research community and in their usefulness to the business community. They suggested that future efforts should be aimed at flexible decision support systems that allow for descriptions of behavioural issues which could include, for example, purchasing negotiation dynamics between buyers and suppliers. Future models should also rely more heavily upon model-based decision support systems that use communication techniques, knowledge discovery techniques and visual aids. They suggested that 'the development of IT driven models will be the wave of the future' (Min and Zhou, 2002).

Analytical research methodologies have included case studies, simulations, dynamic programming, etc., and have been applied to all manner of problems. For example, to *fast moving consumables*, and the retail industry where technological innovations such as electronic data interchange and vendor managed inventory have harnessed the capabilities of information technology to radically alter the flow of information and to be more responsive to changing customer trends (Stock, 1990; Introna, 1991). This flexibility in the chain, in response to customer demand, has been the cornerstone of the *agile manufacturing* concept which is discussed in the next Section 3.3.2.

3.3.2 Distribution: construction

Construction researchers have applied the SCM philosophy to materials flow, seeking to establish a relationship between site productivity and improved materials management.

Research in the United Kingdom has considered materials and components supply and flow with an emphasis on the role of builders' merchant in the supply chain. This work has called for long-term relationships and alliances to be developed between construction companies and merchants (Agapiou *et al.*, 1998). It was also suggested that, during the design phase, merchants should become the party responsible for the flow of information relating to building materials, as this may contribute to significant cost savings and increased productivity. The involvement of the materials supplier/wholesaler at an early stage of the decision making process in, for example, Scandinavia, did lead to cost savings and increased productivity (Agapiou *et al.*, 1988). This study, and those the researchers were referring to, focused on project-based models and considered on-site management through observing material flows and improving individual firm–firm relationships between large contractors and large merchants.

Another approach that relied upon manufacturing distribution theory was offered by those researchers involved in agile manufacturing research. *Agile construction* was taken up by some construction researchers who argued that *lean practices* and benchmarking would be an essential ingredient in achieving the target of a real cost reduction of 30% (Graves, 2000).

The concepts of *agile construction* and *lean construction* are blurred, with some claiming that there is a difference (Barlow, 1998) and others using the concepts interchangeably (Graves, 2000). The key difference, it appears, is that *agile* focuses upon responsiveness and flexibility and to a large extent on business practices (Barlow, 1998). *Lean construction* has been taken up with more enthusiasm by the construction research community with a focus on lean production theory, which is discussed in detail in Section 3.3.6.

3.3.3   *Production theory: mainstream management*

Production theory, particularly *lean* production, has been another major influence on the supply chain movement. The system of production relied upon the just-in-time (JIT) concept developed by the Japanese executive Ohno, the father of *lean* production.

Box 3.22 Just in time

The essence of JIT is 'to provide only the necessary amount of the necessary items at the right time and place – no more, no less' (Nishiguchi, 1997).

One of the seminal texts that promotes lean production and popularized the concept is The Machine that Changed the World by Womack *et al.* (1990), which resulted from an international benchmarking study conducted in the late 1980s by researchers involved with the International Motor Vehicle Program at Massachusetts Institute of Technology. Their study described and analysed the method of production termed lean production, best exemplified by the Toyota Production System (TPS), and pioneered by the Japanese executive, Ohno. There is an extensive amount of literature which describes various aspects of the TPS from a variety of perspectives. For example, Womack, Jones and Roos can be found within the production field; Lamming (1992) can be found drawing upon TPS within the strategic management field; and Nischiguchi (1997) are located within the industrial organization economics field of research.

The epistemology of lean production is posed against *craft* and *mass* production. Craft production was based upon the notion that manufacturers of complex products required skilled labour within a collaborative environment, supported by a system of apprentice-journeyman-master and Craft Guilds, etc. This gave way to mass production whereby unskilled labour could perform tasks designated and instructed by managers. Lean production is often considered a reaction, by the Japanese, to mass production and to have been forced by a need to compete with mass production through producing small and variable runs.

3.3.4   Merging production and distribution

As indicated in the previous section on definitions, prior to the 1960s the fields of production and distribution were fragmented, eventually evolving into two identifiable streams in the early 1980s. In the 1990s, both areas in research and industry practice have been subsumed by integrated logistics management (Coyle *et al.*, 1996).

There has been much debate attempting to distinguish the fields of logistics and supply chain management. The terms *logistics* and *supply chain management* have often been used interchangeably, particularly within the traditional logistics community. As noted earlier, the distinction was clarified by the Council of Logistics Management, a peak international body of industry and academic representatives (Lambert *et al.*, 1998). It was accepted that supply chain management was more than simply logistics and operational issues and that strategic supply chain management subsumed logistics. The distinction between logistics and supply chain management was made in the construction literature in 1998 (London *et al.*, 1998a,b).

3.3.5   Production approach: construction

In the late 1990s, post UK government construction industry investigations (refer to Chapter 2), the UK construction research community reconceptualized the construction industry as a 'manufacturing process', with implications for supply chain research. One approach, the Generic Design and Construction Process Protocol, a normative model, treats the industry as a production process (Aouad *et al.*, 1999). This work described the industry as a single process map for all phases by adopting the manufacturing model of New Product Development. This was quite a comprehensive blueprint; however, there appeared to be little or no empirical work to validate the protocol.

An alternative view of construction production theory was concerned with materials flow and process, and it raised important questions for workflow and resource management. O'Brien in the United States (1998) investigated the production and inventory decisions of multiple firms within the construction supply chain. He indicated that any managerial philosophy, such as Just-In-Time, applied to one site for one project in the construction environment, is problematic due to the temporary nature of project organizations.

O'Brien (1998) makes a further contribution with a systems view of the construction production supply chain, identifying that supply chain management offers the potential to optimize supply chain cost performance (O'Brien, 1998). Borrowing and modifying production manufacturing capacity cost models, he investigated eighteen firms to identify how capacity constraints of subcontractors and suppliers affect the costs associated with construction project schedule and scope changes. This work forms the foundation to develop models for supply chain performance at a project site management level.

The greatest focus of attention by the construction economics and management research community in relation to the supply chain concept has been on the application of lean production to construction – namely, lean construction – and this is now considered.

3.3.6 Lean construction

The *lean* construction movement has, from 1993, led much discussion on supply chains through the International Group for Lean Construction annual conferences. Koskela, Ballard, Howell, Tommelein, Marossezky, Barlow, Formosa, O'Brien, Vrihoelf and Alarcon are key researchers in this community. Formosa, O'Brien and Vrihoelf are particularly focused on supply chain management. *Lean* construction evolved from *lean* production, a developing field that is centred primarily upon a production philosophy for construction. In so doing, key protagonists have explored workflow and conversion processes, waste reduction and efficient use of resources, through *lean* project management, *lean* supply, *lean* design, *lean* partnering and co-operative supply chain management (Alarcon, 1997). The central themes have been eliminating waste and improving workflow in the construction.

To date, much of the construction literature has applied the *lean* concept without contextualization; for example, without the detailed empirical exploration of market structures that underpin the construction environment. Those researchers in the wider lean production field have understood this important issue. The contextualization of *lean* production that supports lean thinking has been provided through organizational and industrial organizational economic descriptions of the automotive and electrical industry supply chains (Hines, 1994; Nischiguchi, 1994; Sugimoto, 1997). This understanding of the organization of the supply chain is an important part of the philosophy of *lean* thinking.

Allied to this criticism and the importance of contextualizing lean construction, in the late 1990s, perhaps just as the lean construction movement was gaining momentum, the value to the construction industry of the *lean thinking* dogma and rhetoric (Green, 1999) was questioned. In his paper, *The dark side of lean construction: exploitation and ideology*, Green challenged the narrowly defined instrumental rationalist approach currently undertaken in the movement. He aimed at introducing literature to the lean community that provided evidence of the human cost of *lean* methods in Japanese industry (Sugimoto, 1997), including repression of independent trade unionism, societal costs (pollution and congestion) and regressive models of human resource management (Kamata, 1982; Sugimote, 1997). Kamata (1982) provides a personal account of life as an assembly-line worker inside the walls of a Toyota plant, discusses the physical exhaustion he experienced meeting impossible production targets, the army-like treatment of employees and the need for conformity and tight surveillance, etc., repression of independent trade unionism, societal costs (pollution and congestion) and regressive models of human resource management. He referred the lean construction community to Kamata (1992), who describes another perspective of the Toyota Production System – one where 'workers were often required to live in guarded camps hundreds of miles from their families and suffered high levels of stress at the workplace as they struggled to meet company work targets.' Green argued that 'whilst the *lean* rhetoric of flexibility, quality and teamwork is persuasive, critical observers claim that it translates in practice to control, exploitation and surveillance.'

He claimed that the lean movement was in reality a system based upon exploitative employment practices by corporate leaders and politicians. It was also maintained that lean thinking is not about customer responsiveness, flexibility and teamwork, as suggested by the lean supporters; rather, it is a system based upon control, exploitation and surveillance.

Green's argument is reminiscent of the dualist theory of subcontracting, of which the core suggests that economic agents located in different segments of the economy are treated unequally, regardless of their objective worth. Traditionally, it has been argued that behind the prosperity of Japanese industries, particularly in the automotive and electronics sectors, lies the sacrifice of many subcontractors. They are characterized as *sweatshops* with cheap labour and labour-intensive technologies (Ito, 1957; Fujita, 1965).

Green also highlighted that construction researchers have notably ignored the extensive literature that addresses the extent to which lean methods are applicable beyond the unique Japanese institutional context (Kenney and Florida, 1993). The notion that management techniques can be applied irrespective of context is in harsh contradiction to the long-established principles of contingency theory (Lawrence and Lorsch, 1986).

Lean proponents then defended the lean movement with the argument that it is based upon a long history of production management thinking, particularly the physics of production (Ballard and Howell, 1999), and that lean thinking simply offers a new way to organize production. Both protagonists are partially correct; *lean* is dependent upon a long history of production management thinking; however, it is also dependent upon a history of economic, technological, political, industrial relations and industrial organizational influences. Green, although using emotive language, was accurate in highlighting political, social, moral and industrial relations evidence that contextualizes the lean movement.

Lean production implementation by large producers would not have been possible had it not been supported by highly organized governance structures in the supply chain. Supply chains were organized into hierarchical clusters of tightly tiered structures of subcontracting firms known as *keiretsu* (Nischiguchi, 1994; Hines, 1994). *Lean* construction researchers, in their quest for production efficiency, in many cases have forgotten that organizing and controlling the market on a very wide and deep scale was instrumental in *lean* implementation. Lean construction proponents typically borrowed the concept and developed normative models for project integration with little regard for the situational context.

Nischiguchi (1994) developed an historical description and analysis of the Japanese subcontracting system. He provides empirical evidence of the development and organization of manufacturing industries from the 1920s through to the present day, highlighting the underlying structural characteristics of markets and the evolution of the subcontracting inter-organizational relationships.

Japanese economists have typically debated the nature of Japan's subcontracting small enterprises from two perspectives. The first position relies upon the dualist theory which holds that 'big businesses accumulate their capital by exploiting and controlling small businesses which have little choice but to offer workers low pay under inferior working conditions' (Sugimoto, 1997). The prosperity of Japanese industries, particularly in the automotive and electronics sectors, lies with the sacrifice of many subcontractors. The core dualist theory suggests that economic agents, either workers or firms located in different segments of the economy, are treated unequally, regardless of their objective worth.

The second position emphasizes the 'vitality, dynamism and innovativeness of small businesses that have responded flexibly to the needs of clients and markets' (Sugimoto, 1997). Nischiguchi (1994), along with Sugimoto (1997), attempts to reject the dualist theory, claiming that Japanese business is more complex. Nischiguchi presented empirical evidence of sustainable growth and high asset specificity of the small- to medium-sized subcontractor firms within the *lean* system. He also showed that union membership in Japan has remained the same and that interscale wage differentials between large and small firms are not as marked as some suggest.

Sugimoto (1997) concluded that both positions exist and that the variation in value orientations and life style of workers is dependent upon the extent of control of the small businesses by the larger companies at the top of the hierarchy. Those who tended to diversify their connections were less controlled and more innovative, participatory and openly entrepreneurial.

Strategic procurement management is an important aspect of the lean production system and, in fact, an important part of supply chain management. It is important to note for the supply chain concept that lean production implementation by the large producers, in many cases, would not have been possible had it not been supported by highly organized governance structures. As discussed in more detail in Section 3.4, the supply chains were organized into hierarchical clusters of tightly tiered structures of subcontracting firms. Lean construction researchers, in their quest for production efficiency, have forgotten that *organizing the market* was instrumental in lean implementation and that the construction environment differs on a number of significant points.

3.3.7   *Strategic procurement management*

A strategic perspective of the supply chain concept emerged in the 1980s which subsequently evolved into strategic procurement (Porter, 1985; Lamming, 1992; Cox and Lamming, 1995). One of the most revolutionary changes in the last 10 years that affects the strategic management of the supply chains is the rethinking of organizational structure to suit the demands of the market (Christopher, 1998). Many organizations have traditionally been hierarchical, vertical and functionally defined and have grown heavy with layer upon layer of management and bureaucracy (Christopher, 1998). To respond to market demands, organizations have restructured and a new form of organization, the network, has emerged, and part of this is the use of the supply chain as an extension of the organization.

Lamming and Cox (1995) identified the importance of supplier procurement and co-ordination through allied business partners and strategic collaborative partnerships to enable *lean* production to take place. He termed this *lean supply*. Underpinning lean manufacturing is an industrial organization of supply chain that supports the concept, which has been well described (Nischiguchi, 1994; Hines, 1994; Lamming, 1992). As mentioned previously, thoughtful and informed descriptions of the Japanese subcontracting in the automotive and electronics industries can be found in Nischiguchi's text *Strategic Industrial sourcing: The Japanese advantage*. Although a clear proponent of the current mode of industrial organization of Japanese subcontracting, he also provides empirical evidence from the 1920s through to the present day of the structural and behavioural features of supply chains. It is a complete picture of the history of the evolution of the inter- organizational relationships. One of the most significant themes throughout his analysis was the description of the industrial organization and subsequent changes to the industrial organization structure. Where Hines

(1994) was concerned with suppliers forming associations, Lamming (1992) described the nature of the strategic partnerships between supply chain participants, and then termed this lean supply.

Lamming and Cox (1995) conducted empirical research on strategic supply relationships in the automotive industry. Lamming's perspective is in contrast to Porter's view of the value chain and suppliers and others which is often the accepted view of control in the lean supply chain. Lamming suggested that achieving lean supply is a complex matter because of the nature of competition in markets as the suppliers are involved simultaneously in several other chains. Jealous guarding of expertise cannot be maintained in the *lean* enterprise as it requires trust between firms.

Strategic procurement is much wider than the *lean* movement. It is a concept applicable to all firms and not simply those involved in production and manufacturing. A significant part of strategic procurement is concerned with business alliances. Co-operation among firms has grown rapidly since the early 1980s as alliances have proliferated in one industry after another (Gomes–Casseres, 1996). Alliances are, however, only one method available for strategic procurement.

In parallel with the development of the concept of supply networks, many different contractual arrangements have evolved, including strategic alliancing, network outsourcing, partnering and joint ventures. These differing arrangements have developed primarily for one or a combination of the following reasons: reduced costs and financial risk; improved innovation (product/process) and reduced risk; entry into new markets, trust and reciprocity in volatile markets.

Typically, this involved positioning a firm competitively in the marketplace by developing appropriate sourcing and management strategies for suppliers. Porter (1985) developed the concept of the value chain as a tool for firms to improve competitive advantage in an industry. The value chain arises from identifying the discrete activities a firm performs and then developing appropriate strategies to optimize these activities to position the firm to achieve competitive advantage. Integral to the concept is that it is desirable to purchase from suppliers who will maintain or improve the firm's competitive position in terms of their own products and/or services. The question is, how to purchase so as to create the best structural bargaining position.

Porter's value chain (1985) is a product of understanding the relationships between supply chain actors for the individual organization's gain. Although his work originates from considering industrial organization concepts, it is firmly located within a strategic management perspective. The strategic management of the supply chain is reinterpreted by Porter (1985) as the management of the 'value chain' to achieve competitive advantage. The arguments tend towards understanding the supply chain as the relative distribution of power between an individual organization and its suppliers (Porter, 1985).

3.3.8 Strategic procurement management: construction

Strategic management was a relatively new field in construction in 1991, with little literature available (Langford and Male, 1991). There has not really been a growth in this area at all, which is surprising; but rather there has been a growth in the research related to the management and procurement for the individual project. The strategic management of firm to firm relationships is still relatively new, with the focus on project alliances. Early research into such concepts as strategic alliances, serial contracting (Green and Lenard, 1999), multiple project delivery (Miller, 1999), organizational design (Murray *et al.*, 1999), vertical integration (Clausen, 1995) and supply chain procurement (Cox and Townsend, 1998), supply chain clusters (Nicolini *et al.*, 2001), supply chain alliances (Dainty *et al.*, 2001), supply chain constellations (London, 2001) and supply chain transaction cost economics (Winch, 2001) are indications of the awareness of strategic organizational management in construction supply chain research.

Cox and Townsend (1998) proposed the Critical Asset and Relational Competence Approach to construction supply chain management which relied upon clients controlling the supply chain. The authors advocated for clients to understand the underlying structural market characteristics of their own construction supply chains and to develop contingent approaches to procurement based upon this understanding.

They considered the UK construction research, based upon *lean* and *agile* manufacturing, inappropriate, because it lacked contextual understanding of the construction industry. They even suggested about unique supply chain properties as follows:

Box 3.23 Call to understand unique supply chain properties

It is our view that if the Latham report, and the somewhat naive research industry into automotive partnerships and lean and agile manufacturing processes that it has spawned, had devoted more time to analysing and understanding the properties of the unique supply chains which make up the complex reality of the UK construction industry a greater service might have been done to value improvement in construction (Cox and Townsend, 1998).

This view was largely based upon their findings reported in the seminal text Strategic Procurement in Construction: towards better practice in the management of construction supply chains (Cox and Townsend, 1998). In this text they reported on the findings of their study where they analysed the management methods taken by six client organizations. The profiles of these organizations differed and included an international Japanese 'design and build' management contractor, one multinational restaurant chain client, two UK international transportation clients, one UK national property developer and a US multinational client organization that is involved in the development of innovative products.

This critical asset and relational competence model relied upon observing the strategic and operational approaches of six organizations that appeared to indicate better practice construction procurement strategies and methods. The six firms were client organizations and were from the private sector, two whose core business was in the construction industry. The criteria for choosing these six were that they exhibited better practice and that they were private sector clients (Cox and Townsend, 1998).

Other researchers have conducted similar case study research on strategic procurement and supply chain management. Olsson (2000), through a qualitative case study on supply chain management of a Swedish housing project driven by Skanska and Ikea, highlighted that a conventional construction approach was found to be too expensive to meet particular client demands. Similar to the work of Cox and Townsend, the conclusion of the research was that, for supply chain management to be effective, the construction industry re-arrangement of existing structures may be necessary. Both studies focused upon strategic supply relationships of project team members.

With a similar business approach, Cardoso (1999) developed a model for comparing entrepreneurial business strategies of contractors based upon two competitive strategies: cost leadership and differentiation. A small study was conducted comparing these variables between six companies in Brazil and France. A methodology developed by Hong-Minh *et al.* (1999), *terrain scan mapping*, aimed to identify the key problems and relevant good practices for each industrial partner. The origins of her research are found in such concepts as business process re-engineering and business systems engineering. The empirical work involved nine companies at various stages of a house-building supply chain involved in action-oriented research activities.

Consistently, researchers have concentrated upon a small group of firms and the supply chain management concept related to an individual project. For example, Clausen's Danish study (1995) also focused upon the key firms in the main construction contract as they evaluated a government programme where the government, acting as a large client, intended on improving productivity and international competitiveness in the construction industry. The central argument to the programme was the 'need for vertical integration of the different actors and their functions in the construction process', with the premise that key actors in the process should be involved in strategic decisions at the outset. Therefore, through a tendering process, four consortia were selected to carry out experimental building projects. The core of the consortia include a contractor, an architect, a consulting engineer and 'manufacturing firms and suppliers of materials and components [who] were associated on a more or less permanent basis'.

Clausen (1995) determined that the programme was much less successful than anticipated, because there was discontinuity in the supply of projects to the consortia; firms were concerned about the financial risk of committing their resources to a single client. The conclusions suggest that the degree of uncertainty in supply of projects and the inherent risk for firms involved is a very important factor in supply chain management.

The interplay between supply and demand, the balance of power or control and incentive, have been considered by others in the form of serial contracting (Green and Lenard, 1999) and multiple project delivery (Miller, 1999). Although many authors suggest the importance of understanding the entire scope of the supply chain (Vrihjhœf and Koskela, 1999), the supply chain is often still perceived as the contractor's supply chain.

Box 3.24 Portfolio and project focused supply chain activities

The client is the more likely proponent and beneficiary for the management of the supply chain (London *et al.*, 1998a,b; London and Kenley, 1999).

Further to this, therefore, as initiator, the client has a greater stake in effective supply chain management, whether they directly manage the chain or abrogate the responsibility to first tier suppliers. Therein lies the fundamental link between Chapter 2 and 'supply chain management'.

Of course, large organizations in any industry can equally take a wider and more long-term perspective of various supply chains which they depend upon – just as we have seen in various industries such as auto manufacturing.

A repetition of project activity, longevity and a strategic perspective of procurement in the construction industry provides an ability for selected customers/clients to go to the next level in an industry and develop a range of portfolio- and project-based activities related to their supply chain in response to a deep consideration of supply markets, an analysis of their own demand, a risk and expenditure analysis and an organizational audit towards developing a supply strategy. Perhaps there are examples where this is being done to some extent.

In 2001, the supply chain field developed further with strategic procurement management concepts such as work clusters and supply chain integration (Nicolini *et al.*, 2001); firm constellation of network of alliances (London and Kinley, 2001b); subcontractors and barriers to supply chain integration (Dainty *et al.*, 2001) and transaction cost economics and supply chains (Winch, 2001) being explored in greater detail.

Nicolini *et al.* (2001) reported on two UK pilot construction projects which were organized on a work cluster basis aiming at supply chain integration. Members of the cluster included engineers, architects, subcontractors, suppliers and the contractor (depending upon which work packages were identified as a work cluster). Traditional roles and hierarchies were challenged as a cluster leader was appointed. An action research approach was undertaken and the benefits of the cluster approach were reported through comments and observations made by participants in the process and researchers involved in the process in both formal and informal data collection situations.

3.3.9 Supply chain constellation

Although not as deeply entrenched in the ethnographic tradition of research as the Nicolini *et al.* study, an equally qualitatively based study on a supply chain constellation indicated the leveraging power that is possible with small- to medium-size enterprises when firm-to-firm procurement relationships are considered outside the boundaries of a single project (London and Kenley, 2001a). I reported on a strategic approach to procuring and managing the supply chain which was identified in a study of very small Australian construction firms which had formed a network of alliances to penetrate international markets. In construction, alliances have been suggested as a form of governance structure to solve procurement issues, as there is a need for improved firm-to-firm relationships.

As mentioned in the previous chapter, project alliancing has found favour in the construction industry in recent years. In the project alliance the contractual relationships between key firms form for a project; then, when the project is completed, the temporary organization is disbanded as there is no contractual relationship binding the parties. The project alliance has a contractual relationship tied to an individual project and is largely short-term in focus and includes the client in the relationship.

The network of alliances identified in that case study (2001b) differs from the project alliance perspective. The perspective of the network of alliances is that the relationships were formed for more than one project and had a longer term perspective. The case study indicated that there were alliances related to the procurement of individual projects; however, they are simply one small part of what was termed a constellation of firms and alliances.

In the constellation, the governance structure is much broader than the usual individual construction project approach. The case study indicated the existence of different types of alliances: learning, positioning and supply (Gomes and Casseres, 1996). In some cases all three elements occur within the one alliance. The focus of the case study is not specifically on the type of alliance or a single alliance; rather, it is on describing the context of the constellation of alliances – that is, the relationship between the network of alliances and the strategy for the group.

In this case study the network of firms joined by various types of alliances evolved over some 20 years. Over the two decades, the key decision makers in the core of the constellation learnt and reacted to the market environment and gradually clustered together more and more firms with a variety of contractual relationships. At the time of the study, approximately twenty firms were involved in the constellation. The firms were typically classed as small or even micro, employing in some cases two or three people. The constellation was structured according to the strategy to penetrate the international market for affordable housing through using an innovative building product system of prefabrication. As the case study revealed, the affordable housing market is closely allied to the search for markets for innovative building products using waste material. The case study demonstrates some of the conflicts, constraints and issues that concerned the actors in small companies that are involved in the process. The following Figure 3.4 is the structure of the constellation in late 2000.

The importance of this study is the focus on describing the variety of firm–firm relationship types, the connection of the relationship types to the business strategy or market and the description of the cluster of firms that are contractually linked outside individual projects.

3.3.10   Barriers to supply chain integration

A study by Dainty (2001) and his co-authors revealed that, regardless of the government initiatives and debate and discussion on the benefits of supply

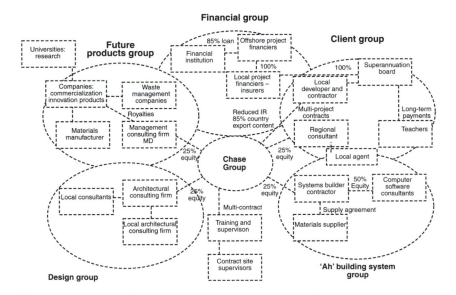

Figure 3.4 Constellation of firms in a supply chain network.

chain management, there are significant barriers to supply chain integration by smaller subcontractors and suppliers. Largely, 'there remains a general mistrust within the SME companies that make up the construction supply chain and a general lack of belief that there are mutual benefits in supply chain integration practices' (Dainty *et al.*, 2001). The researchers suggested that leading clients should take responsibility for engendering the necessary attitudinal change throughout the supplier networks if further performance is to be realized within the sector.

It is noted that Dainty *et al.* (2001) did not describe the methodology explicitly, except to state that there were semi-structured interviews and focus group interviews with suppliers to main contractors. The main contractors provided the introduction to subcontractors working on their projects. The questions in the interviews were not described, nor were any sampling details, except with regard to the method for approaching the subcontractors.

Much of the strategic procurement research is related to governance of contractual relationships. Winch (2001) explored the application of *transaction cost economics* to construction project and supply chains. Transaction cost economics is an economic theory derived to develop the theory of the firm and is concerned with how the boundary of the firm is governed by the attempt to reduce the cost of transacting with other firms. This has connections to industrial organization economics and is explored in greater detail in the following chapter. It is discussed now as it was presented by Winch as a conceptual framework for understanding the governance of construction project processes.

As noted by Winch (2001), there were previous attempts to apply this theory over the years (Eccles, 1981; Reve, 1984; Winch, 1989; Masten *et al.*, 1991; Winch, 1995; Pietroforte, 1997; Walker and Lim, 1999; Lai, 2000). Previously, transaction cost economics had only been applied to the principal contractor procurement, whereas Winch's contribution was to consider its application across all project procurement relationships and further down the supply chain. The paper was theoretical and proposed that empirical work was required to explore this further. This work is discussed in more detail in the next Chapter 4.

In all these models and/or empirical observations, construction supply chains are largely viewed from the perspective of the individual firm, situation or project perspective – perhaps my small case study on the constellation of firms was beginning to move away from this narrower perspective. However, this is quite a limited case study. There is still, however, a dearth of empirical research to address the supply chain across the breadth of the industry towards understanding, describing and analyzing the *structure* and *behaviour* of supply chains.

The types of relationships between firms is central to strategic procurement; however, the type of relationships cannot be fully comprehended without an understanding of the markets. Many authors in strategic procurement have moved the debate regarding supply chains with respect to the need for the development of appropriate relationships; the problems of unreliable supply; the different degrees of control between firms and the difficulties due to the temporary nature of a project-based industry. However, these are all characteristics of the real world construction industry. The approach advocated in this text is to accept the characteristics of the real world in the first instance and the inherent structural characteristics of a project-based industry, as opposed to a process-based industry, rather than attempting unachievable, inappropriate and unrealistic changes to an idealistic model of a supply chain (Cox and Townsend, 1998).

3.3.11 Summary

The first two major themes, distribution and production, are concerned with the flow of information, goods and services along supply chains, which has been heavily influenced by 'hard' systems engineering methods. Much of the management literature focuses upon logistics, and similarly the construction literature focuses upon the flow of materials to the construction site. There are numerous methodologies in the logistics literature used by researchers. It has been noted that, even within logistics research, it has become dominated by two schools of thought: economic and behavioural (Barrett, 1998).

The third major theme, strategic procurement management, is that construction supply chain literature appears to be influenced by organizational theory and, in particular, strategic management. This is quite a broad net to cast and once again this has resulted in a rich, yet fragmented approach to supply chain theory. The developing trend in the construction field is through the use of normative models, based upon the supply chain concept, which will go toward improving relationships and building trust between firms and thus addressing perceived industry problems. The assumption is that integration of project supply chains will address problems including: litigious adversarial relationships, low productivity, lack of innovation, fragmented industry structure and fragmented project structure.

Although seemingly different in approach, the major themes assume that the supply chain is a mechanism that can be structured and controlled to allow better integration of project actors and improved performance. There is a particularly strong sense in the construction research literature of positioning the supply chain as a tangible phenomenon that can be observed and controlled. The most common aim for researchers has been to attempt to develop supply chain theory by proposing that the supply chain management concept is the panacea for widely held beliefs regarding the industry ills. The approach has been from a single organization perspective and has tended to concentrate on productivity, profitability or performance of that organization.

The supply chain research community is in the early stages of theory development in both construction and mainstream management. There is a realization that there is a lack of theory development in the supply chain field (Lambert *et al.*, 1998; Day, 1998; Hines *et al.*, 1998; Barrett, 1998) and rigour in research methodologies (Hines *et al.*, 1998). The lack of depth and rigour may be attributable to the 'newness' of the field, but it has also been attributed to the lack of leadership by academic research. Regardless of the cause, it is acknowledged that a lack of theory underpinning a field may not be a concern for practitioners seeking tools and techniques to apply to their own situation but is more of a concern for the academic community. From past experience in conducting research with industry partners in Australia, I'd have to say that a lack of theory does actually concern the industry partners that I have worked with – perhaps not explicitly, but more implicitly. Very soon after one reports on 'what you found out' when you conduct interviews or similar types of data collection modes you need to interpret what you found out – and by and large you need to have something to say about the data you collected... if for no other reason than to provide useful and perhaps unique interpretations that have some depth and thinking to them. But there is another reason why it is important to understand what theory and problem conceptualizing is 'going on' around the world – it is that, time and again, I have been asked by my industry partners: what happens in the United Kingdom, United States, Europe etc.? I even had an industry partner know about research which was being conducted in South Africa on a particular project – which would have caught me out had it not been that I had just been to a CIB conference in South Africa. Therefore, in my short academic career I have come to realize that the gap between theory and practice – academia and industry – is not as wide as we are often led to believe. I am greatly heartened by this, of course.

But I digress.... Specifically, little research has focused upon identifying and exploring the underlying social, economic, technical, political and legal structures within which supply chains are embedded and the influence that this has had on supply chain actors and thus procurement. These underlying drivers within the industry are what shape how actors can act – the practices and the norms that are acceptable. Conversely, it is the actors that produce these structures. Some supply chain literature discusses industry structures; however, much does not extend beyond describing the size of the industry and summarizing key events, and little explores firm–firm relationship behaviour.

In construction research the impact of a strategic supply chain procurement perspective is not well researched. The strategic identification of supply chains and the location and leverage of relative negotiating power in the supply chain is an emerging area of interest (Cox and Townsend, 1998). The lack of understanding of the structural and behavioural factors that impact upon the realization of project supply chain integration is widespread across the industry (Cox and Townsend, 1998). Deeper issues such as market performance, firm performance, market structure and firm conduct are worthy of investigation.

To date, much of the construction literature has applied the lean concept without the industrial organization contextualization, that is, without the detailed empirical exploration of market structures, chain organization, firm behaviour and firm performance which underpin the construction environment. Industrial organizational economic context is an important issue which has been understood by those researchers in the lean production manufacturing research field. The contextualization of lean production which supports lean thinking has been provided through organizational and industrial organizational economic descriptions of the automotive and electrical industry supply chains (Nischiguchi, 1994). This description of the supply chain is an important part of the philosophy of lean thinking and there is a void in the construction research in terms of supply chain theory, methodology and empirical research in relation to this.

This approach is moving towards the theory of industrial organization economics and industrial dynamics. The work to date by supply chain theorists, who have merged industrial organization and the supply chain concept, has tended to identify supply chain actors, map contractual relationships, identify horizontal and vertical positions and identify relationship types. The rigour of the industrial organization economics methodology may serve to address some of these concerns in construction. The argument being developed along the way and the critique and explanations is, of course, setting the scene entirely for the case studies which are reported on later in this text.

3.4   Industrial organization economics

As noted in Chapter 2, industry analysis has traditionally focused on sectors, which includes groups of firms with similar characteristics, engaged in similar production processes, producing similar goods or services and occupying similar positions (AEGIS, 1999). According to Marceau (AEGIS, 1999), recent attention is on chains, clusters and complexes. This represents a shift from the purely mechanistic conceptions of the nature of industrial organization, as a market consisting of a collection of establishments producing homogenous outputs (Scott and Storper, 1986), to a more complex interconnected and interdependent set of markets and firms.

The industrial organization methodology deals with the performance of business enterprises and the effects of market structures on market conduct (pricing policy, restrictive practices, and innovation) and how firms are organized, owned and managed. The most important elements of market structure in these models refer to the nature of the demand (buyer concentration, number and size of buyers), existing distribution of power among rival firms (seller concentration, number and size of sellers), entry/exit barriers, government intervention and physical structuring of relationships (horizontal and vertical integration). The role of the industrial organization model is to give substance to the traditional neoclassical abstract concepts of market types.

Selected supply chain research, published in mainstream management literature, has studied the complex system of supply chains through inter-organizational structure. These are important models that merge the field of IO and supply chain theory (Ellram, 1991; Harland, 1994; Hines, 1994; Nischiguchi, 1994; Hobbs, 1996; Lambert *et al.*, 1998).

There are two main differences between this stream of research and that previously discussed. First, industrial organization supply chain research tends, in the first instance, to be primarily descriptive rather than prescriptive and is about supply chains rather than supply chain management. It provides interesting grounds for discussion regarding the scope for research of the supply chain and explains particularly the twin dichotomy between research on 'supply chains' versus 'supply chain management'. Second, the unit of analysis is not the individual organization but rather an aggregation of firms within a market.

Having said that, Ellram (1991) took an industrial organizational perspective which was cognisant of the market concept, although from a single organization's ability to manage the supply chain. She suggested types of competitive relationships that firms undertake from transactional, to short-term contract, long-term contract, joint venture, equity interest and acquisition. These involve increasing commitment on the part of the firms. She described the key conditions under which supply chain management relationships are attractive according to an industrial organization perspective. The main thrust was that supply chain management is 'simply a different way of competing in the market' that falls between transactional type relationships and acquisition and assumes a variety of economic organizational forms (Ellram, 1991).

Her paper analysed the advantages and disadvantages of obligational contracting and vertical integration. She continued by describing the key conditions under which supply chain management relationships are attractive according to the industrial organization literature. This was one of the first discussions to explore the implications of Williamson's transaction cost economic theory and industrial organization economics related to supply chain management.

Box 3.25 Transaction types suitable for supply chain management

Situations conducive to supply chain management included:

1 recurrent transactions requiring moderately specialized assets
2 recurrent transactions requiring highly specialized assets
3 operating under moderately high to high uncertainty.

However, as Ellram warned, supply chain management is 'not a quick fix nor is it the best competitive form for every situation'. Such prescriptions should be considered with caution: 'arguments designed to prove the inevitability of this or that particular form of organization are hard to reconcile, not only with the differences between the capitalist and socialist worlds, but also with the differences that exist within each of these' (Ellram, 1991).

Transaction cost economics theory has just as many critics as supporters, though. One of the main criticisms is that it has tended to assume a market and hierarchy dichotomy (Ellram, 1991). Theorists have found it difficult to explain contractual relationships between firms where clearly the transaction costs were high and yet firms did not vertically integrate. There are a variety of institutional arrangements between the two extremes of market versus hierarchy which do not fall neatly into the transaction cost model and clearly demonstrate that markets are not the only way prices are co-ordinated (Alter and Hage, 1993). However, there is potential for future research relating transaction cost economics to the supply chain movement for the construction industry. Transaction cost economics tends to focus upon individual contractual relationships, whereas supply chain theory aims to understand many interdependent relationships as the unit of analysis.

Box 3.26 Network sourcing

Network sourcing became an important area of research in the 1980s (Miles and Snow, 1992). Throughout the 1980s, organizations around the world responded to an increasingly competitive global business environment, moving away from centrally co-ordinated, multi-level hierarchies and towards these more flexible structures that closely resembled networks rather than monolithic traditional pyramids. These networks – clusters of firms or specialist units co-ordinated by market mechanisms instead of chains of commands – are viewed by both their members and management scholars as better suited than other forms to many of today's demanding environments (Miles and Snow, 1992).

Larger primary contractors at the top of the pyramid develop mechanisms to monitor right throughout the chain by ensuring management and co-ordination at each tier.

Hans Hinterhuber *et al.* (1994) categorized networks into four groups based on both intra- and inter-business unit networks.

Box 3.27 Types of networks

Types of networks include:

1 vertical: franchising, subcontracting
2 diagonal: interdisciplinary
3 horizontal: alliances; and
4 internal: profit centres, strategic business units

An alternative typology was proposed by Miles and Snow (1992) that suggested that networks were either

1 stable
2 internal
3 dynamic

dependent upon the extent of linkages.

Hines (1994) and Nischiguchi (1994), who are clearly advocates of the lean system of supply, merging the supply chain concept and industrial organization theory, explored the nature of sourcing in the Japanese manufacturing industry and found it was an example of network sourcing. Some of the more significant contributions of their research were the descriptions of the historical, organizational and economical structure of the Japanese system of supply across automotive and electronics industries. In many ways this has provided a richer picture of lean production and supply chains than other writings which portray an apocalyptic posturing of the field's success.

Suppliers are categorized and organized into either specialized subcontractors or standardized suppliers, based upon the degree of complexity of the supply item. It is within the specialized subcontractors that the pyramidal Japanese subcontracting system or the concept of clustered control lies. As Hines and Nischiguchi used the terms networks and clusters interchangeably for the same industrial sourcing scenario in Japan, for the remainder of this discussion networks and clusters are considered the same.

Box 3.28 Japanese subcontracting system: network sourcing

This system has traditionally been described as a pyramid with an individual assembler corporation at the top and successive tiers of highly specialized subcontractors along the chain, increasing in number and decreasing in organizational size at each progressive stage. More importantly, though, is that each tier would procure, co-ordinate and develop the lower tier through Supplier Associations.

Hines (1994), in his seminal text, *Creating World Class Suppliers*, described this procurement as network sourcing.

In his comprehensive research on suppliers, Hines (1994) considered network sourcing as one example of a type of buyer–supplier relationship with its origins in what he termed the Japanese School. He suggested that the buyer–supplier relationships can be categorized into three groups based upon the origins of the relationship.

Box 3.29 Schools of buyer–supplier relationships

1 *Trust School* relationships are primarily due to a complex mix of social and moral norms with technological, economic and government policies also of some importance. There is, therefore, some suggestion that such approaches may be more difficult, or even impossible, given the set of external and internal factors in the Western world.

2 *Partnership School* relationships are developed on the basis of a one-to-one partnership with individual strategic suppliers – with the emphasis on the formal creation of the partnership. This primarily UK model plays down the potential problems of cultural specificity in following an approach designed to form relationships of the type exhibited in Japan (Ellram, 1991).

3 *Japanese School* which appears to take the middle ground of the above schools; the Japanese school suggests that, although conditions are different in the West, a somewhat modified or developed Japanese-style approach can be translated to other cultures. A number of authors describe the route to developing the desired supplier–buyer relationship in this mode (Burt and Doyle, 1993; Lamming, 1993; Hines, 1994; Nischiguchi, 1994) and it involves supplier grading, supplier co-ordination and development.

The network sourcing model was developed within the Japanese School context (Hines, 1996). A study using data collected on forty Japanese companies within the automotive, electronics and capital equipment industries, through semi-structured interviews and questionnaires, illustrated the relationship between the ten causal factors (refer to Table 3.1 based on Hines, 1995). The results of the study demonstrate that, within network sourcing, supplier co-ordination and supplier development have emerged as the critical causation factors. This work is discussed further in Section 4.4.3.

Table 3.1 Network sourcing overview (reproduced from Hines, 1995)

1	A tiered supply structure with a heavy reliance on small firms
2	A small number of direct suppliers but within a competitive dual-sourcing environment
3	High degrees of asset specificity among suppliers and risk sharing between customer and supplier alike
4	A maximum buy strategy by each company within the semi-permanent supplier network, but a maximum make strategy within these trusted networks
5	A high degree of bilateral design, employing the skills and knowledge of both customer and supplier alike
6	A high degree of supplier innovation in both new products and processes
7	Close, long-term relations between network members, involving a high level of trust, openness and profit sharing
8	The use of rigorous supplier grading systems increasingly giving way to supplier self-certification
9	A high level of supplier co-ordination by the customer company at each level of the tiered supply structure
10	A significant effort made by customers at each of these levels to develop their suppliers

Box 3.30 Supplier co-ordination and supplier development

Supplier co-ordination refers to the activities made by a customer to mould their suppliers into a common way of working, so that competitive advantage can be gained, particularly by removing inter-company waste. This type of co-ordination would involve areas such as: working to common quality standards, using the same paperwork system, shared transport and employing inter-company communication methods such as EDI (Hines, 1994).

Supplier development refers to the activities made by a customer to help improve the strategies, so that suppliers could plan their processes more effectively, as well as the customer offering specific assistance to the suppliers in areas such as factory layout, setup time reduction and the operation of internal systems (Hines, 1994).

Associated with each tier are supplier associations, which are a 'mutually benefiting group of a company's most important subcontractors, brought together on a regular basis for the purpose of co-ordination and co-operation as well as to assist all the members'. Industry associations are quite common in various sectors and are becoming more and more active – particularly in construction and related to training. However, they tend to be industry-based and less focused on an individual company. Major materials-supplier corporations are sourced directly by the large assembler corporation, through strategic procurement. Sourcing alliances are dealt with separately from the pyramid system. The supply of the material is provided to appropriate subcontractor tiers for the manufacture of components.

It was suggested that to achieve a lean supplier network both of these activities must be undertaken simultaneously. It is interesting that Hines (1995) commented that, in the West, the best organizations generally endeavour only to address one or other of these areas, with the majority of firms failing to address either. Nischiguchi (1994) explored strategic industrial sourcing through his exhaustive empirical investigation as part of the internationally acclaimed MIT Motor Vehicle Research Program (MVRP). He suggested the concept of clustered control (refer to Figure 3.5).

However, this represents a single company network encompassing all the relevant tiers necessary to produce the end product, suggesting a closed system. However, in reality, first tier suppliers supply to many assemblers across the whole industry. This conflict led to the Alps Structure of supply chains, a series of overlapping pyramids resembling mountain alps across

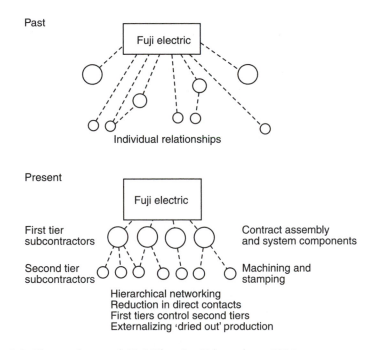

Figure 3.5 Clustered control: Fuji Electrics Tokyo plant, 1986.

Source: Nischiguchi, 1994.

an industry, which enlarged the industry-specific view to look at the wider economy and suggested that, rather than this closed rigid system, the Japanese subcontracting system was moving more towards a structure of interlocking supplier networks. In this system, many firms supply more than one industry sector and potentially operate in different tiers; for example, the electronics suppliers operate in a number of sectors (refer to Chapter 4 for a more focused discussion).

Lambert *et al.* (1998) also provide insights for mapping supply chain structure. They claimed, quite simply, that there are three primary structural aspects of an organization's supply chain structure:

Box 3.31 Supply chain structure

- members of the supply chain
- structural dimensions
- types of process links.

Lambert *et al.* (1998) developed, for the supply chain structure of an organization, a generic map of a complex network of suppliers and customers arranged in successive tiers from the focal organization. In many ways this model suggests methods of strategic procurement. However, the importance of this model for the industrial organization debate is the inclusion of a number of empirical case studies indicating the structure of different supply chains and the interconnection between a number of focal organizations' supply chains and the resultant networks of supply.

Harland (1996) widened the industrial organization of supply chains debate, suggesting the term, *the supply network*, as a means for capturing the full complexity of the firms involved through a more holistic view of the process.

Further research on organizational networks has been conducted by the ION (inter-organizational networks) project, a collaboration between three UK universities. The ION team has divided the research into three areas, including supply networks, learning networks and innovation networks. It has been difficult to find publications of their work – but it would be worthwhile, if you were interested in this area, trying to track some down. The output of this project to date included a literature review (Callaghan, 1998); however, there are no empirical studies yet published. Callaghan (1998) discussed a number of concepts related to supply networks, including environment, strategy, structure, process, net-work evolution and product/service dimensions. He concluded that little existing research has examined these in detail. The majority of research on supply networks examines the structural and strategic issue of vertical integration, but this has been on a general level, non-specific to particular circumstances. There are few empirical cross-comparisons of supply networks between industries.

Much of the supply and industrial network literature builds upon the industrial networks movement of the 1980s (Piore and Sabel, 1984). This body of research has tended to suggest that close-knit inter-organizational networks produce superior economic performance and quality, and that there should be a move away from the large, vertically integrated firms (Alter and Hage, 1993).

To change track a little – Tombesi (1997) conducted a case study on the networks of architectural firms involved in a single construction project in the United States. It was significant in that it developed an architectural design model based upon the theory of flexible specialization brought to us by Piore and Sabel (1984) and described in their Italian industrial networks study. This model has important implications for understanding the industrial organization of the supply chain of firms involved in design; however, it did not concentrate upon the contractual relationships between firms and the wider contextual market environment. The network of firms was, however, studied many many years ago by Eccles in the US residential construction industry when he first coined the term, the 'quasi-firm' (Eccles, 1981).

The AEGIS (1999) study in Australia was perhaps the closest step towards understanding the construction industry supply chain using an industrial organization economic perspective. This study was discussed in the previous chapter as part of the government-led investigations where the significant limitations were also outlined. The AEGIS model for the Building and Construction Industry Cluster also contributes to the development of the wider industrial organization perspective. It discusses the industry as a chain of production and conceptualizes the industry through five main sectors: onsite services; client services; building and construction supplies and products and fasteners, tools, machinery and equipment. To remind you – existing statistics were used to describe the sectors in terms of industry income. However, the researchers themselves note that this is contrived, as sufficiently detailed data is not available. The major firms are organized and listed according to groups: suppliers, project firms and project clients. The discussion focused upon general information about size and turnover and addressed a market view of some key markets and the major players in the key markets, but does not seem to address the firm or project level of supply chains. It certainly didn't explore in practice the realities of procurement; how customer firms the market, negotiates contracts and eventually engages suppliers.

There is a need to develop this further and explore the explicit inter-firm supply chain relationships on projects within the context of the firm and market. The firm and market level of analysis lies within the field of industrial organization economic theory.

3.5   A final word

> ### Box 3.32   Further reading
>
> A compilation of scholars' work has been in the following set of Readings; Handbook of Construction Supply Chain Management, an edited volume by O'Brien, W., Formosa, C., London, K. and Vrijhoef, R., and is currently in print. This collection of chapters represents authors from Australasia, United States, South America, Europe and the United Kingdom. It will be published by US Taylor and Frances.

Chapter summary

1   The role of the supply chain concept in construction is moving beyond the rhetoric that it is a management tool towards improving the performance of the industry. As the field develops, research may include optimization of supply chains and will enable more credible discussions of advantages of different types of networks, clusters or chains.

2   Some studies have widened the perspective and introduced industrial organizational concepts; for example, vertical integration (Clausen, 1995; Tommelein and Yi Li, 1999), design specialization and fragmentation (Tombesi, 1997), subcontractor/contractor dependence and the 'quasi-firm' (Eccles, 1981) and buyer concentration or pooled procurement (Taylor and Bjornsson, 1999). There is no shortage of construction supply chain research that is action, applied or case study in orientation. Much of this empirical work is oriented to the project as the unit of analysis. There has been in the past a lack of work that approaches the research problem from a wider industrial context. A deeper and more detailed understanding of industrial organization theory and supply chains is established in this text.

3   Government public policy, particularly competition policy, should be informed by observing the current state of the supply chain. Until we are able to describe the vertical and horizontal relationships between firms and understand interdependencies at a firm level in relation to the market level, it is difficult to compare the long-term impact upon changes to the relational position between firms. In a global economy this may also have implications for competitiveness, sourcing, monitoring and traceability of products and materials.

4   Boundaries between sectors are blurring when we think of how our business processes and the various supply chains. Specifically, we are

now seeing new players in the chain as e-business perhaps becomes more and more significant; for example, dedicated supply procurement managers or transaction organizing companies. This has begun to evolve in the IT world. A dissertation by one of my students on outsourcing strategies used by five of the IT 'giants' in the AsiaPacific region is worthwhile reading. Sheila Wang identified a new wave of thinking in IT whereby procurement is being outsourced using strategic alliances. In Australia we have seen a large IT portal consortium involving major civil and building contractors align with a telecommunications company to develop the first construction portal. To date the portal has not been a success and the barriers and drivers for adoption of e-business throughout the supply chain has been the topic of a major national study which I completed in 2006. The social, cultural and economic context plays a major part in the diffusion of new technologies and the structure of the different markets provides different types of technology adoption pathways. Three main pathways were identified in practice.

5   Development of an industrial organization model specifically for construction supply chains also has implications for designing co-operative associations across markets for purchasing. It will assist in locating innovative supply clusters and make transparent roles of co-ordinators and controllers in the chain. Important questions are:

> What is the overall nature of the organizational relationships along the supply chain?
> What is the nature of the competitive environments that organizations operate within and how will this affect performance of firms in that market?
> How do firms source their suppliers?
> How does a supply chain form?
> Who actually supplies to whom?
> How is sourcing organized?
> What are the power relationships between firms and their suppliers along the chain?
> How do we analyse such fundamental structural and behavioural properties in the supply chain?
> What does the supply system look like as you move away from, say, the building or civil engineering project environment, and begin to unravel the delivery of commodities that feed in to various sectors that are involved in systems of construction?
> What inter-sector chains exist and how does this cross-fertilization impact sectors?

In many cases the developing case studies on the construction industry to explore these was not that easy – as the industry is often secretive about methods of industrial sourcing coupled with the problem that the number of transactions on projects is so vast. These types of questions are relevant to all supply systems which involve **constructed systems**; that is, *people who make things.*

6   The following chapters seek to answer these questions by specifically addressing the relationship between the industrial organization concepts of market structure, firm conduct, market and firm performance in relation to sourcing. The theory related to industrial organization economics will assist in procurement modelling for the constructed systems industry supply chains based upon industrial organizational economics theory.

4 Industrial organization economics methodology and supply chain industrial organization approaches

Can I really manage something if I don't understand the economic environment?

A discussion bringing supply chain and economics together

4.0 Orientation

> **Box 4.1 Chapter orientation**
>
> **WHY:** Chapter 4 provides a brief overview to the historical development of the industrial organization economics field of research and then focuses on the work that has related industrial organization to the supply chain. It provides more detail than in the last chapter.
>
> **HOW:** The strong division of the field into two main schools of thought, namely the Chicago School and the structure-conduct-performance school, is described. Some fundamental principles relevant for understanding industrial organization concepts are discussed, including: market structure, firm conduct and market performance. This provides some detail on economic structural and behavioural concepts and is the background for understanding existing supply chain industrial organization approaches and the case studies in the later chapters of this text.
>
> **WHAT:** This chapter highlights that those models that have merged the supply chain concept and the industrial organization methodology have not addressed markets orientated towards projects and short-term production scenarios. Much of the empirical work in the field which validates methods and techniques is associated with auto manufacturing and retailing and not construction – not large constructed projects either in civil, building, shipping nor aerospace. This chapter develops the principles for us. To develop the model the procurement relationship between two firms is identified as the key concept that ties the industrial organization and supply chain fields.

4.1 Introduction

Industrial organization economics has been identified as a field of research that has contributed to the *practical* organization, integration and management and the *theoretical* understanding of supply chains in a variety of industries including retail, auto manufacturing, information technology and electronics engineering. In particular, it has contributed to the description and understanding of the structural and behavioural characteristics of these supply chains. These industries were not project-based industries and therefore we do need to think about what other implications there are for a *project-oriented industrial organization economics supply chain procurement model*. Before this empirical model of supply chains in construction environments can be developed, an understanding of the components that would make up that model needs investigation. Therefore, the fundamental principles of the industrial organization economics field are explored in this chapter.

This chapter outlines the origins of the industrial organization economics field and then discusses the key concepts. This allows for the examples which are discussed in this chapter to be placed in context. There have not been any substantial empirical studies or theoretical models that have attempted to merge the two concepts within the construction management and economics supply chain research community. This chapter seeks to do this by identifying the difficulties inherent in applying the industrial organization economics methodology to the construction industry which is a project-based industry. This is achieved first through an examination of procurement in the supply chain as evidenced by empirical studies from other disciplines and industries. Second, it is achieved by a critical examination of the procurement as it relates to a project-based industry in terms of structural and behavioural characteristics of the chain. The chapter is therefore organized as follows:

- overview of industrial organization economics field, including its origins and development
- description of key concepts of market structure, firm conduct and firm and industry performance
- procurement relationships in the supply chain
- supply chain organization; and
- a summary of the issues for IO procurement modelling for construction: a project-based industry.

4.2 Industrial organization economics overview

The theory of industrial organization economics, often simply termed industrial organization or industrial economics, is a useful framework to understand the economic structure of markets, the economic behaviour and performance of firms in those markets and the performance of the market as a whole. People have been interested in the market structure, firm behaviour and performance of markets since the beginning of the industrial revolution. These three elements of market structure, firm conduct and firm and market performance have come to be the cornerstone elements of the field. However, the delineation of a specific area of economics under the title of industrial organization economics only emerged in the early 1940s.

4.2.1   *Industrial organization economics field delineated*

We don't want to get too caught up in the economic theory and all the various nuances, but it is worthwhile to be acquainted with some of the key issues and the history is always a good place to start; therefore, the next section maps key economists who have been instrumental in industrial organization economics.

Industrial organization economics and microeconomics, another branch of economics sometimes termed price theory, are quite similar in subject matter. In particular, the *theory of the firm* and the associated *transaction cost economics theory*, which form part of microeconomics, are often confused with industrial organization. The nature, scope and methodology of industrial organization economics is marked by confusion and conflict. My first reaction when I began to dig in to this body of knowledge when I uncovered disagreements was that this doesn't really make me feel that comfortable – I am coming to this field for answers, not more problems. However, what it does serve to illustrate is the contested nature of *knowledge* – in any field, if we didn't have debates, conflicts and general discourse, we probably would all agree in the first hour and then go home – I wonder where we would be then? After a while you get to feel more and more comfortable with knowing this about academic literature and that this is the scholarly *modus operandi*. The following discussion serves to explain the differences between various economic fields and theories.

A really brief economic lesson. ... Two major factors have been identified that distinguish the industrial organization economic and microeconomic fields. Neoclassical economics identifies four market types, based primarily upon the degree of competition, including: monopoly, monopolistic competition, oligopoly and perfect competition. The term, 'market structure in microeconomics', refers to characteristics of a market which influence the relationship between the firms in a market (Terry and Forde, 1992). This typology of market structure is developed through the following three criteria: the number and relative size of firms in the market, whether firms supply identical or slightly different products and the ease with which firms can either enter or leave the market. Microeconomics primarily focuses on simple market structures, perfect competition and monopolies. Perfect competition has many firms, and they are insignificant; whereas in monopolistic markets the single supplier is very significant (Terry and Forde, 1992).

> ## Box 4.2 Industrial organization economics versus microeconomics
>
> The first factor which distinguishes industrial organization economics from microeconomics is that it focuses upon the oligopoly – the type of market in which firms are neither monopolists nor perfect competitors and there are a few suppliers who are generally significant (Martin, 1993). Generally, oligopolies are the types of markets of the real world and perhaps that is why I think it is a most useful framework for us to work within.
>
> The second factor that distinguishes microeconomics from industrial organization economics is concerned with policy questions; that is, government policy towards business. According to Martin (1993), 'industrial economics, in contrast to microeconomics, is profoundly and fundamentally concerned with policy questions' (p. 1).

The differences between the approaches has been explained by Hay and Morris (1979) through an historical discussion that traces the roots of the various theories through two main schools of thought. A more detailed literature review of the history is located in Appendix A in my PhD dissertation, but perhaps you don't want to go there! So, a 'potted' view is now provided.

There have been two schools of thought in the analysis of industrial organization economics and in many ways the divide relates to this fundamental issue of government intervention and the epistemological differences. The two schools of thought are discussed in detail in the next Section 4.2.2.

4.2.2 Origins and development of industrial economics

The origins of the industrial organization economics field have been traced back to Adam Smith and his seminal writings in 1776 in the *Wealth of Nations* on the two prices of a product, the natural price or value and its market price (Hay and Morris, 1979). Hay and Morris (1979) suggested that from Smith two quite distinct paths have evolved which can be categorized by fundamental epistemological and then ensuing methodological differences. Figure 4.1 indicates this historical development of the field (Hay and Morris, 1979).

The two paths are the deductive theoretical and the inductive empirical observation. The emphasis on logical deduction from precise assumptions to determinate conclusions has been a powerful epistemological approach in economic analyses of firm behaviour. The deductive researchers attempted to analyse market competition and in particular tried to establish the specific conditions under which competition would result in the equalization of prices and costs. Various researchers built upon this thread which

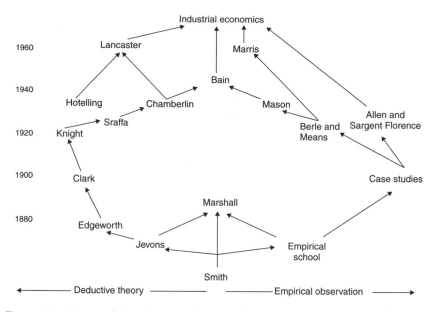

Figure 4.1 Historical development of industrial organization economics field.
Source: Hay and Morris, 1979, p. 6.

culminated in the 1920s in the work of Knight (1921), who refined the perfect competition model. As an aside to this discussion – I find it particularly heartening that I can go in to another discipline and find some terms that mean the same thing in my own discipline – interdisciplinary work is difficult at the best of times as you wander through a minefield of concepts that when you find someone interpreting something in the same way that you would it makes it easier. Much of my work is interdisciplinary – at the moment I do find it quite time-consuming as it simply takes time to understand the histories and traditions within a discipline. Before I go any further...it would be presumptuous of me to say that I fully understand this discipline – but perhaps enough to borrow key concepts and help to make sense of our own industry.

Hay and Morris (1979) suggest that the 'so-called Theory of the Firm [was] concerned almost exclusively with price and output decisions and their impact on firm efficiency, resource allocation and economic welfare'. The approach was deductive and the firm was representative and bore little relationship to the reality of actual firms in industries. The study of the behaviour of firms in the traditional approach to the theory of the firm assumes that producers aim to maximize profits. Much of this work has involved presenting a theoretical firm's cost and demand curve and attempting to explain the question: what will the long-term relationship between them be?

At the same time, further to the work of Smith (1776), the empirical school developed and much of this work observed the historical development and actual behaviour of particular firms in industries. There were also studies of the current structure and behaviour of one or more industries. A wide range of industrial organization issues was considered in these studies, although it was noted that many of the studies were not rigorous and few could predict generalizable results (Hay and Morris, 1979). They were largely a reaction to the deductive approach, the abstract idealized way of viewing the world, and involved case studies, descriptions of firms and people and practices in the real world.

Empirical techniques have progressed over the ensuing years and become even more specialized. They have become heavily influenced by econometric techniques and this is discussed in detail later in this chapter. According to Hay and Morris (1979) Bains' and Chamberlin's work represents the beginning of the market structure and market performance model that has become so prevalent as an underlying framework in industrial organization economics today. Chamberlin's work is claimed to be the catalyst that generated how industrial economics was practised in the 1970s (Hay and Morris, 1979).

4.2.3 Schools of thought

The distinct development in two paths explains, to some extent, the two schools of thought that characterized the industrial organization economics field from the 1970s onwards. The two schools of thought have become formally known as structure-conduct-performance and the Chicago School.

Box 4.3 Structure-conduct-performance school

The structure-conduct-performance school 'argued that the private exercise of monopoly power is a persistent feature of many markets'. This means that the impediment to the effective functioning of markets is strategic behaviour by some firms, which prevents other firms competing on the basis of merit. From this perspective, firms can have a great deal of power in the market and therefore governments should implement competition policies that moderate this. Market structure determines the behaviour of firms in the market and the behaviour of firms determines the various aspects of market performance. This shall be explored in more detail in the Section 4.3.4 on performance.

The Chicago School, which takes its philosophical roots from the deductive lineage, has always been theoretical. Much of the work by the Chicago School after the 1970s has criticized or critiqued the structure-conduct-performance research. The following comment is indicative of these critiques:

> ## Box 4.4 Criticism of structure-conduct-performance school
>
> Casual observation of business behaviour, colorful characterizations... eclectic forays into sociology and psychology, descriptive statistics, and verification by plausibility took the place of the careful definitions and parsimonious logical structure of economic theory. The result was that industrial organization regularly advanced propositions that contradicted economic theory (Posner, 1974, p. 929).

The distinction between the two schools of thought is often regarded through the structure-performance and performance-structure relationships. The structure-conduct-performance school focused originally upon the industry as the central concept or unit of analysis in an empirical manner; thus, market structure was central to how firms behaved and markets performed. Therefore, governments should intervene to alter market structure to alter performance. The Chicago School is much more theoretically based and focuses on the Firm as the central concept and therefore does not advocate market intervention.

Industrial economists had come to the point in the 70s that both schools were lacking and that each had something to offer the other (Martin, 1993). It has been largely accepted, in recent years, that one approach is not more valid than another but that one can be more appropriate to a particular investigation and this is the philosophy that underpins the *new* industrial economics field. This then allows us to understand the new industrial economics field. However, before that discussion takes place, it is particularly relevant to reflect upon the approach to government policies described in Chapter 2 in the light of this discussion on industrial organization economics.

4.2.4 Commentary on normative versus positive economic models: government policies

Whether explicit or not, the construction policies underpinned by positive models identified in Chapter 2 are heralding the Chicago School approach to industrial economics. It should not go unnoticed that the US and Australian models are primarily positive. More and more Australian economic policy has followed the lead of the economic rationalist approach of the United States. Economic rationalism simplistically means that our economic resources are better allocated through market forces than by government intervention (Stilwell, 1993).

Australian construction industry policies support freedom of entry, through various mechanisms: for example, by easier access to markets through electronic procurement. Other policies aim to support or encourage technological innovation. These policies are aimed at market performance and free market policies that are based upon deductive theories which 'purport that a competitive market economy will generate equilibrium outcomes in which resources are efficiently allocated' (Stilwell, 1993).

The policies impact upon the full extent of the construction supply chain. Another open market policy that impacts every market in the supply chain is that which relates to tariffs in particular industries. For example, the lifting of tariffs to ensure Australia's metal and fabrications industry (DSRD, 2000) is internationally competitive, impacts upon the construction supply chain downstream from the client and ultimately back upstream to the project.

Box 4.5 Import impact upon supply chain

Victorian imports in the metal fabrication sector have now grown to 41% of total Australian imports in the metal fabrication sector. In other words, the Victorian industry is becoming more reliant on imports and potentially at the expense of the local manufacturing industry.

Despite the significant promise that more open access to global markets offers to the industry, the commodity-based nature of the sector's products creates an additional set of challenges (DSRD, 2000).

The economic philosophy supporting tariff policy is that protectionism for inefficient industries has reduced the capability of that industry to compete successfully in international markets and is worthwhile to explore. Ultimately, the federal and state governments seek to address Australia's balance of payments problem as tariffs on imported goods have protected these industries.

Contrary to this is the assertion that most of Australia's current account deficit is not due to imbalances on merchandise trade, but due to interest and dividend payments on overseas borrowings and payments for foreign shipping and insurance. Stilwell (1993) further comments on the lack of logic of this policy:

Box 4.6 Economic rationalism

Ignore the reasons why tariffs were introduced in the first place to provide for the development of industries which would otherwise never have been established at all. Ignore the array of non-tariff barriers to trade which other countries have developed as part of their programs of national industrial development. All these considerations must be set aside because we know that the nation's economic problems are all the legacy of excessive government intervention with market processes. Economic rationalism is the solution (Stilwell, 1993).

Economic rationalism, as a theory, is based upon deductive reasoning and at the industry level of analysis finds its roots in the Chicago School of industrial organization economics.

Box 4.7 Chicago school – non-interventionist public policy

Chicago economists have always been openly and strongly anti-statist... they have been opposed, resolutely, to government engaging in the regulation of private business, fixing of prices, or direct production of goods and services with the usual grudging exceptions... and the result has been that Chicago school of economics has been successful in providing a rationale for political conservatism (Reder, 1982, p. 13).

At the level of public policy, many have suggested that, now, economic policy is wide open for debate (Stilwell, 1993). It appears that such policy should be grounded in more detailed understanding of industries and economies and more enlightened policies of government intervention. This raises the question: does the new industrial organization economics hold any promise in this respect? In particular, does it hold any promise or guidance for the construction sector and its policy makers?

4.2.5   The new industrial economics

Industrial economics for many years has been a dialogue between two groups of researchers who had different methodologies and different epistemologies. Perhaps one of the major changes is that, in the past, the linear structure-conduct-performance model assumed causal relationships and the new industrial organization model assumes greater *interactions* between the elements of the model. The relationships are not so simple; they are complex and interactive, as indicated in Figure 4.2 (Philips, 1974).

Box 4.8 Interactive structure-conduct-performance framework

Structure and conduct are both determined, in part, by underlying demand conditions and technology. Structure affects conduct. Structure and conduct interact to determine performance. Sales efforts – an element of conduct – also feed back and affect demand. Performance, in turn, feeds back on technology and structure. Progressiveness moulds the available technology. Profitability, which determines how attractive it is to enter the market, has a dynamic (intertemporal) effect on market structure (Martin, 1993, p. 7).

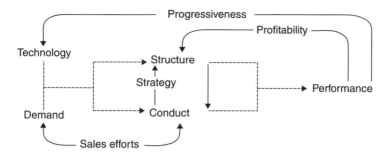

Figure 4.2 The interactive structure-conduct-performance framework.
Source: Martin, 1993.

4.2.6   Inadequacies of firm theory

Both schools of thought have been preoccupied with the notion of firm behaviour, albeit from different world views. Many researchers make the mistake of assuming that industrial organization economics is primarily an extension of the Theory of the Firm and the associated Transaction Cost Economics theory. In reality, the 'development of industrial economics can partly be seen as a consequence of several important inadequacies and faults of analysis in the theory of the firm' (Hay and Morris, 1993).

Hay and Morris (1979) outlined their problems with the Theory of the Firm after consideration of a number of contributors to the field; for example, Smith (1776), Jevons (1880), Edgeworth (1881) as cited by Hay and Morris (1980), Clark (1899) and Knight (1921). Those problems are summarized:

Box 4.9   Inadequacies of theory of the firm

- it had little regard for empirical support (forerunner to the deductive Chicago School)
- generally ignored historical and institutional aspects as factors affecting firm behaviour; and
- the firm was indivisible and being representative did not embrace differences between actual firms.

The theory encompassed a perfect competition model which was far from reality of firms' activities and a monopoly model which was equally unrealistic.

In more recent times, the Theory of the Firm has come to greater significance through the concept of Transaction Cost Economics. Fundamental questions such as why are there firms and what determines what production takes place within the firm and what takes place external to the firm were the starting point of Coase's seminal paper in 1937, 'The Nature of the Firm'. According to Coase (1937), 'there are costs to carrying out transactions and these transaction costs differ depending on both the nature of the transaction and on the way that it is organized. The tendency is to adopt the organizational mode that best economizes on these transaction costs' (Milgrom and Roberts, 1992). There are inherent problems in the work that has adhered to the Theory of the Firm. A fundamental problem is that the organizational modes have been considered to be dichotomous, either purely market or purely hierarchical – that is, external to the firm or internal to the firm.

It was not until the 1970s that further work specifically on the transaction costs concept emerged. Two key papers are worth mentioning here. Coase raised many unanswered questions regarding the origin and nature of these transaction costs. It was not until the now widely cited Williamson's paper, 'Transaction Cost Economics: the Governance of Contractual Relations' (1975), that the characteristics of transactions were described across five dimensions:

Box 4.10 Five dimensions of transactions

- the *specificity* of the investments required to conduct the transaction
- the *frequency* with which similar transactions occur and the *duration* or period of time over which they are repeated
- the *complexity* of the transaction and the *uncertainty* about what performance will be required
- the *difficulty of measuring performance* in the transaction; and
- the *connectedness* of the transaction to other transactions involving other people.

The second important piece of research at this time was a challenge by Ellram (1972), who pointed out that in reality an either/or distinction between firm and market is simplistic and that the firm enters into many different relationship types.

Others have equally indicated that the transaction cost economics approach is limited although appealing (Dore, 1976; Ouchi, 1986; Milgrom and Roberts, 1992; Alter and Hage, 1993). For example, Milgrom and Roberts (1992) indicated that the 'approach can not be correctly applied to all problems in economic organization because without additional conditions its fundamental argument – that economic activity and organizations are arranged so as to minimize transaction costs – is problematic,' because it is difficult to empirically identify transaction costs.

Further to this, conveniently, transaction cost economic theorists reduce the level of analysis to the microanalytical. The set of transactions are distinguished from the broader set of issues at the macro-analytical, or institutional, level which set the *rules of the games* in the national business system and the broader socio-economic context. Transaction cost economists typically exclude not only the macro- but the meso-economic conditions as well, that is, the market or sectoral set of issues that impact upon contractual governance structures. The market and sector conditions are quite dynamic in some industries and one suspects should not be excluded.

Many of the proponents of transaction cost economics have heeded the criticisms and adapted their approach and even Williamson (1996) now acknowledges that there are a variety of relationships between the dichotomous market and hierarchy. He also elaborated his original framework to include contingency factors, behavioural factors and context.

Nevertheless, even with these problems transaction cost economics has found favour in construction research. It is a seductive theory for construction researchers as it focuses upon the costs of negotiating and the costs associated with carrying out transactions. Since the construction sector is primarily driven by the constant negotiating and carrying out of transactions, it appeals as a useful conceptual framework for reduction of tender costs. In recent times a number of attempts to apply the theory of TCE to construction have been undertaken, but with limited success.

It is important to remember, though, it is the minimization of the costs of these activities that is central to the theory. Transaction cost economics indicates that firms organize their activities and structure their firm for the sole purpose of reducing transaction (tender) costs. This is challengeable and it is suspected that firm behaviour in a project-based industry is also directed towards increasing access to projects and competing in the market as well as reducing tender costs. The behaviour then gives rise to a variety of forms of contractual governance. It is this variety of behaviour that has been ill-considered by the transaction cost economics theorists in the past.

However, when applying transaction cost economic theory, construction researchers have not explicitly related the market structure and the nature of competition to the variety of governance structures. Transaction cost economics applied to construction began with the pioneering work of Bowley (1966) to the more recent exploration by Winch (2001); however, the problems or richness of contextualizing the types of governance structures are still not very well addressed by construction researchers at the market level (Winch, 2001). For example, the nature of market competition is still not considered as an explicit characteristic of the transaction. Winch (2001) does explore the idea, however, that the choice of governance structure is primarily driven by power that the upstream supplier has over the downstream supplier. This concept has merit and was explored further in the case studies which are reported later in this text. For more detailed discussion on transaction cost economics as applied to construction, refer to

Professor Winch's papers in the *Journal of Construction Management and Economics* (Winch, 1989, 2001).

It is quite understandable that, in approaching this topic of research, industrial organization economics, one tends to become confused with the vast array of research studies scope or purpose, theorems, assumptions, methodologies and methods. It is multidisciplinary as it draws from economics, law, management and sociology and various combinations of these. Within the core field there is disagreement and therefore attempting to borrow the concept as a framework to apply in another field remains problematic. The multiplicity of its origins characterizes the field even today as it has absorbed and inherited the different approaches of the past. There appears to be quite distinct approaches to industrial organization economics in different countries, but this discussion is left to another day. The following is a more detailed explanation of the three key core concepts of market structure, firm behaviour and performance.

4.3   Key concepts of structure, conduct and performance

In Section 4.3, the key concepts of market structure, firm conduct and market performance are explored. An understanding of these components will assist in understanding your own supply chains. The firm conduct is particularly relevant to the supply chain concept in terms of governance and firm boundaries; that is, industrial sourcing and procurement.

4.3.1   Market structure

To understand the term market structure, a clarification of what constitutes a market is required. The terms, 'industry' and 'market', are often used interchangeably in the industrial organization economics literature.

Box 4.11 Market definition

Chamberlin (1933) proposed that groups of closely substitutable products would form an industry or market. Similarly, some four decades later Porter (1985) suggested that industries are a group of firms producing products (or services) that are close substitutes for each other; this relates to produce differentiation and industry segmentation. He also considered that drawing industry boundaries is always a matter of degree. He later suggests that an industry definition should encompass all segments for which segment interrelationships are very strong (p. 272).

The definition of an industry or market indeed raises difficulties because at what point is a product a close substitute or not?

Box 4.12 Market structure

Market structure tends to deal with the characteristics of markets across the following dimensions:

- seller concentration: that is, existing distribution of power among rival firms, measured by number and size of sellers
- buyer concentration: that is, nature of the demand, measured by number and size of buyers
- degree of product differentiation and
- entry/exit barrier.

The nature of supply and seller concentration has been well considered in the literature; however, the nature of demand – that is, the number and size distribution of buyers as an important element of firm conduct and market performance – has not been considered at such length in the field.

Box 4.13 Countervailing power

The earliest work that progressed the importance of the number and size of buyers was the theory of countervailing power (Gailbraith, 1952). This theory suggests that the concentrations of power in one part of a market will evoke balancing concentrations of power in other parts of the market. For example, when a few large buyers bargain with a few large sellers (as when automobile manufacturers purchase steel or rubber tyres) it will be more difficult for sellers to hold the market price above the production cost. The number and size distribution of upstream buyers is an element of market structure that affects firm conduct and market performance.

Product differentiation is quite complex in the real world of economics and yet the simple model of competition would suggest that all rival firms sell a standardized product. In reality, this is hardly ever the case. Products are always differentiated in some way if only by location of the supplying firm.

The following dimensions are now considered explicitly in relation to the construction supply chain concept: market definition, market structure measures and differentiation.

4.3.2   *Construction market definition*

First, the definition of the construction industry is explored as this formulates the boundaries of markets and an understanding of which firms constitute the market. There has been much debate regarding the definition of the construction industry and a divide exists as to whether or not it is a single industry or multiple industries. The industrial organization economics literature assists in providing a framework to understand this debate.

Construction is often regarded as one industry whose total product is durable building and infrastructure projects. Industry definition based upon the total product has been fairly common and is used in most countries by national data-collecting agencies to describe statistics about the construction industry. The product-based approach has been quite useful to differentiate sectors of the industry and typically has included the following three broad and general categories:

- residential (dwellings, houses for long-term residential purposes)
- non-residential (commercial and social infrastructure, including shopping centres, tourism, health care and educational facilities, etc.); and
- engineering (civil infrastructure and major plant, including bridges, roads, utilities) (ABS, 2003).

Further support of the one industry concept, consider that the contracting section of the industry undertakes to organize, manage and assemble the materials and components so that they form a whole building or other work. This is basically a service to manage the whole process for a client. Hillebrandt (1982) suggested that, in these abstracted terms, this service is largely similar across various building and infrastructure projects and to this extent the industry can be regarded as a single industry.

However, it has also been argued that the service and management will vary according to the technical process involved. Therefore, in reality, there is not one industry but many sub-industries, or it is a construction economic sector system made of numerous segments (Carrusus, 2001).

However, we can look at this a little differently. Even this product-based delineation is problematic when considering the industrial organization economic approach; which relies upon the firm as the unit that defines the commodity that is supplied. There is no one firm that provides the product such as the shopping centre, the road, etc.; rather, it is a chain of firms. Even if we consider large developers, they do not in reality supply the *product*; instead, they supply the management services and are one firm involved in one of many chains of firms.

Firms may even supply products and/or services to each of these categories. For example, practitioners such as general contracting, project management and design firms may specialize within each of the above project product categories, or sectors, although this may not be the general rule. Many firms provide design or management services with equal ease to both the residential and non-residential categories or non-residential and engineering categories.

Some materials suppliers typically would supply to all three categories and are themselves often referred to as a separate industry or are involved within another industry. For example, the upstream materials supplier for an aluminium window manufacturing firm could be supplying bauxite and then smelted aluminium to a range of component manufacturers in a variety of industries. Likewise, the upstream concrete suppliers could typically supply to all three categories.

Box 4.14 Construction as a multi-industry

The multi-industry approach has found favour with many (Groak, 1994; Jennings, 1997; London *et al.*, 1998; AEGIS, 1999; Andersson, 2001; Carassus, 2001; de Valence, 2001; London, 2001; Lopes, 2001). Groak (1994) took up this argument and questioned the existence of a single industry and more importantly the validity of such a perspective. He claimed that the assumptions such a perspective brought with it affect research analysis and policy debate on construction activities within the industry. He observed that from the late 1950s onwards in the United Kingdom a number of research studies were based upon the assumption that a single construction industry existed as a feasibly coherent and responsive organism. Groak (1994) disagreed with the assumption that it was a single industry and instead that the phenomenon of construction is better represented by a multi-industry model where there are several overlapping industries, each with its set of unique characteristics. Each of these submarket industries is either directly or indirectly linked to the core activity of construction. Unfortunately, he did not provide any further theory or methodology to support his observations or progress any empirical work.

It is still ill-defined, but perhaps the supply chain concept becomes quite useful, at this point, as it is a means of providing a framework or ordering principle for the multiplicity of markets that arise. Within the property and construction sector, firms may operate within a single market for a product and/or service or diversify into other markets and compete with another set of firms. In so doing, they may locate themselves within a number of supply chains as a response to market demands.

On an individual project, firms compete within a particular market for the particular transaction. Within the larger boundaries of the construction project, there are many firms operating in different market types related to different products and/or services. The different characteristics of each market type may affect the way in which an individual chain is formed. For example, the firms involved in a supply chain can rely upon the degree of vertical integration and extent to which one firm provides more or less of the production activities along the chain. A specialist subcontractor may have to provide all the façade components, including manufacture of aluminium components and glazing processing, or alternatively may only construct the components and conduct very little of the manufacturing activities. These two scenarios provide for very different structural organizations of supply chains which could have further impact upon pricing or timing of activities during construction and thus the performance of the chain. If we can at least map that these scenarios exist and in what circumstances they exist, then we begin to really comprehend the structure of the industry.

This also introduces the relevance of the countervailing power theory suggested in the previous section, which relies upon the power relationship between the seller market and the buyer market. The number of firms in the market and their relative sizes and the degree of product differentiation are key issues in this scenario. The nature of competition in the various markets may provide information and assistance in knowing when production cost meets market price.

Market concentration is the market structure measure and is concerned with size of firms and number of firms in markets. The method of measurement of market structure has experienced a growth in the industrial organization research field and is now briefly considered. Measurement of market structure is an attempt to make comparisons between different markets in terms of performance.

4.3.3   Market structure measures

The structure of a market is often presented in terms of the degree of product differentiation and market concentration. Market concentration is related to the number of firms in the markets and their relative sizes. The previous discussion has considered the dilemmas in using product differentiation in defining the industry and this section now considers the difficulties reflected in the empirical work related to attempts at analysing market concentration. In the case of market concentration, the construction industry has often been presented in a static manner using the available statistical data collected by national agencies. The difficulties with this approach in practice have been long identified by industrial organization economists. The construction research work in this regard is divided. Some recent studies in construction research have highlighted the problems with this as a data source (AEGIS, 1999), whilst other quite recent studies have not fully comprehended these problems (Andersson, 2001; Carassus, 2001; de Valence, 2001; Ruddock, 2001). Although construction studies have been completed and attempts have been made to benchmark across countries, this work is extremely problematic because of data definition and the industrial classification of sectors.

The most recent construction industry comparative analysis that explored the use of secondary statistical data did not address these issues

(Andersson, 2001; Carassus, 2001; de Valence, 2001; Kaklauskas and Zavadskas, 2001; Lopes, 2001; Ruddock, 2001). This study across six countries attempted to use a similar methodology to analyse each country's construction industry. In this analysis, structure was described but there was no consistency about which concentration measure was used and no consistency in tackling this data source dilemma. It was quite a problematic study in that there was different terminology used, different data sources, different methods for placing boundaries around market sectors and different data analysis. Terms such as a 'highly concentrated market' were used, but this was not defined explicitly and not consistently used. Two of the country studies used a similar approach in defining the size of companies related to number of employees and the percentages in those categories. However, even this was not consistent and so comparisons could not really be made as the size categories differed.

Conclusions could not be any more meaningful than the following summary:

> The structure of the (UK) contractors' sector is similar to that in most industrial countries with a small number of large firms, a relatively less significant band of medium-sized firms and then a mass of small firms, which are either specialists or work in extremely local markets.
>
> (Ruddock, 2001)

This study did not identify which industrial classification system was being used and at times some data was used that was collected by national agencies and some data was sourced from various trade magazines. It seems that some countries were analysed using the one-digit classification and at times a two-digit classification system. Market concentrations were described for materials suppliers for one country in the study but not for any other country. Concentration was confused at one point in one country and was related to employment figures rather than market share. In summary, it was a laudable attempt; however, an extremely problematic study that probably served to illustrate unwittingly what a difficult task it is to compare market structural characteristics across nations.

Even within the industrial organization economics field, concentration ratios have been criticized for various reasons (Hay and Morris, 1980). If market performance is then related to market structure it may become meaningless. This impacts upon describing market performance and such notions as productivity, growth and efficiency, if these concepts are then related to markets with a particular concentration measure. If governments rely upon the output from their statistics to make conclusions about market structure and also decisions about construction industry policy, then it is evident that this could become a flawed process. The degree of rigour in construction industry policy development related to market structure understanding is problematic.

Most discussion on the structure of the construction industry has relied upon examining the first element in structural analysis of industries that is, the number and relative size of firms in the industry. The construction industry is populated by a large number of small- and medium-sized firms. This has led to a perception that the industry is fragmented. The discussion on industrial organization rarely proceeds further than this generalized statement.

A number of construction researchers motivated by their own research agendas have clearly identified these problems for construction research and have found the statistics lacking (Rosefielde and Mills, 1979; Bon, 1990; AEGIS, 1999; Ruddock, 2002).

There is a trail of work which still highlights the problems of how construction data is collected. For example, Rosefielde and Mills (1979) challenged the statistics collected in the US construction sector that established a technologically stagnant industry with low productivity and limited economic growth. Their systematic questioning of the statistics accuracy, relevant measurement and interpretation of results, was developed through an analysis of data definitions, data collection and analysis. Their analysis suggested that construction is not a technologically stagnant sector and did not have a low productivity. These are important findings as they have several implications for public policy in the United States.

Likewise, Bon, much later in 1990 through his use of the macroeconomic technique of input–output modelling, identified problems with data. The study commissioned by the federal government of Australia in the late 1990s and conducted by the Australian Expert Group for Industry Studies (AEGIS) attempted to use an industrial organization approach and highlighted this problem with statistics. This study, which was discussed in Chapter 2, is one such example where the federal government commissioned a holistic view of the industry based loosely upon an industrial organization approach (AEGIS, 1999). The industry was conceptualized as a chain of production and described through five main sectors: onsite services; client services; building and construction supplies; products and fasteners; and tools, machinery and equipment. Existing statistics were used to describe these sectors in terms of industry income. However, the authors note that this is contrived, as sufficiently detailed data is not available.

Ruddock, in 2002, with his strategies for better macroeconomic data to allow international comparisons, has progressed the issue further. The Council of CIB saw that the problem was significant enough to ratify a task group to explore the development of macroeconomic information on the construction sector worldwide.

4.3.4   Fragmented industry versus specialized chain of markets

Different market models operate at different times in the industry related to the individual characteristics of the construction project. In a project-based industry the structure may be constantly changing because each project market can be unique. Once again, the factors affecting which firms compete can depend upon the complexity and type of project, the contractual value, procurement methods and project location. Manufacturing industries are characterized by long-run production and markets do not tend to change so quickly; therefore, statistics collected annually reflect a closer representation of the nature of market structure.

Hillebrandt (1982) identified that the different market models are sensitive to the various tender selection models used for individual projects in the industry. The conclusion was that, prior to a tender being awarded, there are a number of different types of markets operating, dependent upon how the client chooses to approach firms to provide a bid for the project. The theoretical discussion only focused upon the role of one key participant and the major transaction in the supply chain: the contractor and the construction tender. There are numerous other transactions that take place on a project outside the procurement methods and market characteristics for these transactions are largely unknown.

Box 4.15 Multi-market dynamic project supply chain

The construction sector is complex and varied, with numerous chains of firms located in multi-markets. As a whole, there is a large number of small- to medium-sized firms giving the impression of little specialization. In reality, when each subsector for particular commodities is examined along the chain, there will be considerably fewer firm numbers in individual markets. A multi-market chain model suggests that there will be considerably fewer numbers of firms in specialized markets. Within particular project markets, there may even be smaller markets as firms compete to differentiate themselves and provide specialized services and/or products.

For example, not all architectural firms are able to design hospitals, educational institutions or art galleries and museums, just as not all structural engineering firms are able to design systems for buildings, bridges, towers and roads infrastructure. Similarly, there may also be considerable overlap in the markets that firms supply. For example, not all window manufacturers supply standardized windows, simple curtain walls and complex façade systems, but selected firms may be active within each of these markets to varying degrees.

It is suspected that there may not be differentiation amongst this final group but, certainly, there will be differentiation to arrive at the group at a project level that tenders for a particular project. There are similar scenarios for much of the consultant, contracting, subcontracting and component supplier markets that are project-oriented. There are smaller groups of firms competing against one another for tenders, rather than the larger pool; there may be little differentiation between firms in the project market and price could well be the final selection criterion. Figure 4.3 is a graphical representation of the multi-market model for the construction project-based industry.

4.3.5   Market–market relationships

Another important aspect of market structure for the supply chain is the power *relationship* between the buyer's market concentration and the

> ## Box 4.16 Case study: multi-markets aluminium construction supply chain
>
> To illustrate that different market models can operate along the supply chain, a small empirical study was conducted by London *et al.* (1998). Various characteristics of markets related to the aluminium window frame supply chain in a remote construction market in Australia were observed. The characteristics of the market included: number and relative importance of firms, differentiation, barriers to entry and models of competition. This study sought to establish the range of market types and the project supply chains were not described.
>
> The characteristics of individual projects may impact upon the type of market. The study highlighted probably the simplest scenario for a supply chain for the design, fabrication, supply and installation of aluminium framed windows. It is noted that not all the firms supplying to the materials and component suppliers have been included in the study. The data for the interpretation of the market features was obtained from both the upstream customer and their general perceptions of the market and suppliers within the market.
>
> There were ten different suppliers identified in the chain, including; engineering consultants, cost consultants, architectural design, primary contractor, aluminium subcontractor, aluminium extrusions supplier, glazing supplier, seals supplier, hardware supplier and aluminium supplier, displaying characteristics of various market types of models of competition ranging from monopoly to monopolistic to oligopoly.
>
> The results represent a snapshot of a single supply chain for a specific group of products and services; however, there are large numbers of supply chains involved in any construction project. The results do not give us any understanding of firm conduct in relation to procurement nor the markets on particular projects.

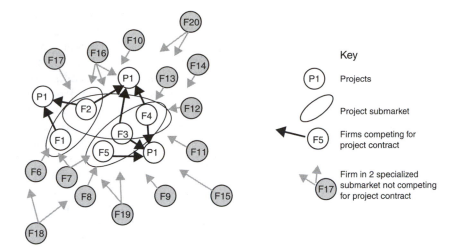

Figure 4.3 Differentiated project markets.

seller's market concentration. This has been explored previously in terms of countervailing power. This market power relationship can be affected by the project or the upstream customer/client characteristics and what each of these represent to the supplier. For example, the type of client can be varied, the client may be a regular customer and therefore *feeds* many projects to the firm. Alternatively, they may not be a regular customer and in turn the project becomes the significant factor; that is, the firm has *won* the project and it is significant as it either represents a new market or in itself it represents a significant proportion of the firm's yearly turnover for the firm. The *significance* of the transaction to both the supplier and the customer is an important indicator.

4.3.6   *Product and/or service differentiation*

Product differentiation is quite complex in the real world of economics and yet the simple model of a competitive market would suggest that all rival firms sell a standardized product. In reality, this is hardly ever the case. Products (and services) are always differentiated if only by location of the supplying firm. This is perhaps the greatest point of departure by the construction industry participants. There seems to be a perception that all customer firms in the industry are procuring all suppliers based upon price and that there is no consideration of supplier characteristics at all. The assumption then is that all purchasers in construction markets make decisions between a product and/or service from various suppliers based on a price criterion alone and that commodities are homogenous.

This belief that there is little differentiation between products and/or services offered by firms is reinforced by the research community. This belief also forms the basis upon which the majority of construction projects are procured. The industry practice of awarding construction project contracts through tendering, and the ensuing competitive bid process for cost leadership (Runeson and Raftery, 1997), assumes a perfect competition market. This practice ripples through the various markets and provides the framework upon which other contractual relationships along the construction supply chain are based. It follows, then, that across the entire industry the competitiveness of firms is based upon cost leadership alone and not differentiation. In this environment, firms have no real power and are price-takers and, as such, contractual relationships are based upon the arm's length philosophy.

If the vast majority of construction work is procured in this manner, it suggests that the construction sector is composed of all arm's length supply relationships. In other industries such arm's length supply relationships are normally only suited to non-strategic, low-value and infrequent purchases, where there is a great deal of choice from a market of expert and capable suppliers (Cox and Townsend, 1998).

It may be that all construction firms are making decisions about procurement using this mindset – yet this remains to be demonstrated and it is a problematic method to underpin an industry that seeks performance improvement. The construction sector is more varied than this. It is postulated that there is product differentiation and that purchases are strategic, high-value and frequent as well as non-strategic, low-value and infrequent. Therefore, it is illogical for all relationships to be treated in the same manner. This philosophy leaves little room for consideration of the relevance of firm-firm relationship types in the construction process. If this is true, then it also follows that relationships are not varied and can all be arm's length and categorized as spot contracting; that is, that each transaction has no context of history or past performance. This leaves little room for the value of shared knowledge of systems and product and/or process innovation across firms. This highlights a real dilemma in construction research as little is actually known in a systematic manner about the extent of the types of relationships and on what basis they are formed and what is the nature of the purchases that are being made.

As differentiation increases, the products of different producers become poorer substitutes for one another; furthermore, each producer becomes more like a monopolist. Therefore, product differentiation makes competitive industry performance less likely. Martin (1993) suggests that this might be simplistic and that there is a trade-off between market power (the power to control price) and variety. Customers are willing to accept some market power for variety.

The arm's length approach to procurement impacts dramatically upon our conceptualization of the supply chain and the theoretical position of strategic procurement, which suggests that there is differentiation within the supply chain or that it can be created. Construction clients typically appear to naively frame their actions towards purchasing a single product without understanding the chain of events that lead to the purchase. There is some evidence to suggest that some clients have orientated towards thinking that they are purchasing a supply chain rather than a single product or service (Cox and Townsend, 1998).

The premise for managing the supply chain is that it should be managed for competitive advantage rather than to reduce costs (Hines, 1994). Almost two decades ago it was noted that the interdependencies between customer firms and suppliers is the largest remaining frontier for gaining competitive advantage and that nowhere has such a frontier been more neglected (Drucker, 1992). In the last 20 years, the recognition of an altered competition model for many industries, whereby supply chains compete rather than single organizations, has prevailed (Christopher, 1998). The altered competition model may be operating in parts of the construction industry; little research has explored this. The following Figure 4.4 summarizes the

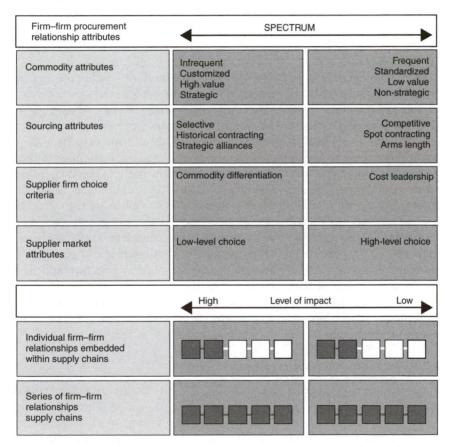

Figure 4.4 Commodity differentiation versus cost leadership supply chain spectrum.

key concepts in relation to the idea that firm–firm procurement relationship within supply chains and across an entire supply chain are perhaps on a spectrum of product differentiation versus cost leadership type attributes.

This is a complex issue for the construction sector as there is a wide variety of suppliers on projects operating with so many different levels of technical and managerial expertise. Much of the theoretical discussion of the construction supply chain would be more meaningful with descriptions of the types of market structures that exist. Differing levels and types of differentiation may occur and it is the knowledge and understanding of the degree and location of where this occurs, along the supply chain, that may inform firm conduct and construction sector performance as a whole. Both firm conduct and firm and market performance are integral to the industrial organization economics literature and firm conduct is now explored in detail.

4.3.7   Firm conduct

Firm conduct, in industrial organization economics literature, has often been described in terms of how firms seek to control price and their strategic behaviour towards achieving this. A critical part of the subtext of strategic behaviour is simply how do firms structure their own firm in order to operate in the market?

Firm conduct has typically involved discussion on such concepts as collusion, strategic behaviour and advertising/research and development. Collusion refers to competitor firms co-ordinating actions and as a group of firms (cartel) restricting the output and raising the price of product. Strategic behaviour has referred to how established firms may be able to discourage the entry of new firms by various mechanisms, including research on mergers, joint ventures and internal structure that explore controlling production output. Finally, research and development impacts primarily upon product differentiation (by product or process innovation).

Of particular importance to supply chain industrial organization economics is a detailed consideration of something more fundamental than these concepts on firm conduct – which is the firm boundaries and governance structure. In terms of the relationship between market structure and firm behaviour (in terms of boundaries and the supply chain), there are three key areas of interest, including: horizontal, vertical and conglomerate structure. In terms of the supply chain concept, it is vertical and conglomerate integration that is most relevant and these are now discussed.

4.3.8   Firm governance structure

There can be a number of stages during the production and distribution process of a product. There are ways to divide this process which can range from a fully vertically integrated individual firm taking on all stages to the other extreme whereby individual firms carry out each activity in the chain separately. The most infamous early case of vertical integration was tried by Ford Motor Company before World War Two; it was later abandoned. These days it is suggested by Milgrom and Roberts (1992) that few firms ever approach the pattern of complete vertical integration. Although integration is of primary interest to construction researchers, it is project integration across the project team participants that has been the focus of research and not vertical integration. Project integration appears to be an idealized concept of a harmonized project environment with little understanding of the real world nature of markets; the nature of risk which businesses experience and profitability expectations.

As discussed earlier, the governance structure has been tied closely with the transaction cost economics concept. As noted by Winch (2001), there have been attempts to apply the transaction cost economics approach to the construction industry (Eccles, 1981; Reve, 1984; Masten *et al.*, 1991; Winch, 1995; Pietroforte, 1997; Walker and Lim, 1999; Lai, 2000). All this work explored the costs of transaction in relation to the transaction between client and principal contractors.

Winch (2001) developed a conceptual model that merged the transaction cost economics approach with the entire supply chain. He described the construction project value system in terms of vertical and horizontal transaction governance and suggested that there was a project chain and a supply chain. The vertical transaction governance involved design services offered by the engineer and the architect. He also introduced the concept of trilateral governance whereby a third party is used to facilitate the governance of the transaction. Third party control actors in the UK construction system involve the 'architect or engineer for quality of performance and the principal quantity surveyor (PQS) for programme and budget' (Winch, 2001). The vertical transaction governance involves the project chain where there may be first and second tier suppliers. There is some confusion in the description of the model as the graphical representation indicates that the project chain involves contractors as well as designers, although the text description clearly excludes the contractor. The diagram is also misleading in that the first tier suppliers are arranged hierarchically and yet they are actually linked directly to the client. The following Figure 4.5 represents this model of the project and supply chain. The second tier suppliers are those that are not involved in the main project contract with the client. In some ways this is problematic conceptually and empirically, as the influence of the project is acute at the second and even third tier suppliers.

Horizontal governance structure refers to the manner in which the construction supply chain is organized – the make or buy decision; that is, the structure of the firm and employment versus the decision to subcontract and external governance mechanisms. The external governance structures options for the management of the supply chain were then described in a matrix relating the two dimensions of asset specificity and transaction frequency. For example, the reason why firms choose to enter into different

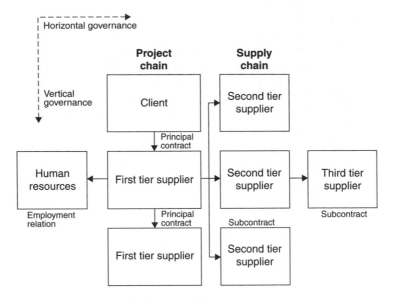

Figure 4.5 Project and supply chain model.

Source: Winch, 2001.

types of relationships is based on how frequently a transaction occurs for them and then the degree of asset specificity required. Asset specificity is loosely described as 'potential market difficulty' (Winch, 2001), whereas the definition that is typically accepted in the literature is that it is the nature of investments that the parties to a transaction must make. It is noted that transaction complexity and uncertainty were excluded from the discussion. The primary concern of this model is the relatively low uncertainty transactions later in the project life cycle, the subcontractor-to-contractor transactions.

The transaction cost considerations are a useful and integral part of the industrial organization economics of the supply chain. The Winch (2001) model does not provide any empirical work to support the discussions; however, it is an important conceptual model. Although he has stepped down the supply chain, it is somewhat restricted in terms of widening the perspective on what makes up the construction sector and really does not address the problem of complexity inherent in the real world of the construction supply chain. For example, the following issues are not addressed: the wide variety of different commodity sectors and segregation within the sectors; the various types of market types in terms of market structure, firm conduct and particularly firm–firm conduct in regard to types of procurement relationships. The subcontractor governance is only considered for low uncertainty transactions and yet it purports to understand the variety of different external governance modes for different trades. This deficiency perhaps relates to the lack of empirical work to ground the theory and progress its development. However, the model is an important critical step for considerations of application of transaction cost economics theory to construction.

The lack of empirical work to further TCE is not only endemic to construction research. One of the greatest criticisms that has plagued the transaction cost economics approach since its inception in the late 1930s has been the lack of empirical work to substantiate or develop the theory and even to this day this problem still persists. The lack of empirical work can be traced to the difficulty of obtaining good quality data as firms do not routinely collect the costs of doing business (Winch, 2001). As noted earlier in Section 4.2.6, Milgrom and Roberts (1992) have suggested that the difficulty is even deeper than that – it is difficult to separate costs of firm into the two distinct categories: costs associated with production only and costs associated with transaction only.

4.3.9 Performance

Industrial organization deals with the performance of the market. The key industrial organization economics concepts related to market performance are profitability, efficiency and progressiveness, and these are briefly considered in this section. This completes the three major components of the industrial organization economic model: structure, conduct and performance.

In a competitive market, the quantity demanded equals the quantity supplied at a price equal to the marginal cost of production (Martin, 1993). Production is efficient and all firms have access to the same technology and firms unable to use the technology efficiently lose money in the short term and in the long term leave the market. In a competitive market, firms are able to earn only a normal rate of return. Economic profit is the profit above the normal rate of return, and the reason firms seek to acquire and maintain market power is to be more profitable. However, in an imperfectly competitive market, firms will earn economic profit. The more concentrated the market, the greater the tendency for higher individual firm profit but lower market performance on the whole. The closer profit is to the normal rate of return, the closer price is to marginal cost and the better is market performance. Empirical evidence suggests that market structure explains a considerable amount of variance in profitability but that the relationship is not precisely as simplistic as the theory suggests (Hay and Morris, 1979).

Efficiency refers to the extent of a firm's use of their resources and the degree of waste. A firm that is insulated from competition may be slow to reorganize production and therefore market power will sometimes be associated with inefficiency. High market power *can* signal low firm efficiency and low market efficiency.

Progressiveness refers to the rate of technological progress. In recent times this has been equated with innovation. New technologies, which may become evident in either product or process innovation, lead to greater product differentiation. It is assumed that, the less competitive the market, the less the likelihood for firms to innovate.

The relationships between structure, conduct and performance are not so straightforward and, according to Phillips (1974), are actually interactive, as highlighted earlier in this chapter in Section 4.2.5. Past studies of the industrial organization economics of the construction sector have assumed the same type of linear causal relationship; that is, structure determines conduct and conduct determines performance. Past studies of the construction sector have relied upon a discussion of the market structure based upon describing existing statistics on firm size, firm numbers and turnover. As highlighted earlier and discussed, this is problematic as there has not been any further exploration of firm conduct nor of market performance and conclusions that are drawn are fairly simplistic. It has come to be fairly well accepted in the industrial organization economics literature that this is an outdated view of the relationships between the various components and that they are interactive and complex. In an industry such as construction, given the significant role of the chain of firm–firm relationships, it might be more useful to consider performance elements of profitability, efficiency/productivity and innovation as criteria of supply chain performance, as well as market performance.

In the construction industry, firm–firm procurement relationships are constantly changing for each project. Procurement relationships link one market to the next on projects. Procurement relationships give an indication of structural characteristics of both the sector and project markets. They also give an indication of the structural characteristics of the industry, as a whole, as these are the physical links between firms in the supply chains; they explain what constitutes the supply chain and map relationships. They also indicate the conduct of firms in markets – that is, the behavioural characteristics of firms in supply chains – as they interact with the market to determine governance strategies. Therefore, firm–firm project procurement relationships, which are representative of both upstream client sourcing strategies and downstream suppliers' bid strategies, are a fundamental component of the project-based industrial organization economic model. The procurement relationship is an entity that reflects both structural and behavioural characteristics and can provide information that describes structural and behavioural characteristics of the construction industry's industrial organization through the supply chain concept. The nature of the procurement relationship and methods of industrial sourcing are integral to the industrial organization model of the construction industry.

4.4   Procurement relationships

One of the significant contributions by those merging concepts *derived* from the industrial organization economics literature and supply chain research is the attempt to describe and analyse the structural and behavioural features of supply chains. These are important models that draw from both the field of industrial organization and supply chain theory (Hines, 1994;

Nischiguchi, 1994; Bowersox and Closs, 1996; Harland, 1996; Hobbs, 1996; Lambert *et al.*, 1998) and they have typically described industrial sourcing methods, types of contractual relationships, the nature of the commodity being transacted and the manner of supplier management. All of these attributes are important parts of firm–firm relationship. To capture all these attributes between a customer and supplier the term 'procurement relationship' is now used. This procurement relationship is the foundation of the supply chain and its characteristics provide an understanding of the underlying structural and behavioural characteristics of an industry. For this reason, theories and industry examples related to the following are now explored:

- relationship types
- modes of subcontracting
- industrial sourcing

4.4.1 Relationship types

The types of competitive and collaborative relationships that are available to link organizations can range from acquisition through to transactional. Ellram (1991) suggested a continuum of competitive relationships as the following:

Box 4.17 Continuum of competitive relationships

- acquisition
- equity interest
- joint venture
- long-term contract
- short-term contract
- transaction.

In each relationship type there are a number of criteria that affect the choice of relationship type, including:

Box 4.18 Criteria for choice of relationship type

- the purpose of relationship: the reason why the exchange is taking place – that is, the nature of the commodity (service and/or product) being transacted and the associated transaction complexity and significance, in terms of immediate financial return and longer-term economic value, and the extent of differentiation between commodities offered by suppliers

---

- the governance structure: how the relationship is formed, co-ordinated and controlled
- the duration: the length of time and frequency of the anticipated contract (Ellram, 1991).

Although identifying some similar attributes of transactions to the TCE theorists, Ellram pursues the impact that the dynamics of markets has on the transaction and the real world that the transaction is located within. It is supportive of an inductive empirical approach where it is acknowledged that a firm may choose another firm as its supplier for a variety of reasons. The TCE approach to understanding the dimensions of transactions is concerned almost exclusively with price and output decisions and their impact upon efficiency, resource allocation and economic welfare (Hay and Morris, 1980). This employs a deductive approach with little real regard for empirical support and generally ignores historical or institutional aspects. The 'Firm' is indivisible and representative and does not embrace differences between actual firms. These are the problems that have plagued the 'Theory of the Firm' for over 80 years and there is still little real advancement in the theory relating to transactions costs. However, the dimensions of the transaction provide a useful framework along with Ellram's criteria of choice of relationship type with supplier.

Where does all this sit with the construction industry? It is suspected that a variety of relationship types proliferates in the construction industry. In recent years such concepts as partnering, alliancing and public–private partnerships have begun to emerge in the construction industry between the client and tier one suppliers, and have been proposed as a panacea for the proliferation of short-term adversarial relationships. A number of the alliance contracts are based upon risk/reward contracts aligned to a particular project; in Ellram's continuum this would be similar to a joint venture. There is also evidence of many international joint ventures between firms at other tiers for individual projects which are of a similar nature (Ellram, 1988).

These types of relationships are typically individual project contracts between contractors, owners, financiers and sometimes selected specialist subcontractors, suppliers or consultants. They impact upon the project supply chain and, for that particular project, impact directly upon the unique markets that these firms operate within. The purpose of many of these contracts is linked to improving communication, problem solving and minimization of litigation on projects between tier one participants. Public–private partnerships have been primarily used as a risk and financial management strategy. However, construction contracts are made up of no less than twenty or thirty trade packages and these other sectors may remain largely unaffected by such arrangements if they are not a part of the relationship. Also, there are different types of contractual relationships along the supply chain that are not contracts for individual projects.

It is probably less known that long-term contracts for national price agreements have supported relationships between manufacturers' agents and contractors or specialist subcontractors for many years. There is evidence to suggest that firms in the residential sector have developed such agreements that are outside the boundaries of single projects which have relied upon volume purchasing to negotiate reliability of supply in terms of timing and price (Horman *et al.*, 1997; Barlow, 1998).

At this point it might be worthwhile remembering the industrial organization descriptors for distribution of power, namely seller and buyer concentration. Much of the power in the construction supply chain is derived from actual volume and purchasing power. The greater the volume and the greater the need of the downstream firm to win the work, the weaker their position in the market. This vulnerability is evident during tendering, negotiation and during the life of the contract. Alternatively, in some cases there are only a few materials, component suppliers, consultants; and it is then that the downstream firm is able to exert a higher degree of negotiating power in the relationship. These are important considerations in understanding the organization of the supply chain and the different types of relationships that may develop. The political economy or nature of power inequality between upstream and downstream firms has underpinned much of the research associated with the theories of modes of subcontracting. This theory is useful to consider in brief as it is particularly relevant to the construction industry.

4.4.2 Theories of modes of subcontracting

Five different theories of subcontracting have been identified in the literature, including: dualism (Berger and Piore, 1980), obligational contracting (Williamson, 1985), goodwill and benevolence (Dore, 1987), flexible specialization (Piore and Sabel, 1984) and collaborative strategic industrial sourcing (Nishiguchi, 1994). It is not the intention of this text to support one form or another but simply to uncover the theories that may help to develop generic characteristics of procurement relationships.

The project is central to the understanding of modes of subcontracting in the construction sector. Time is an important factor to relationships that develop on projects. Even though the project is considered as an extremely temporary organization, the project contractual relationships may be embedded in longer-term industrial environments. This longer-term perspective and explicit strategic view to industrial sourcing has been examined (Hines, 1992; Lamming, 1992; Nischiguchi, 1994).

4.4.3   Collaborative strategic industrial sourcing

Nischiguchi (1994) challenged each of these previous four modes: dualism, obligational contracting, goodwill and benevolence and flexible specialization, in his study in his attempt to explain the contemporary practices of subcontracting found in Japan. In doing so, each was found lacking and he suggested an alternative which could partially, although not completely, be explained by some of these concepts. For example, he argued that the most telling problem with interpreting subcontracting only through the previous mode's philosophy is the fundamental assumption of a dichotomous division, and even confrontational positioning, between one production mode and economic organization and the other. He proposed that the reversal of roles of the large and small firms within the economic system takes place – that both co-exist in supply chains, that sourcing relationships are not always about small firms clustering around large firms in an unequal economic subordinate position.

Box 4.19 Collaborative strategic industrial sourcing definition

The text, *Strategic Industrial Sourcing*, largely defines this type of subcontracting as inter-firm collaboration aimed at production problem solving. The mode of sourcing has been described variously as clustered control (Nischiguchi, 1994), lean supply (Lamming, 1992) and network sourcing (Hines, 1994). The most significant difference in this system of subcontracting is that, at times, the system includes more than the immediate subcontracting relationship and extends down the supply chain.

The empirical work by Hines (1994) and Nischiguchi (1994) involved descriptions of the historical, organizational and economical structure of the Japanese system of supply across automotive and electronics industries. They provided much of the context of the Toyota Production System that has come to be known as Lean Production. However, they did actually conduct research wider than the one firm, Toyota, and therefore they developed wider insights than those related to one dominant market leader. This section continues to extract some key characteristics of procurement relationships within the network sourcing system.

Nischiguchi (1994) traced the historical evolution and reorganization of the subcontracting system from the 1920s through to the 1990s, with a particular focus upon the relationship between subcontractor asset specificity and successful problem solving. Asset specificity can be understood by the attributes of a transaction; that is, complexity and significance. The more significant a transaction is to a subcontractor, the more likelihood there is of a higher degree of asset specificity. The more complex, the higher the likelihood that problem solving is required and therefore the greater the likelihood of asset specificity in the relationship.

The actual commodity involved in the transaction plays an important role in our understanding of the nature of the procurement relationship. Typically, suppliers can be categorized into a commodity supplier-type based upon the attributes of the commodity. For example, a manufacturing firm is faced with a simple choice for all of the products and their individual parts it can either make them in-house or buy them in. Of the *bought-in* parts, there are two different types: made-to-order parts and off-the-shelf parts. Customized parts are specific and unique to a customer and this type of component is termed a subcontracted part by the Japanese. Subcontractors can be further subdivided into those that have equity links with their customers (partially integrated subcontractors) and those that do not (independent subcontractors).

Typically, suppliers are categorized and organized into either specialized subcontractors or standardized suppliers, based upon the degree of complexity of the supply item (refer to Figure 4.6). It is within the subcontractor group that the network sourcing subcontracting system occurs.

The sourcing strategy taken by the upstream customer is a key characteristic of the procurement relationship. In the section on relationship types we saw that there was a continuum of types. The relationship type would range from long-term contracting to equity to joint venture to equity-based arrangements. In the Japanese situation there is a highly organized, hierarchical and ordered system of sourcing strategies. This system is at all tier levels in the supply chain. Each tier sources down the chain from a small group and through a highly structured method. Typically associated with each tier are supplier associations, which are a 'mutually benefiting group of a company's most important subcontractors, brought together on a regular basis for the purpose of co-ordination and co-operation as well as to assist all the members'. Major materials-supplier corporations are typically sourced directly by the large assembler corporation through strategic procurement sourcing alliances and are dealt with separately from the pyramid system.

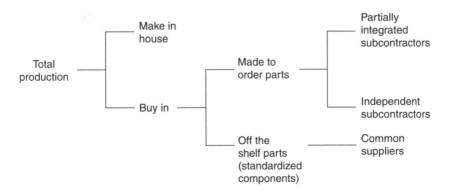

Figure 4.6 A classification of subcontractors and common suppliers.
Source: Hines, 1994, p. 53.

4.4.4   Network sourcing characteristics

The subcontractors, whether equity linked or not, are the key to network sourcing. The structural and behavioural features of this subcontracting network, described by Minato (1991), are as follows in Figure 4.7.

4.4.5   Summary

Firm–firm procurement relationships between customers and suppliers provide clues to the firm conduct, the market structure and the market performance. The relationship is typically a result of the interaction between

Structural features	Behavioural features
High dependence on made-to-order goods in the production process	Preference for dealing with a limited number of firms when placing or filling orders
Large numbers of subcontractors of high quality	Preference for long term ongoing relationships
Large number of specialist subcontractors that depend on firm to purchase more than half of their output	Dealings regulated through tacit understanding of log rolling rather than documented contracts
Single as well as multiple level of subcontracting (widespread use of secondary subcontractors)	Preference for close relationships between core firm and subcontractor which allow extensive exchange of information
Organizations of subcontractors (kyoryokyu) or a single firm	Close attention by the core firm to the quality of management when selecting subcontrators
	Acceptance by subcontractors of partial control of their internal management by the core
	Sharing in the fruits of growth

Figure 4.7 Structural and behavioural features of subcontractor network sourcing.
Source: Minato, 1991.

market structure, firm conduct and performance. There are numerous empirical studies that have identified attributes of relationships between firms in supply chains. The previous discussion has served to clarify a collection of key attributes that can assist in differentiating between different scenarios of relationships and assist in describing the nature of the relationships. Those key attributes can be grouped by supplier market environment or customer demand environment and are as follows in Figure 4.8.

Supplier market environment	Customer demand environment
Suppliers project market: that is, number of supplier tendering on a project	Sourcing strategy: method of approach to market open tender pre-registration, negotiate
Supplier location: proximity to client and competitors	Supplier choice: decision making cirteria including price, quality, past performance, etc.
Transaction complexity: ranging from standardized to highly customized commodities	Transaction type: joint venture, long term contract, equity, short term contracts
Transaction significance to supplier	Transaction frequency: degree of interaction during project
	Supplier management: coordination and expectation of asset specificity
	Number of parties to contract
	Payment method: indicates degree of closeness of parties
	Financial value: indicates significance to customer

Figure 4.8 Key attributes of supplier market and customer demand environments.

Coupled with characteristics of the individual relationships between firms is the manner in which these relationships across the entire chain are organized. Past studies on chain organization give an indication of the context for the individual firm–firm procurement relationships, thus providing a wider understanding of both structural and behavioural characteristics of supply chains that impact upon chain performance. The chain organization is a method for drawing together market structure and firm conduct. This contextual background is serving to provide a clear picture so that we have ways of interpreting behaviours in the supply chain and thus are more open to different mechanisms to change those behaviours if we are so positioned to be able to do so. The earlier high hopes for integration of business processes between firms across the entire supply chain as one of the cornerstones of supply chain management is perhaps achievable if we have a deeper understanding of the complexities involved. Various methods for describing chain organization are now discussed.

4.5   Chain organization

Although not entirely comprehensive, Hines (1998) developed a typology that is a useful model to locate the research that concerns the supply chain. More importantly, it is also a useful framework to locate the industrial organization literature to the supply chain concept. The typology was intended for researchers within the field of logistics and supply chain management and was an attempt to categorize the type of supply chain they are describing and the level of analysis. A significant contribution of the supply chain framework model was the acknowledgement that there are different levels at which to approach supply chain research, a situation which brings with it varying degrees of complexity (Figure 4.9).

The model was based upon 'increasing complexity and holism' of the supply chain that was being described rather than being an evolutionary or mutually exclusive typology. Hines (1998) suggested that there were five types of supply chain research, including the intra-functional, inter-functional, inter-organizational, network and regional clustering supply chains.

Network supply chain typically focuses upon the networks of supply for a single customer. The fifth element of regional cluster supply chains broadens the previous element of network supply to include the competitors within the markets which is often at a regional level. Hines (1998) suggested that this is where the least amount of research had been done with regard to the supply chain concept and yet this level of research is actually the 'closest to the real world'. However, it also reveals the most complex arrangements of relationships and is possibly the most difficult to investigate or model. The fourth and fifth elements of the model, network supply chain and regional clustering supply chains, are the most closely aligned with the aim of describing and classifying supply chains within the industrial organization context that this text seeks to address.

1. Intra-functional supply chain

Purchasing   Material control   Production   Sales   Distribution

2. Inter-functional supply chain

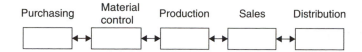

Purchasing   Material control   Production   Sales   Distribution

3. Inter-organizational supply chain

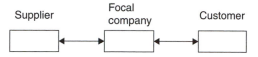

Supplier   Focal company   Customer

4. Network supply chain

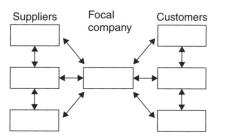

Suppliers   Focal company   Customers

4. Regional clustering supply chains

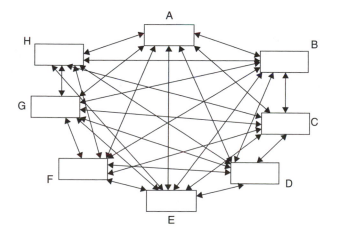

A   H   B   G   C   F   D   E

Figure 4.9 Supply chain framework model.
Source: Hines, 1998.

This section outlines various methods that have been developed to describe and classify chain organization primarily at this level. According to Hoek (1998), the study of supply chain channel configurations goes back to the 1960s and there are five general approaches used to study and describe supply chains, which are: descriptive institutional; channel arrangement classification; graphic; commodity groupings; and functional treatments. This method for categorizing the approaches is derived from a logistics perspective. In the 1990s, two other methods for describing chain organizations were developed, including the structural mapping of Lambert *et al.* (1998) and Nischiguchi's (1994) tiers of pyramids. The five approaches noted above are now discussed, followed by the two developed in the 1990s:

Box 4.20 Supply chain channel configuration categorization methods

Descriptive institutional

Channel arrangement classification

Graphic

Commodity groupings

Functional treatments

Structural mapping

Pyramid tiers

The descriptive institutional approach focuses on the identification, description and classification of middlemen institutions. Such institutions are grouped with respect to the services they perform in regard to the level of risk of ownership that they take on (refer to Figure 4.10). For example, merchant middlemen buy and sell on their own initiative, thereby dealing with the risks of ownership. Functional middlemen do not take ownership and therefore do not assume the risks of inventory ownership; however, they do provide some necessary service to both client and customer. At the second level, the middlemen may provide a range of services. The third level represents descriptive criteria commonly applied to the various categories of wholesalers specified by the first two levels. In construction, the different classes or types of suppliers has not really been addressed in this way except by large class descriptions such as consultants, contractors, subcontractors and materials suppliers. It is suspected that within each of these types there would be further categorization based upon the range of services rendered, type of customer, ownership, type of product and method of operation.

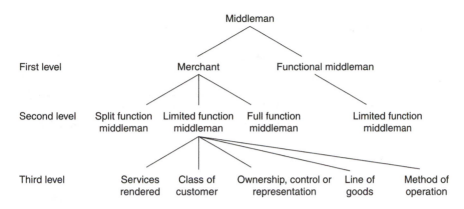

Figure 4.10 An analytical framework of middlemen in the structure of distribution.
Source: Bowersox and Closs, 1996, p. 119.

A graphic approach is a useful technique to identify the flow of commodities between the various ranges of alternatives in firm types as they are grouped by the similar service they provide in the marketplace. This approach indicates general patterns or the general channel structure. The most useful aspect to this approach is that at an industry or sector level the structure is presented, although the simplicity sometimes belies the multiplicity of different channel structures that can occur at the firm level. The following Figure 4.11 is an indication of typical channel structures in consumer goods and industrial goods distribution. The major difference between the two chain organizations is that the incidence of functional middlemen such as selling agents, brokers and manufacturing agents is much greater in industrial goods than in consumer goods chains. The advantage of this approach is the ability to show, in a logical sequence, the variety and positioning of firms that participate in ownership transfer in general patterns. The disadvantage is that it tends to understate the complexities of chains for individual firms and their immediate competitors.

The channel alignment classification maps channels based upon the relationships in terms of acknowledged dependence, which, according to Bowersox and Closs (1996), is the prime indication of channel solidarity. This simply means that firm–firm relationships are classified according to commitment levels. Three channel classifications are identified, ranging from least to most open expression of independence: single transaction channels, conventional channels and voluntary arrangements. Within the voluntary arrangements, some four different types are identified, these being: administered systems, partnerships and alliances, contractual systems, and joint venture. Single transaction channels indicate unique transactions whereby the relationship is a one-time event. Conventional

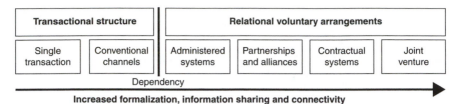

Figure 4.11 Classification of channel relationships based on acknowledged dependency.

Source: Bowersox and Closs, p. 119.

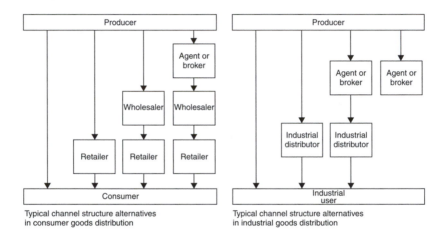

Figure 4.12 Channel structure: graphic approach.

Source: Bowersox and Closs, 1996, p. 118.

channels is a classification best viewed as a loose arrangement or affiliation of firms that buy and sell products on an as-needed basis. Voluntary arrangements are those whereby firms acknowledge dependency and develop joint benefits by co-operating. This forms part of the suite of relational contracting arrangements and involves such inter-firm behaviour as joint ventures, alliances and partnerships which have become well known to the construction sector. The following Figure 4.12 illustrates the classification of supply chain structure based upon relationship types. It is important to remember that we have now gone beyond the classification of the individual relationship and are trying to move towards that which I suggested previously in Figure 4.4 Commodity differentiation versus cost leadership supply chain spectrum.

Figure 4.13 is a map of alternative arrangements to deliver products to different market segments. This gives more of an indication of the

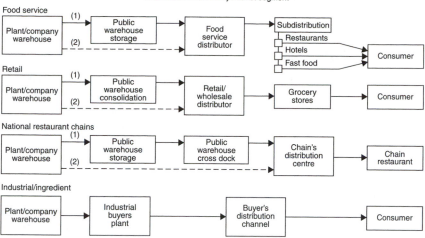

Figure 4.13 Channel alignment of one manufacturer.
Source: Bowersox and Closs, 1996, p. 118.

complexity that may arise in relationships. This is a more detailed empirical example for an individual manufacturer.

The commodity groupings method was developed in an effort to define channel structure in detail for specific commodities. These are useful for specific instances and several studies have been completed and are empirical in nature. They are largely the same as the graphic approach but specific to individual firms. Another variation is the functional approach which tends to map the channel arrangements by market segment of a single firm. This particular approach is similar to that of Lambert *et al.* (1998), which is described in the next Section 4.5.1.

4.5.1 Industrial organization-derived models

This section discusses two methods for describing chain organization at the firm level within a sector. The first method is more empirically based and describes the subcontracting system that underpins many of the manufacturing supply chains in Japan.

Nischiguchi (1987) described the manner in which the procurement relationships between suppliers and customers had become organized through descriptions of the number of suppliers in the market and the size of the supplier firms in Japanese manufacturing. The fundamental building blocks of Japan's manufacturing are its small- and medium-sized enterprises (SMEs). It is important to note that the size of an SME varies considerably and, in the research by those considering the manufacturing subcontracting networks, the SME is usually considered to be those firms employing fewer than 300 people. There can be differences in sectors across countries.

Box 4.21 SMEs and economic growth

Japan has twice as many small companies as the United States – and nearly ten times as many as Britain. For the last 30 years, they have been the critical first stage of the economic rocket that has made Japan a by-word for industrial competition.

The structure into which SMEs in Japan fit has traditionally been referred to in terms of a pyramid. Such a pyramid, demonstrating the tiers within the system, is illustrated below in Figure 4.14. There are multiple layers, or tiers, delineated by the size of the firms. At the apex of the traditional

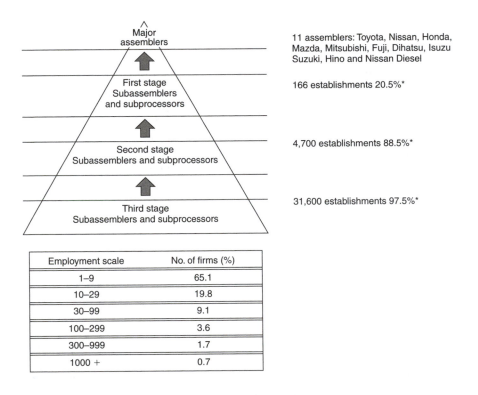

Employment scale	No. of firms (%)
1–9	65.1
10–29	19.8
30–99	9.1
100–299	3.6
300–999	1.7
1000 +	0.7

Figure 4.14 Japanese subcontracting: pyramidal structure.

Source: Hines, 1994.

Note
* Small medium sized establishments as proportion of total.

pyramid sits the final assembler. With the automotive industry, this is one of the eleven giants – such as Toyota or Nissan – which employ thousands of people each.

These market leaders are supplied by first tier firms typically employing 300–1,000 employees, although there are a number of significantly larger firms such as Nippondenso. In this industry there may typically be around 200–300 such suppliers per final assembler providing sub-assemblies or systems. These companies in some instances are owned, partly owned or have a minor equity stake, from one or more customers. They have been classified as affiliated (where no one customer owns more than 30%) or independent (Hines, 1994).

The first tier firms are supplied by a larger number of second tier suppliers providing them with sub-assemblies. Each of the first tier firms has 25–30 of these second tier suppliers, which are typically small, employing 10–300 people. The percentages associated with each tier indicate the number of small/medium-sized establishments as a proportion of the total. Clearly, there is a relationship between the position in the chain and the size of establishments. The second tier suppliers have their own subcontractors who provide specialist processing. At the third tier, there are a large number of SMEs, typically with fewer than ten employees. There also may be fourth and fifth tier suppliers depending upon the type of product; however, little research has explored these situations. This traditional pyramidal structure is a representation of the individual firm network of supply.

The common conception of the Japanese subcontracting system has been that it was a simple tiered pyramid structure; however, in recent times, this has been challenged. The Japanese system of subcontracting is no longer the closed, highly integrated, pyramidal and hierarchical structure it used to be. The type of relationship and the forms of collaboration have diversified. In the early 1990s it had been identified that the Japanese structure of subcontracting had become a network system and it was noted that before long it will become quite an extensive network system.

This system has traditionally been described as a pyramid with an individual assembler corporation at the top, and successive tiers of highly specialized subcontractors along the chain, increasing in number and decreasing in organizational size at each progressive stage. From these studies, Hines (1994) enlarged the industry-specific view to look at the wider economy and has suggested even further that, rather than this closed rigid system, the Japanese subcontracting system has moved more towards a structure of interlocking supplier networks. In this system, many firms supply more than one industry sector and potentially operate in different tiers. Electronics suppliers are an example of this. Nischiguchi (1987) aggregated the pyramids that were described in the previous section and suggested that the structural formation was similar to a series of mountain peaks. He called this the Alps structure. The Alps structure of supply chains represents a series of overlapping pyramids resembling mountain alps across an industry where each mountain represented a large assembler (Nischiguchi, 1994). The most significant finding in this assessment was that the structure was not as simple as previously claimed and that, far from the Japanese system of subcontracting being a closed keiretsu as some Western researchers had observed, it was more open. Subcontractors supply to many final assemblers and at higher levels of independence compared to Western automotive firm counterparts.

The network sourcing model as indicated in Figure 4.15 (Hines, 1994) mapped the structural organization of the supply chain as a way of providing clarity of the systems supporting the industrial sourcing strategies.

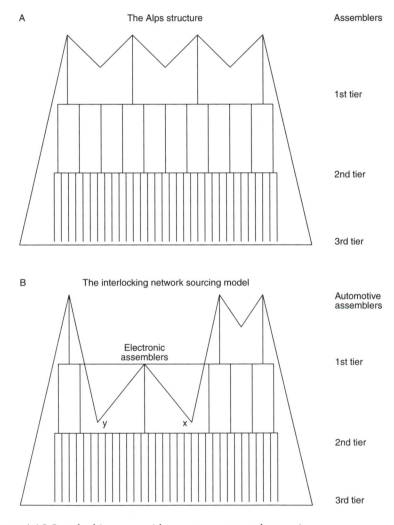

Figure 4.15 Interlocking pyramids structure: network sourcing.
Source: Hines, 1994.

Box 4.22 Supply chain aggregations

This mapping of supply chain organization can be interpreted even further at the following levels:

- individual firms supplying to a single large assembler
- firms aggregated across tiers and across commodity types supplying a single assembler
- firms aggregated across tiers and across sectors supplying to an aggregated tier of customer firms; and
- procurement relationships aggregated across tiers.

The following supply chain mapping model (Lambert *et al.*, 1998) also provides insights for mapping supply chain structure and behaviour. The supply chain structure is described as 'the network of members and the links between members of the supply chain' and provides descriptions similar to the previous approaches. However, the mapping model is more concerned with providing an abstract mapping model which could be a method applicable to providing descriptions of any chain.

Box 4.23 Supply chain network elements

The three primary structural elements of a company's supply chain network are:

- members of the supply chain
- structural dimensions; and
- types of process links.

Lambert *et al.* (1998) suggested there were three dimensions of the supply chain network to consider when describing, analysing and managing.

Box 4.24 Supply chain network structural dimensions

With regard to structural dimensions, there are three critical dimensions:

- horizontal structure refers to the number of tiers across the supply chain which is, in effect, the number of different functions that occur along the supply chain and indicates the degree of specialization; chains may be long with numerous tiers or short with few tiers

- vertical structure refers to the number of suppliers and customers represented within each tier. This reflects the degree of competition amongst suppliers. A company can have a narrow vertical structure, with few companies at each tier level, or a wide vertical structure with many suppliers and/or customers at each tier level; and
- horizontal position is the relative position of the focal company within the end points of the supply chain.

Various combinations of these three structural variables were found in a study conducted by Lambert *et al.* (1998). Findings from case studies, involving 11 companies and 80 in-depth 1–3-hour interviews, suggest as follows:

Box 4.25 Wide horizontal structure and lower tier 2 active management

Supply chains that...burst to many Tier 1 customers/suppliers will strain corporate resources and limit the number of process links that management of the focal company can integrate and closely manage beyond Tier 1. In general we found that companies with immediately wide vertical structures actively managed only a few Tier 2 customers or suppliers.

Lambert *et al.* (1998) developed the generic map for the supply chain structure of an organization as a diagram of a complex network of suppliers and customers arranged in successive tiers away from the focal organization. The members of a supply chain include all companies/organizations with which the focal company interacts directly or indirectly through its suppliers or customers, from point-of-origin to point-of-consumption.

Box 4.26 Primary and supporting supply chain members

A **primary member** of a supply chain is:

[A]ll those autonomous companies or strategic business units who actually perform operational and/or managerial activities in the business processes designed to produce a specific output for a particular customer or market.

In contrast, **supporting members** are:

[C]ompanies that simply provide resources, knowledge, utilities or assets for the primary members of the supply chain.

In this mapping model the focus tends to be on primary members. However, I would say that it is all members and all suppliers which need to be on our radar...it is perhaps wise to remember the words '...it is the weakest link in the chain...'.

Lambert, Hines and Nischiguchi importantly took us further down the supply chain. The focus on the relationships at the first tier has been a problem, with much of the past research on the construction supply chain. London and Kenley (1999) highlighted that Cox and Townsend's perspective of the supply chain was narrow and focused only on the relationships at the first tier; that is, between the client and consultants and contractors and in rare cases a specialist subcontractor. An attempt to widen the traditional perspective of the construction industry resulted in Figure 4.16. The production suppliers are explicitly considered as well as the suppliers that are involved when the building is in operation. After the Lambert *et al.* (1998) method for mapping supply chains, the diagram represented the various firms, competitors and the markets. Each tier would have numerous groups of different types of suppliers and within each group numerous firms. The horizontal structure is considered to be the arrangement of the tiers of firms away from the focal firm. In the construction industry, the focal firm is considered to be the client (London and Kenley, 1999).

This method for mapping the supply chain has been used in an exploratory study of the construction industry by London and Kenley (1999). This method has serious limitations in that when more than one client is being mapped in this manner, the multiplicity of links and cross-links between firms makes it difficult to comprehend. It is useful perhaps for mapping simpler scenarios-clusters of chains around an individual focus organization; for example, a specialist subcontractor or the primary suppliers for a project. A more sophisticated methodology that allows for multiplicity between firms, multiple projects, multiple clients and multiple markets is required. In construction it seems that we are lacking in largely both; in particular, we need to build credible aggregated project supply chain organization maps which would provide a picture of sector-specific supply channels.

4.6 Issues for procurement modelling using an industrial organization economics approach

It is important to remember that industrial organization and supply chain theory both have evolved from the tradition of understanding the world through the *permanent organization* framework rather than the *temporary organization*. The empirical studies to date are derived from manufacturing and retail sectors. The form of a supply chain for the temporary organization is flexible and adaptive, can be designed with an exact purpose, and can also be easily redesigned. Relationships that link firms are traditionally re-formed for each project and it is often claimed that the instability of demand necessitates this. Perhaps there is a continuum and variety of relationship

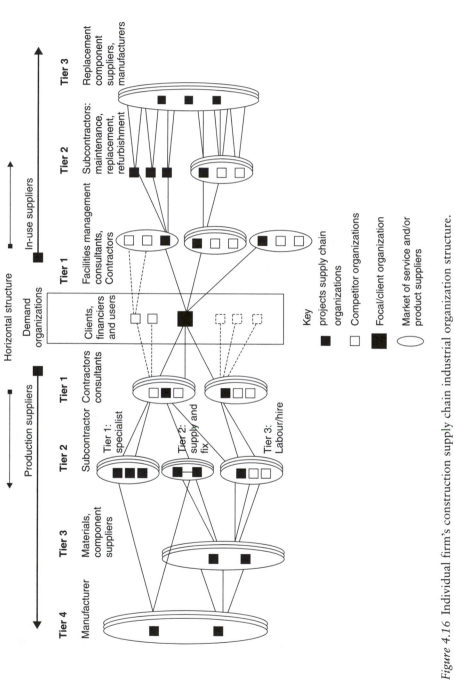

Figure 4.16 Individual firm's construction supply chain industrial organization structure.

Source: London and Kenley, 1999.

types between the extremes of flexibility and rigidity and associated with it a variety of structural organizational maps. At the moment we are left with the impression that in project-based industries all relationships are quite atomized and not embedded within any historical context.

Some construction studies have already widened the perspective and have introduced industrial organizational concepts; for example, vertical integration (Clausen, 1995; Tommelein and Yi Li, 1999), flexible specialization (Tombesi, 1997), subcontractor/contractor dependence and the *quasi-firm* (Eccles, 1981) and buyer concentration or pooled procurement (Taylor and Bjornsson, 1999), SME constellation of supply (London, 2001) and transaction cost economics and project governance (Winch, 2001). There is no shortage of construction supply chain research that is action, applied, or case study in orientation as any glance through the IGLC conferences will indicate. Much of this empirical work is oriented to the project as the unit of analysis and is not approaching the research problem from an industry or market perspective. The results are difficult to generalize and are quite focused. There is a lack of examples in construction that provide an understanding from a wider industrial context.

The established industrial organization theory is an extremely useful theoretical framework to model firms, markets and supply chains in industries. This is developed further in the following chapters, where we explore the explicit firm–firm supply chain relationships and various attributes and properties of these relationships on projects within the context of the firm, industrial and project market. The firm and market level of analysis lies within the field of industrial organization economic theory. A neo-industrial organization economic or project-based industrial organization economics approach to understanding industries uses the traditional concepts of industrial organization coupled with the project as a key construct. What this might entail is explored in detail in Section 4.6.1 Nature of construction projects, and then taken up explicitly in Chapter 5.

4.6.1 Nature of construction projects

The behaviour of firms in supply chains in the construction industry is influenced by the way in which the industry operates as a project-based industry. It is unknown how far the influence of the project environment extends down the supply chain, but it is suspected it at least impacts directly on the subcontractor and their suppliers and it is suspected that the nature of the project has a degree of influence at the materials and manufacturer level.

The initiators of the whole construction process are the clients of the industry. How to procure the construction project is one of the most significant first decisions made by all clients. Clients are often, though not always, dependent upon expertise outside the client's organization; that is, products and services are generally not in-house and are sourced from the industry. Vertical integration of all the specialized products and services required for the construction project is typically not possible as they are generally not available within the one firm. It is noted that in some cases in the residential sector there has been known to be a degree of vertical integration; however, in many cases there is a high level of subcontracting in this sector too. There are a variety of project procurement mechanisms available to the client and the decision upon which strategy to choose largely relies upon role definition, risk allocation and budget and time considerations and, it is suspected, advice given by close confidantes either from the construction industry or other clients.

Following the decision to procure a project, there are a number of identifiable phases: conception, inception and realization. The principal purpose of the conception phase is to assess the strategic need of the project, which typically includes financial feasibility studies, future growth and market expectations. The purpose of the inception phase is to clarify specific project objectives which involve determining the design brief and the method of procurement, developing a financial model and producing conceptual designs. Finally, the realization phase involves resolving the detailed design, construction planning, tendering and construction. There is often considerable overlap between phases.

The simplification of conception, inception and realization belies the variety and number of interdependent firms involved in the entire process. These firms, as we know, form temporary organizations to provide specific productive capacity for a given project to satisfy client demands. They can be categorized into the specialist roles of consultants, contractors, subcontractors and suppliers. Within each of these classes or types of firms there are various subclasses of firms who internally structure their own firm and then organize their required suppliers to respond to delivering a product and/or service in order to fulfil their contractual obligations for the realization of the project. Each of these firms has a role to play for the duration of the procurement process that is interdependent on the other firms within their own network of supply.

During the early phases of conception and inception, services are purchased by various tendering mechanisms, ranging from open tender, selective tender, to varying degrees of negotiation amongst a small group, two suppliers or single sourcing. The criteria for selection of services could be based upon any of the following or a combination of the following: lowest prices, quality, past historical relationship, trust in performance, long-term personal relationships, or there may not be any substantial competition. At some point there is enough information to commit to a tendering process to manage the construction process and to realize a building or infrastructure project. Again, the various mechanisms to purchase the service, in order to manage the construction process, range from open tender to negotiation. Similar processes are typically repeated for both the procurement of the services for constructing or assembling of the component parts on site and also the procurement of products/components.

The relevance of a supply chain metaphor begins to be realized at this stage of the discussion as the transactions take place typically between firms along the chain. Supply chains do exist once contracts are agreed upon; however, they are of a very different nature to the static industrial organization of manufacturing and long-run production supply chains.

To summarize, the construction industry is a set of projects initiated and capitalized by clients to which firms allocate resources according to the terms of their individual contracts. Contracts are not necessarily made between the client and all firms participating. Each project will have different governance structures and each firm on each project may differ in its project governance structure; that is, the number and type of suppliers which they procure to complete the contract. The importance of the concept of the supply chain procurement firm–firm relationships to the construction industry is realized as there are a multitude of these transactions occurring at progressive tiers. These transactions on individual projects and within one supply chain are being repeated simultaneously by the firm in other supply chains that may be in different stages, with different firms and of a different form.

The significance and extent of these supply chains in the construction industry is only realized when this complexity of procurement along the chain is revealed. The modelling of the underlying fabric of the industry is important for government policy as the imprint upon the construction sector and various other sectors is significant. One of the fundamental premises to industrial organization economics is its relationship to policy, and this is now considered.

4.6.2 Government policy

In recent years, a great deal of research has been reliant upon econometric modelling of markets to explain market structure and to describe changes to market structure. The purpose of the time series type of perspective of market changes is to provide information to assist governments to make sector-wide policies and decisions. As noted previously, governments play a dual role in many countries in the construction sector; they act as both a regulator of the industry and a major player in the industry as a large client. Strategic procurement, including supply chain management, is of interest to certain governments as they grapple with supplier development because they are a major procurer in the industry.

Government public policy, particularly competition policy, can be informed by observing and distilling certain properties of supply chain structural organization and structural and behavioural characteristics. Until we are able to describe the vertical and horizontal relationships between firms and understand interdependencies between firm, market, project, multi-project and sector level, it is difficult to compare the long-term impact upon changes to the relational position between firms. In a global economy this will also have implications for competitiveness, sourcing, trade agreements, monitoring and traceability of products and materials. More and more the property and construction industry has become internationalized at all levels in the supply chain. We will also be able to understand new players in the chain, such as dedicated supply procurement managers or transaction-organizing companies, as e-business and building information modelling becomes more and more significant. If we can in the future access reliable data, we are able to borrow and apply numerous econometric models from industrial organization economics. Much of this could inform the reason firms behave in a certain manner and the use of supply chain management in certain circumstances.

4.7 A final word

Box 4.27 Further reading

Ellram, L. (1991) Supply chain management: the industrial organisation perspective. *International Journal of Physical Distribution and Logistics Management*, **21** (1), 13–22.

Harland, C. M. (1996) Supply chain management: relationships, chains and networks. *British Journal of Management*, 7 (Special), 63–80.

Hines, P. (1994) *Creating World Class Suppliers: Unlocking Mutual Competitive Advantage*. Pitman Publishing, London.

Nischiguchi, T. (1994) *Strategic Industrial Sourcing: The Japanese Advantage*, Vol. 1, 1st edn Oxford University Press, New York, NY.

Reve, T. (1990) The firm as a Nexus of internal and external contracts, Ch 7 in *The Firm as a Nexus of Treaties* (eds) Aoki, M., Gustafsson, B. and Williamson, O., Sage Publications, London.

Chapter summary

1   Although the traditional industrial organization economics perspective is a useful framework, more important to the supply chain concept are those industrial organization economic concepts that assist in understanding firm–firm relationships and firm-market behaviour. Various models that enable supply chain descriptions have emerged in the last two decades in the manufacturing industry. Current supply chain techniques in the construction industry have focused on project-based models and little work has considered the entire context of the supply chain. Theoretical supply chain procurement modelling for construction needs to consider that the underlying structure involves many

firm–firm relationships within a temporary project scenario against a background of a sector underpinned with multiple projects and multiple transactions.

2   In the construction industry there are various types of supply chain scenarios and numerous different paths of supply chains. Each supplier market provides a range of different firms from which customers may choose, and therefore a pool of potential quite varied firm–firm linkages prior to contracts being established. This supplier market is not associated with an individual project and includes all competitor firms. Modelling structural and behavioural characteristics of supply chains allows clients and customers at each tier in the construction industry a mechanism to understand the impact of their decisions when choosing particular firms. It also allows the possibility for clients to develop a variety of different contractual arrangements with firms along critical supply chains as it provides clients with information to make strategic decisions about supply chain management.

3   There is a need to develop this further and explore the explicit firm–firm procurement relationships on projects within the context of the firm and market. The firm and market level of analysis lies within the field of industrial organization economic theory.

4   Mapping construction supply chains should ultimately inform government public policy. Until we are able to describe the vertical and horizontal relationships between firms, and understand interdependencies at a firm level in relation to the market and project level, it is difficult to compare the long-term impact upon changes to the relational position between firms. In a global economy, this may also have implications for competitiveness, sourcing, trade agreements, monitoring and traceability of products and materials. Specifically, it will assist in understanding new players in the chain, such as procurement managers, as e-commerce becomes more and more significant.

5   Development of a project-based industrial organization model specifically for construction supply chains also has implications for designing co-operative associations across markets for purchasing. It will also assist in locating innovative supply clusters and make transparent roles of co-ordinators and controllers in the chain.

6   Important questions to consider:

> What is the overall nature of the firm relationships along the supply chain?
> How do firms source their suppliers?
> How does a supply chain form?
> What firms actually supply to whom?

How is sourcing organized?

What is the structure of project markets?

What are the power relationships between firms and their suppliers along the chain?

What is the relationship between firms, projects and markets?

What are the differences/similarities in supply channels across projects/sectors?

7   The first research question for this study is now posed:

What are the structural and behavioural characteristics of the key objects associated with procurement in the construction supply chain?

In many cases, the collection of this data will not be an easy task as the construction industry is often secretive about methods of industrial sourcing, but it will provide some fundamental knowledge upon which to develop construction industry policy. The following chapter explores the development of a project-oriented industrial organization economics-based supply chain procurement model that is concerned with describing structural and behavioural characteristics.

5 Project-oriented industrial organization economics supply chain procurement model

How can we describe the industrial structure of our industry?
A methodology to describe project-oriented sectors

5.0 Orientation

Box 5.1 Chapter orientation

WHY: Many aspects or our industrial society are about projects and about making things. As such, we work in environments where we design, procure and construct. Although this study is using the built environment, building and civil projects, as the intellectual, theoretical and practical framework, many of the principles are relevant to those involved in industrial construction systems of various major projects, including mining, aerospace, facilities/asset management, buildings, civil structures, pipelines, industrial design. Perhaps what is common is the language and fusing of project environments, design, construction, procurement relationships and markets. Supply chain literature has not really addressed this – that is, the economics of the supply chain – and is perhaps a reason why supply chain management has not really diffused throughout the sector. This chapter intends on explicitly building the model to create a language for supply chain economics.

HOW: Chapter 5 defines a system for a new project-oriented industrial organization economic model for procurement in the construction supply chain. It develops the model through the synthesis of the principles within the industrial organization literature and the supply chain literature.

WHAT: The principal components of the model include: project attributes; firms: their commodities and their market structure; attributes of firm–firm procurement relationships; structural organization

---

of firms and events in the formation of the chain. Each construction supply chain is composed of a contractual chain connecting firms which relate to a construction project. The contractual chain is formed by firms that are providing services and/or products along the chain. The product and/or service is termed a *commodity*. A construction supply chain forms in response to a construction project which has particular characteristics; has firms with various attributes that provide commodities that may or may not be homogenous that reside within different types of markets. Firms are linked through relationships that have certain attributes. The forming and re-forming of firm–firm procurement relationships for individual projects occurs within unique project markets which are embedded within industrial markets.

There are two parts to the system: the structural elements and then the behavioural characteristics. The methodological framework for the development of the system is logical argumentation.

WHO: The system can be appreciated as a way to describe project-oriented supply chains and as a way forward for supply chain management on a much larger scale than appreciated before. It can also be appreciated as an information model as it begins to amass the key data which would support descriptions of supply chains and would be of use to numerous construction stakeholders seeking to improve or evaluate the performance of parts of the industry. The information model is designed to be supported by the computer information sciences object-oriented methodology; however, this is not explicitly described in this text.

5.1   Introduction

Interest in the supply chain management concept by the construction research community arose from the successful implementation by manufacturing sectors to resolve firm and industry performance problems. Construction industry policymakers have also appropriated the concept (London, 2004). Researchers tend to develop normative models to improve industry performance through supply chain integration (Barrett and Aouad, 1998; Cox and Townsend, 1998; Saad and Jones, 1998; Taylor and Bjornsson, 1999; Tommelein and Yi Li, 1999; Vrihjhoef and Koskela, 1999; Olsson, 2000; Nicolini *et al.*, 2001). Typically, such models are based upon the assumption of a homogenous industry which is fragmented and composed of numerous small- to medium-sized enterprises. However, policymakers are seeking positive economic models (London, 2004) and yet current policies are not based upon an explicit understanding of the nature of the industry nor an explicit model of firm and industry performance. The positive economic model, as defined by economists (Scollary and John, 2000) accepts that the industry is specialized and heterogeneous with varied structural and behavioural characteristics across individual markets. The widespread implementation of supply chain management has proven difficult in construction (Dainty *et al.*, 2001; London, 2004; Briscoe *et al.*, 2005) and one of the greatest difficulties with supply chain management in terms of construction research theory and practical application is that it relies upon both interdependent *management* of firm to firm relationships and corresponding holistic *information* about the characteristics of these relationships along the chain by large market leaders. Currently, too little is known about the characteristics and how to describe them (London, 2004) and also no-one really has a 'big picture' view of the industry. Perhaps such a model is too information-intensive. I find that hard to believe though, and believe that there is a way to refine and develop the findings presented in this text – but more on that in the final chapter.

Procurement modelling across the supply chain is fundamental to describing the underlying structure and behaviour of the industry. Such a model would merge the elements of the accepted industrial organization economics model of market structure, firm conduct and market performance; the concepts of supply chain structure (Minato, 1991; Hines, 1994; Nischiguchi, 1994; Bowersox and Closs, 1996; Harland, 1996; Hines, 1998; Lambert *et al.*, 1998) and behaviour and the characteristics of the project-oriented industry. The rationale for this approach to a procurement model is that government construction industry policy is being developed within a vacuum of appropriate economic models. Uninformed and/or simplistic assumptions are being made with regard to the structural and behavioural characteristics of the construction industry.

This chapter proposes a new model to describe the structural and behavioural characteristics of procurement along the construction supply chain using a hybrid project-oriented industrial organization economic approach and does so by assembling the components of the model from first principles. The model is described in terms of three key elements: projects, firms and firm–firm procurement relationships and the various associated entities that link these elements. The model abstracts the construction supply chain to an object composed of firms and industrial relationships brought about by construction projects. Each construction supply chain is composed of a contractual chain connecting firms which respond to a construction project. The contractual chain is composed of firms that are providing services and/or products along the chain. The product and/or service is termed a commodity. Any construction supply chain:

- forms in response to a construction project which has particular characteristics
- has firms with various qualities that provide commodities that may or may not be homogenous that reside within different types of markets; and
- has firms that are linked through relationships that have certain attributes.

Structural characteristics – supply chain entities:

- project attributes
- firms, their commodities and their market structure; and
- attributes of firm–firm relationships.

Behavioural characteristics – mapping relationships between the supply chain entities:

- organization of firms; and
- firm procurement events.

The following discussion describes each of these dimensions in more detail and serves to highlight that each of the supply chain entities – namely: project, firm, commodity, market and procurement relationships, structural organization and firm procurement events – have certain characteristics that distinguish them. The model assists in developing a framework to describe industrial organization of project-based industries supply chains and, in particular, the construction supply chain. The following section assembles the structural elements of the model in detail. The underlying idea to the development of the model is the assemblage and definition of the elements and the various associations between the elements which are then central to the attributes of the contractual procurement relationships associated with a project. The methodology is logical argumentation (Groat and Wang, 2002).

5.2   Assembling the structural elements of the model

The construction project represents the catalyst for construction supply chains. At the most basic level, a project represents a market opportunity for a firm to supply its commodity for a return. There is an association between a project and the firm; a firm mobilizes its resources and *works on* a project. Figure 5.1 illustrates this relationship graphically.

Figure 5.1 Project-firm: one-to-one relationship.

However, construction projects are much more complex than this and, on any individual project, there are numerous firms that work on the project. Each firm has a relationship with the project (refer to Figure 5.2).

This simple abstraction, although a useful start, masks the depth and breadth of the construction industry, as it is a narrow and limited view on the number of projects that are occurring concurrently. In reality, there are multiple projects in various stages taking place simultaneously and, although there is one project that has many firms working on it, there are also many other projects in the industry. Despite these multiple connections, each firm usually only forms one relationship that connects them to the project; which is typically through their upstream client on each project. In some cases, there can be different stages to projects, and firms are engaged for different contracts. Firms also have multiple suppliers on projects and so an individual firm may have many firm–firm relationships on an individual project. These firm–firm relationships are discussed later in Section 5.2.3 Firms: members of supply chains.

Figure 5.3 indicates the underlying structure of the construction industry in terms of projects and firms; that is, the industry is composed of many projects and numerous firms working on these projects.

However, this is simplistic, as firms typically supply to more than one project simultaneously. To fully appreciate the structural and behavioural characteristics of the industry and the impact that this has on the formation of supply chains and ultimately the performance of the industry as a whole, we need to explicitly account for *multiplicity*. The project is generally not a single entity that occupies all the resources of a firm; in reality, firms work simultaneously on projects and manage their resources and relationships across many projects and with the same or other upstream clients. The construction industry is composed of layers upon layers of individual projects,

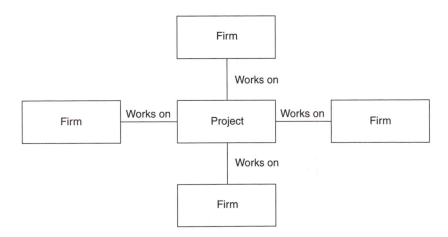

Figure 5.2 Project-firm: one-to-many relationships.

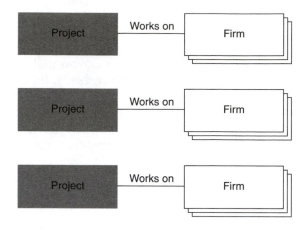

Figure 5.3 Associations between isolated multiple projects and multiple firms.

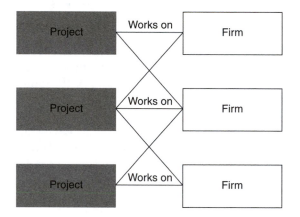

Figure 5.4 Network of many firm-project associations.

each populated with numerous firms. Therefore, each firm has many project relationships on multiple projects (refer to Figure 5.4).

This is similar to Hines' (1994) representation of the supply chain in regional clustering, except that the network is made up of many projects and many firms associations rather than simply between many firms; or alternatively Reve's (1990) idea as the firm as a 'nexus of contracts'. The construction industrial organizational structure has to contend with both firm–firm networks and firm-project networks and even Figure 5.4 belies the real complexity of the network of multiple firm–firm associations related to multiple projects. Figure 5.4 illustrates the firm-project networks and Figure 5.5 illustrates a simple firm–firm and firm-project

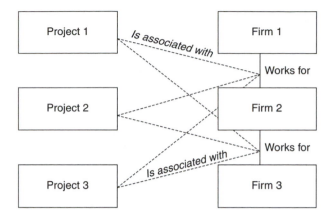

Figure 5.5 Network of many firm–firm-project associations.

network. This is, of course, the crux of the project nature of the industry, which is the core attribute of our industry and thus our response to this attribute, contributes to the difference between the built environment professional and research disciplines and many other disciplines.

5.2.1   Project attributes

The types of firms that are drawn to a project are both typical and unique to that project and the attributes of a project can alter the type of suppliers required. For example, a project requires a firm that supplies glazing and this is typical to many projects; however, the project may require a particular type of glazing or a particular production process which makes the supplier firm more specialized and perhaps unique. There are many product- and service-related firms that will be repetitive across projects; for example, few buildings are constructed without some form of steel, concrete, glass, aluminium and timber-based products, and without some form of design and construction service specialists. The attributes of the project will help to define which commodities and, correspondingly, which firms are within its project market environment. We, of course, would refer to this as customized and standardized commodities.

After we move beyond the obvious project attribute which defines the commodity required and, therefore, defines the supplier market, there are other project attributes that may define more accurately the boundaries of the project market environments. For example, the construction value, complexity and duration of a project can differentiate the supplier's markets. The following Table 5.1 summarizes key project attributes and an associated explanation which differentiate supplier's markets.

Table 5.1 Key supply chain project attributes

Project attribute	Explanation
Construction value	Construction value locates a project for a firm in relation to resources required. It is suspected that many firms determine their ability to supply to a project by the size of the contract that they are able to manage effectively given their resources, which has implications for market segmentation.
Project complexity	Complexity is often associated with construction value; for example, more complex air conditioning systems are typically associated with larger projects. However, this is not always correct; for example, a large residential apartment complex may not necessarily mean an air conditioning system that was as complex in design as a hospital and yet the contract value may be similar.
Duration	Duration can impact upon the type of suppliers; for example, the length and time-frame of the project may be a consideration for suppliers. They might normally be able to supply to a given project; however, not within the given time-frame. One possible reason is that the firm may simply not have enough resources to cover the project.
Project sector	Project sector is the range of classification schemas; for example, civil versus residential, new versus repair and maintenance, public versus private. The construction sector is typically categorized by the following sector descriptors: residential, non-residential and engineering (ABS, 2003). A project can be located to a sector; however, firms may supply to a range of sectors or may specialize and supply to one. Therefore, it is important to locate the firm to a sector and the project to a sector.
Location	Location is significant as all product-related commodities require transportation to site. The location attribute is linked to the market and its characteristics.
Project procurement strategy	Project procurement strategy has been well documented in the construction literature. It is the organizational structure adopted by the client for the management of the design and construction of a project. The project procurement strategy does establish the structural and behavioural characteristics of the supply chain relationships of those tier 1 firms who are connected to the client. There is a common understanding of various project procurement methods worldwide with typically similar understandings of classification schemas. The project procurement method may have less and less significance for supplier firms downstream in the chain away from the client.

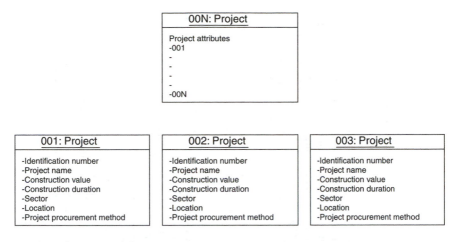

Figure 5.6 Project attributes.

This discussion is beginning to suggest the variety of scenarios possible when representing the firm–firm procurement relationships in the construction supply chain. Figure 5.6 graphically summarizes the project attributes; each of the objects can be represented in this manner. Neither the projects, actual firms supplying to projects, their commodities and the various markets, nor the actual associations that link firms together have been considered in detail.

5.2.2   *Firms, commodities and markets*

The following sections define the attributes of the model entities: firms, the commodities they provide and their commodity markets. It also describes the associations between these entities. It then provides a description of the attributes of the object which is at the core of these associations – the firm–firm procurement relationship – and concludes by defining the associations between all the entities.

5.2.3   *Firms: members of supply chains*

Firms may be categorized by various means. Key descriptors are now considered in the light of the literature review on industrial organization economics and supply chain industrial organization models. The key descriptors include: allocated firm numbers, location, scope, size, project types, specialism, market segmentation, firm differentiation and workload and are described in the following Table 5.2.

Table 5.2 Key firm attributes and definitions

Attribute	Definition
Firm number(s)	Allocated firm numbers include national business numbers or commercial database numbers. Firms are typically allocated a number by public or private sector so that they are identifiable; these numbers are unique identifiers. Information is attached to that number which can vary from country to country, and database to database depending upon the purpose and nature of the database. Many of these numbers are context- and region-specific. The information can be highly variable and therefore as global recognition their merit is questionable; their limited universality and usefulness across international borders is limited. The most universal method is the ISIC number, which is based upon commodity groupings and is largely an international system (sectors/sub-sectors). Typically, the national statistics only classify a firm in its primary sector, which can be misleading. Since firms have a number of different commodities, it would be appropriate to allocate more than one ISIC number to a firm. The number would be allocated through the associations of commodity and market entities as well as a primary ISIC number allocated to the firm. The number is not a unique identifier as more than one firm can have the same ISIC number – it is a categorization firm number.
Location	Location is a useful descriptor of a firm because there are so many associated pieces of information in the various entities that are co-dependent upon understanding location. The relationship between upstream client and downstream supplier can be better described and understood through physical proximity. The location attribute can provide context for logistics (Christopher, 1998), interdependent problem-solving (Nischiguchi, 1994) and flexible specialization (Piore and Sabel, 1984). Every project has a unique location and so too does a firm.
Scope	Scope is related to location and refers to how firms operate across regional and national boundaries. Scope involves grouping firms into the following classes: local, regional/state, national, international and multinational. Local refers to firms who operate with one office within the region that they supply to; regional/state refers to firms that have more than one division but still within the region/state; national refers to firms that operate in more than one region/state; international refers to firms that operate in more than one country; and multinational firms are those that operate in many countries (more than 3). Firm scope gives an indication of the level of operations by grouping firms according to how many different localities they operate within by region and nation. A location attribute then comes under scrutiny. Location can refer to the head office or could refer to the division or unit which is making the transaction.

Resources	Resources can be defined in numerous ways, including firm income (annual turnover); cash flow; profit; employee numbers. The most common descriptor in industrial organization is employee numbers (Martin, 1993). In many cases, when the firm is large and has more than one division and they become accountable at this level, then the resources related to the units might also be a useful descriptor, as the division acts much like a firm; that is, an independent decision-making and financial unit.
Specialism	The construction industry has accepted classifications of firm types and these are generally universal and generic categories such as consultants, contractors, subcontractors, manufacturers and materials suppliers. There may be subtle variations across countries – for example, primary contractor is often used in the United States whereas in Australia the term builder or general contractor is commonly used. In some cases firms may specialize and supply more than one commodity type and this is discussed in the next section on Commodities. Typically, they suggest the following definitions: clients, who will own and manage the facility; consultants, who are responsible for the design to enable construction of the facility to take place; contractors, who are responsible for co-ordinating and enabling the construction of the facility to take place; subcontractors (secondary or specialist contractors), who are responsible for delivering individual trade packages through installation onsite or prefabrication offsite. As we move further downstream in the chain, other firm classifications are not as well defined in the construction research literature. Such classifications of these types of suppliers who are in the construction supply chain include materials and component suppliers, who are responsible for supplying products to enable subcontractors to fabricate or install; wholesalers, who are agents in the chain who distribute raw materials or components to any number of parties at various stages in the chain (wholesalers are often involved with importing and exporting); manufacturers, who include firms/primary manufacturers or second order processors who are responsible for supplying products that eventually find their way to subcontractors to fabricate into something or install directly; and raw material suppliers, who extract raw materials from the earth or sea and supply to manufacturers. The commodity attributes define the commodity types within these broader firm classifications.
Segmentation	This attribute distinguishes firms that are located in markets that are segmented or not; that is, the market of firms supplying wooden buckets may contain 20 firms. However, not all 20 firms compete against each other; smaller groups of firms compete against each other in segmented markets. Segmentation relates to the primary sector that the firm considers it to be located in and it relates to the ISIC sector. Each commodity may also be located in a market that is segmented and this is described through the commodities' market attributes and the attributes of individual firm–firm procurement relationships.
Firm differentiation	This attribute classifies how the market is segmented; whether it is by product, price differentiation, scope, size and/or service.

5.2.4   *Commodities*

Each firm supplies at least one commodity and may supply more than one. There are potentially numerous attributes that can describe a commodity; for example, a product can be described by many physical descriptors such as weight, size and colour. However, for the supply chain procurement model the attributes of a commodity have been narrowed to those that are primarily concerned with describing the type of commodity that is being transacted at a higher level. There are various Building Information models and Product Classification models which are largely inter-operable which ultimately the model described in this text would connect with. The commodity attributes are described in the following Table 5.3.

The previous section on firm attributes enabled simple classifications of firms by generic groupings. The groupings of subcontractor, consultant, etc., describe little except that consultants supply a service, subcontractors supply a product and a service and manufacturers supply products. These generic groupings may also bring with them various stereotypes that are ill-defined. At times these terms may also be too broad and become meaningless. The commodity attributes allow more specific descriptions of what is supplied by the firm for each transaction.

The firm supplies a product, a service or a product and a service within a transaction. Firms may supply more than one commodity type (refer to Figure 5.7) and, therefore, firms may be competing in a number of markets. Each firm may have a set of commodity objects; each commodity object has its own attributes.

Each firm also has an association to a project by virtue of the commodity it supplies. Each project requires a generic commodity which a group of firms may supply; however, the project and the upstream customer may then demand unique attributes for a particular transaction. It is the characteristics of the commodity supplied that is important to the project and the firm–firm procurement relationship. It is a combination of the firms' commodities and market attributes and the project attributes that are critical to the customer's demands and the ultimate composition of the supply chain that forms for each project. Each commodity is located in one sector market, although it may be associated within more than one project market, and this is discussed in the next section on associations between various elements. The uniqueness of each commodity supplied for each project is related to the firm–firm procurement relationship object, which is also described in detail in Section 5.2.5.

5.2.5   *Industrial markets*

Commodities are located in' markets; and four key attributes have been identified to describe the markets. All markets can be classified according

Table 5.3 Key commodity attributes and definitions

Commodity attribute	Explanation
Commodity type	The commodity can be categorized by whether or not it is a service, product or a product and service.
Product type	Products can be categorized by trade element. There are various systems of classification; for example, in Australia the nation-wide accepted system for specification relies upon a trade elemental breakdown which includes the following: demolition, groundworks, piling, concrete, masonry, structural steel, metalwork, woodwork, glazing, hardware, access floors, partitions, roofing, suspended ceilings, windows, doors, finishes, painting, furniture, drainage, electrical, mechanical and preliminaries. This system also aligns with costing and the national building code regulations.
Service type	There is no universal classification system for services. Each service is provided by different groupings of consultants; various specialist engineers, architects, quantity surveyors, land surveyors, building surveyors, project managers, construction managers, landscape architects, interior designers, subcontractors, manufacturers, etc. Each of these professionals is typically governed by various professional associations and, in some cases, legal instruments in relation to definition. A generic list of services includes: conceptual design, performance briefs, detailed design, design/project management, scheduling, inspection, installation, production/ shop drawings, construction management, commissioning, transportation, labour, etc.
Product and service type	Subcontractors typically provide a service and a product; for example, brick supply and installation, aluminium window design, supply and installation, equipment hire and operation. The classification system includes; design, supply and install; supply and install; and hire and operate.

to the International Standard Industrial Classification (ISIC) system or corresponding system for that country (for example, in Australia and New Zealand, ANZSIC). The ISIC system is based upon commodities. Coupled with this is the market structural attribute which is denoted by the number of competitor firms and/or the concentration ratio if applicable. Markets may also be segmented.

The industry classification numbers and concentration ratios are useful to a certain extent; however, they do not describe completely the competitive

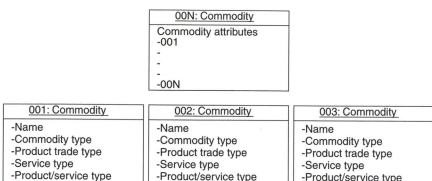

Figure 5.7 Commodity attributes.

nature of the industrial market. In simple terms, concentration ratios are a measure of the market share of the leaders of the market (note there are numerous permutations and fine-tuning of definitions of ratios). Similar to how a firm locates itself in a generic sector which may be segmented, correspondingly, each commodity may be located in a market that is segmented. Although not considered at all in the construction supply chain literature, it is suspected that there are levels within these markets. The levels are related to how a firm differentiates itself. This is difficult to capture through the national industry classification systems and this is what is really interesting in relation to the nature of competition in the market.

According to ANZSIC, a commodity and a firm could be classified as E4222; which would mean that the firm supplies a commodity related to *brick-laying services*. This immediately locates the commodity within that industrial market; however, industrial organizational theory suggests that firms perform in certain ways as they compete in the market which produces various levels within a market. There are many degrees of service levels of brick-laying services possible in the construction industry. A further attribute: segmentation is proposed for the industrial market entity of a commodity which accounts for this quite dynamic attribute.

The nature of competition within the markets can be more specific. In the market attribute for segmentation, this specificity of competition is accounted for. Describing industrial market competition characteristics in a more responsive manner than the national statistical system is important to describing the context of the relationship between upstream and downstream suppliers in the chain. It is a fundamental property of the structural and behavioural characteristics of the construction supply chain. Segmentation identifies the following: façade fabrication for special projects, façade fabrication for standard projects, architectural design service major projects; architectural design services small scale residential projects, etc. The segmentation criteria can be based upon, for example, product complexity, service scope, firm scope, price, etc. The same sort of categorization occurs in the subcontracting group; for example, there are many steel fabricators and yet a more select group may be focused upon less complex projects such as small-scale residential projects or more complex projects such as a national stadium. We can think of segmentation in terms of tender invitations and expressions of interest. Refer to Figure 5.8 for a summary of the industrial market attributes.

The firm–firm relationships that form for individual projects have not yet been considered. The firm–firm relationship brings with it an even greater clarity of commodity market and our understanding of supply chain industrial organization, as there are often project-specific markets. Project markets may provide a further level of segmentation according to the individual characteristics of the specific project. In the construction industry, firms may supply commodities that vary according to the project requirements and, thus, the accepted concept of vertical integration is challenged; because, within one firm, the level of integration up and down the production chain changes with each project. This high level of variability and uniqueness that arises with each project may be a unique characteristic of the industrial organization of the construction industry. For example, a firm may typically supply a commodity that is a product and a service involving the design, fabrication and installation of a façade. This particular firm may also manufacture the aluminium extrusions and conduct second-order glass processing. This represents a degree of vertical integration. If the firm also supplies these extrusions to its competitors, then this represents another commodity and can be clearly identified. The project, the firm, the commodity and the industrial market objects are all interlinked and these associations are now explicitly considered. The association manifests itself in the firm–firm procurement relationship, which is discussed in Section 5.2.7, after the associations are considered.

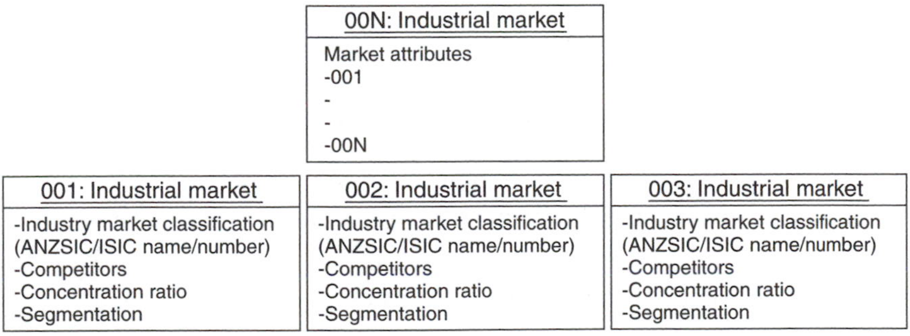

00N: Industrial market
Market attributes
-001
-
-
-00N

001: Industrial market	002: Industrial market	003: Industrial market
-Industry market classification (ANZSIC/ISIC name/number) -Competitors -Concentration ratio -Segmentation	-Industry market classification (ANZSIC/ISIC name/number) -Competitors -Concentration ratio -Segmentation	-Industry market classification (ANZSIC/ISIC name/number) -Competitors -Concentration ratio -Segmentation

Figure 5.8 Market attributes.

5.2.6   Associations: projects, firms, commodities and markets

The discussion in this chapter has focused upon describing the entities that make up the model. This section describes how each of those entities is associated to one another. The following Figure 5.9 graphically describes the project, firm, commodity and market entities and their associations.

Each firm *works on* a project. Each firm *supplies a* commodity on a project. Each commodity *competes in* an industrial market. Figure 5.9 represents these relationships as simple, linear and static associations that occur between these entities. Each entity has a number of attributes and each attribute has descriptors; for example, the commodity type can either be a service, a product or a product and a service, and the firm scope can either be local, regional, national, international or multinational. Each attribute requires a set of criteria that assists in establishing distinctions between the objects; for example, commodity type can be service, product or product and service; firm scope is local, regional, national, international or multinational and industrial market class can be sourced from the national statistic classes and subclasses.

The next stage of the model development is to explicitly focus on the objects that tie the supply chain together; that is, the firm–firm procurement relationships; this is discussed in Section 5.2.7. Before this is done, however, it is possible to consider to some extent the dynamic nature of these relationships and the multiplicity of relationships across many projects.

There are numerous projects that make up the construction industry and firms typically work on one or more projects at any one time. On each of these projects, firms supply a commodity. The commodity may differ from project to project. Figure 5.10 illustrates the multiplicity that underpins the industrial organization of the construction industry. The complexity and high volume of transactions that take place in the industry are now beginning to be realized through this model.

The supply chain organization is reliant upon the firm–firm procurement relationships that arise on projects. This entity draws together the following connexions between the:

*   upstream and downstream firm objects as they are linked in the supply chain
*   supplier firm object and the commodity object; and
*   firm–firm procurement relationship object and the project object.

5.2.7   Firm–firm procurement relationships

The development of the model thus far has considered that the supply chain is composed of the basic entities or objects: projects, supplier firms, their

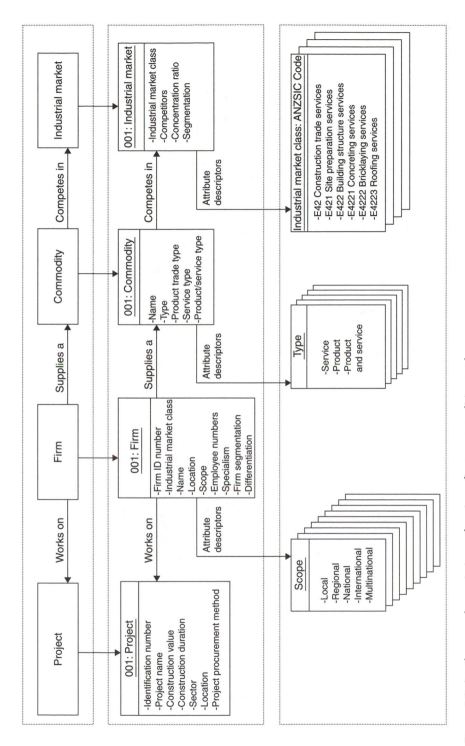

Figure 5.9 Attributes and associations of projects-firms-commodities markets.

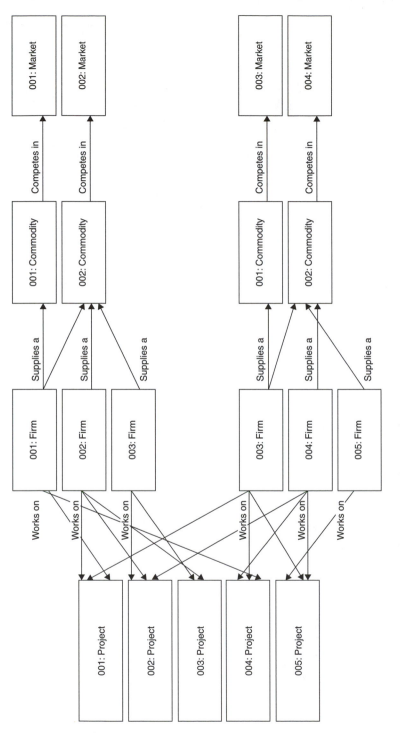

Figure 5.10 Multiplicity of projects-firms-commodities markets.

commodities and the industrial markets. The commodities and markets were considered in relationship to the broad industrial market sector. Now the upstream firm–downstream firm procurement relationship is examined in detail to bring this association directly to the project. Firms don't typically transact with a project; instead, they transact with a 'client' or a 'customer'. However, firms do have an association with a project. Supplier firms are associated with a project through their upstream relationship. The next Section 5.2.8 will deal with discussing first the procurement relationship and its attributes from the perspective of the *supplier market environment* and then will identify the attributes from the *customer demand environment*.

The types of competitive and collaborative relationships that are available to link organizations is a dual system and is defined by what the supplier brings to the relationship and what the customer brings to the relationship. Together there is a firm–firm procurement relationship that has attributes that are derived from the supplier market environment and the customer demand environment. The relationship of the seller and their market leadership and the buyer market and their market leadership and the subsequent power dynamics is referred to as countervailing power.

5.2.8 Supplier market and customer demand environment

The supplier market environment includes the following attributes: project market, supplier location, transaction complexity and transaction significance. The firm–firm procurement relationship also has attributes dependent upon the customer behaviour, including: sourcing strategies, supplier choice, transaction type, transaction frequency, supplier management, number of parties, payment method and financial value. The following Table 5.4 provides descriptions of these attributes.

At this point it might be worthwhile remembering the industrial organization descriptors for distribution of market power; namely, seller and buyer concentration. The distribution of power in the construction supply chain should consider that much of the power is derived from actual volume and purchasing power. The more work that is won by the firm, which translates into more projects and/or higher contract values, then they are in turn able to exert more influence when purchasing commodities downstream. The greater the need of the downstream firm to win the work, then the weaker their position in the market. Let us call this project market vulnerability and it is evident during tendering, negotiation and during the life of the contract. It is a particularly dynamic factor – project market vulnerability and need-to-win work can change within a month, so that when a firm submits a tender to when it begins a contract can be a very different level of market vulnerability.

Table 5.4 Firm–firm procurement relationship attributes and definitions

Attribute	Description
Supplier market environment	
Project market	The IOE theory suggests that the number of competitors is a key element of the market structure (Hay and Morris, 1979; Martin, 1993). The industrial structure of the construction industry needs to reflect the nature of competition in both the industrial market and the project market; as they may differ. There may be a variety of reasons why they differ; for example, the characteristics of the project may dictate that only a select number of firms can supply the particular commodity or that a select number of firms can supply at the particular time. The project market may be referred to as the tender market for onsite construction firms, even though consultants do not consider that they tender. Like all other construction-related firms, they often undergo some form of selection process. There may be numerous suppliers in the market sector who are able to provide the commodity; however, for expediency, a smaller project market may be used. Alternatively, the larger market sector may not provide enough commodity differentiation and therefore a smaller select group is required unique to the project. The characteristics of the supplier are critical to the size of the project market. If product/service differentiation is less significant, then the project market approaches the market sector size.
Supplier	Differentiation may arise through the supplier location and location can apply to firms supplying products or service (Piore and Sabel, 1984; Tombesi, 1997).
Transaction complexity	Transaction complexity can arise from the degree of complexity inherent in the commodity or in the supply of the commodity. Transactions can range from a high degree of complexity to a low degree of complexity; for example, subcontractors may supply off-the-shelf parts (standardized components) which would mean that the transaction has a low degree of complexity, or made-to-order parts (problem-solving contracts) which would mean that the transaction would have a high degree of complexity (Hines, 1994; Nischiguchi, 1994).
Transaction significance	The significance of the relationship to the supplier is an important characteristic of the relationship. Transactions can range from high significance to low. A supplier's perception that a transaction is highly significant is often related to high asset specificity (Williamson, 1975). The following are examples of highly significant transaction scenarios: recurrent transactions requiring highly specialized assets; transaction represents a high proportion of annual turnover and uncertain nature of transaction through either new market or highly technical (Ellram, 1991). Low significance can simply mean that the transaction is one of many with a customer who is regular and who would not be likely to move to a competitor for supply.

Customer demand environment

Sourcing strategies	A firm–firm procurement relationship can be considered from the perspective of its governance structure; that is, transaction formation and management (Ellram, 1991). Formation and management are two activities that are inter-related. A customer typically approaches the market of suppliers with a particular strategy, that is, a sourcing strategy. The strategies can include options that range from transactions that are reminiscent of spot contracting through to those that require significantly more investment by both parties, for example, strategic alliances. Each of these strategies involves a different approach to the market and can includes such options as: open tender, pre-registration of tenderers, tender and negotiate with a small group of suppliers or strategic alliances/single sourcing. Each type of strategy would typically correspond to a smaller project supplier market.
Supplier choice	Choice of supplier can range from lowest price, quality, past performance or trust. In some cases there may also be no choice decision required as there is only one supplier. Once the decision is made then the transaction type is determined.
Transaction type	The literature has identified a range of transaction types from acquisition, equity interest, joint venture, long-term contract, short-term contract and transaction (Ellram, 1991). Particular to construction are partnering and alliancing transaction types, with clients and joint ventures between contractors/consultants. In construction the transaction type includes: project-only contract, a project alliance or joint venture, a purchasing or licence agreement or a firm-wide strategic alliance. In many cases purchasing agreements involve a certain volume of spend and may be a firm wide agreement rather than an individual unit within the firm. Other firm-wide agreements may also arise that are not strictly related to purchasing volume.
Transaction frequency	The degree of interaction is an important firm–firm procurement relationship descriptor. The degree of interaction can also range on a continuum from high to low, depending upon commodity type. Small regular purchasers are classified as high and one-off large irregular purchases are classified as low.
Supplier management	The manner in which the customer manages the supplier outside of the boundaries of the contract. Supplier management strategies range from high levels of supplier co-ordination to low levels (Hines, 1994). Higher levels typically relate into higher degrees of asset specificity upon the part of the supplier (Williamson, 1975).
Number of parties	The number of parties to a contract reflects the significance of the relationship. More complex transactions would require more than one party to a contract.
Payment	Payment method set up by the customer gives an indication of method the closeness of the relationship between the two firms; for example, account payments can be an indicator of a high degree of trust between the two firms and/or that there are numerous small transactions taking place between the two firms. These types of arrangements also reduce transaction costs and are part of supplier co-ordination (Williamson, 1975).
Financial value	This is simply the cash value of the contract and is an indicator of the degree of power that the upstream customer has over the downstream supplier. This can be viewed against firm turnover; a low supplier firm turnover with a high transaction value would indicate that the transaction is highly significant for the supplier.

Alternatively, in some cases there are only a few materials, component suppliers, consultants, and it is then that the downstream firm is able to exert a higher degree of negotiating power in the relationship. These are important considerations in understanding the organization of the supply chain and the different types of relationships that develop. It is speculated that much of what preoccupies the chain participants is their behaviour in relation to this strategic positioning for market power. Figure 5.11 is a graphical representation of the procurement relationship attributes and the associations with the previously discussed entities.

5.2.9   Summary

This first part of the chapter has sought to consider the building blocks of the construction supply chain and the various information required to model the system. The components of this model suggest that we need to

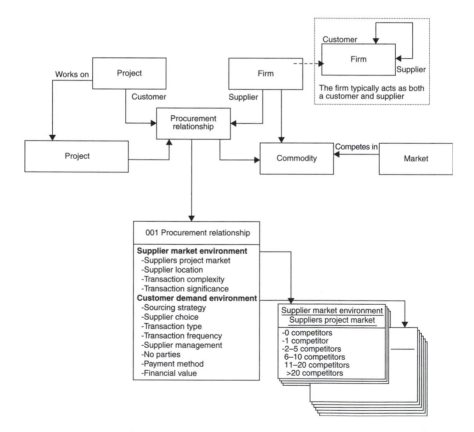

Figure 5.11 Procurement relationship attributes.

understand the following details about supply chains: project attributes, market attributes, customer and supplier firm details, commodity and procurement relationship. The associations between these components have been identified and the key information about each of these components in order to define the system.

Markets typically include an industrial market and an individual project market, which may have different structural characteristics and this is at this stage accounted for in the procurement relationship. One of the key considerations of firm conduct in the immediate environs of the construction project sector is that firms organize themselves in response to the project nature of the industry; that is, their governance structures are typically developed as a strategic response to those commodities for particular markets and often for particular projects or at least project types. As we move down the tiers, the suppliers are still very intrinsically supplying to the construction market and as such are part of the supply chain, but it is interesting to speculate on the degree of influence the project has on the chain.

Both market structure and firm conduct are determined in part by underlying demand conditions and available technologies. Demand in the construction sector ultimately relies upon the number of projects. Firms respond to the number of projects as either individual project contracts, or as aggregates of individual contracts (supplier accounts). As firms move further away from the project environment downstream from the client, they tend to view demand as a series of supply contracts and less as project contracts. It is apparent that when a sector level perspective multiplicity is critical, there are many projects, many firm-to-project associations and many firm-to-firm associations.

Even though there are many project suppliers, they have some common characteristics which assist in understanding the industrial structure of the sector and the generic manner in which firms relate to each other in the sector. Firms supply commodities which are located within an industrial market. Firms may supply more than one commodity. On a project, firms typically supply one commodity. It is suspected that there is a project market operating as well as the industrial sector market; the project market is a unique characteristic of the firm–firm procurement relationship. Each firm is linked to a project by virtue of its association with an upstream customer which has been described as the firm–firm procurement relationship. Procurement relationships may differ from firm to firm, but they all have the same type of underlying attributes.

The aggregation of procurement relationships forms supply chains and the importance of the various linkages as the basic building blocks of a model to describe a construction supply chain. An important aspect to the model has been the understanding of multiplicity of projects, firms, commodities, markets and relationships and the rich-layered and complex nature of the industry and its various procurement relationships which form the fabric of supply to projects. The model described in this chapter thus far is a way forward to begin to describe the industrial structure of the construction supply chain. There are still difficulties encountered when using the industrial organization economic methodology as a framework to represent a model of supply chain procurement. These difficulties have been defined as the concepts of *multiplicity*, *interaction* and *types*:

- multiplicity of associations between entities
- interaction between structural and behavioural characteristics of entities; and
- types of entities which have common characteristics.

The industrial organization economics methodology has limitations when trying to represent the dynamic nature of a project-oriented industry. I also developed a methodology for describing and representing the concepts of multiplicity, interaction and types. I used a methodological framework that allows for this complex real world representation and thus modelling of the supply chain and the many individual project scenarios is provided through the information sciences object-oriented modelling methodology. The industrial organization economic model though was an important first step towards the development of the object-oriented model. The ongoing research in relation to the information model is introduced in the final chapter of this text. Section 5.3 though describes ways to think about some of the behavioural dynamic aspects of supply chains.

5.3 Description of the behavioural characteristics of the model

The next stage in the development of the model is to begin to describe the behavioural characteristics of the *objects* in the system. One way of understanding at a chain level the behavioural characteristics of the model can be understood by mapping the overall structural organization of the firm–firm procurement relationships and by examining closely the individual events that lead to the formation of those firm–firm procurement relationships. The structural organization mapping of firms across supply chains represents the way in which firms conduct themselves to organize production to fulfil contracts. It is their strategic behaviour and firm governance which is reflected; however, it also represents the underlying market structures. These are key concepts that the industrial organization economic literature has provided us. As well, the mapping of the events that occur to form the firm–firm procurement relationships which form the supply chain give an indication of the interactions between various components of the industrial organization model, namely: commodities, market structure, firm conduct, demand and firm and market performance. The structural organization of supply chains is discussed, followed by the firm–firm procurement relationship events. So, by describing the structural characteristics, we are able to describe the behavioural characteristics. This duality of objects having both structure and behavioural characteristics is not new – it pervades not only the industrial organization economic theory but does actually surface in numerous other theories.

5.3.1 Supply chain organization

Firms generally require a number of suppliers to assist in their ability to fulfil contracts. For example, a concrete subcontractor may require contracts with a concrete manufacturing firm, a steel reinforcement supplier, a waterproofing specialist supplier, a labour contracting firm and a concrete formworker. Each of these suppliers then in turn may have a number of suppliers that they require to fulfil their contract to the concrete subcontractor. Each firm appears to be a node for a cluster of firms supplying various commodities. The chain has often been represented primarily as a linear entity when, in reality, it may be a series of clusters of firms. Each firm acts as a node in a network of its own suppliers or as the nexus of contracts (Hart, 1989; O'Brien, 1998; Reve, 1990).

Figure 5.12 indicates one scenario of the arrangement of firms supplying on an individual project from the focal firm, the client. Each connecting line represents an individual firm–firm procurement relationship between firms. This graphical representation maps firms and firm–firm procurement relationships and gives an indication of the transfer of commodities from one firm to another. Past representations of the supply chain in construction literature have not attempted to indicate this volume of transactions that takes place, nor the number and variety of firms involved in the supply chains. For example, the impression of the supply chain that is often given is that there is a homogenous group of firms supplying at tier 1 and they are subcontractors who behave in the same manner and who are located in the same type of markets.

The firms represented in Figure 5.12 are those that have been selected. At any one position in any of the multitude of chains, a different firm may be chosen for the project, which would alter the composition of the supply chain. A permutation of one firm could dramatically alter the supply configuration. The following Sections 5.3.2 and 5.3.3 discuss the vertical and horizontal chain structural organization in more detail. It is followed by a consideration of the role that multiple projects plays in affecting the behavioural characteristics of the construction supply chain, since much discussion in the literature does little to address this added complexity.

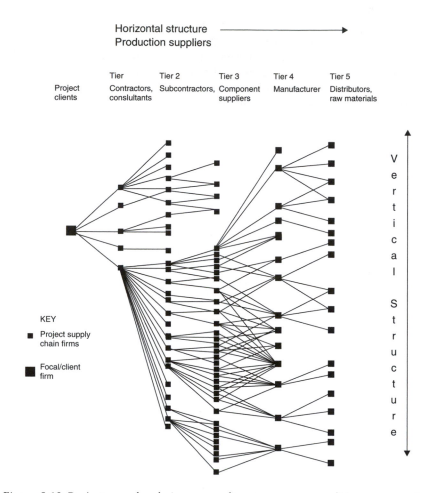

Figure 5.12 Project supply chain: many firms-many commodities-many project markets-many procurement relationships.

5.3.2   *Vertical chain structure*

The vertical structure of a tier reflects the number of different types of commodities required to satisfy the upstream demand. It indicates the degree of commodity specialization across that tier. It also indicates the degree of horizontal market integration in the chain as it reflects the cluster of firms that upstream firms require to contract with downstream in order to fulfil the obligations of the contract. The greater the number of firms, the greater the degree of specialization and diversity of markets or less degree of horizontal integration.

To recap a little – the supply chain begins when a client, that is focal organization (refer to Figure 5.12), determines the need for a new facility. Project inception involves a demand for a facility or infrastructure by client organizations. There may be more than one client organization; however, there is generally only one contractual entity. The decision to build a new facility is then, typically, followed by designers developing sketch designs. Architectural (if a building project) and engineering consultants are typically contracted to the client. There are numerous project procurement strategies and possible project organizations (Walker, 1996) and this affects whether the client contracts directly to the consultants or a project manager does so on their behalf. This attribute of a project – namely, the project procurement strategy – can alter significantly the firm–firm relationships and therefore the structural organization of the chain at the higher levels in particular.

In the automobile industry, the customer is typically not considered in the chain when the structural chain organization is discussed. The chain begins at the major assembler. This represents a critical difference from our project-based industry. The project is where all chains converge – the nexus of contracts. The client draws chains to the end product; the facility. For the model developed in this chapter, the construction demand organization is equivalent to the large assembler in the Hines' and Nischiguchi models in Section 4.4.3. This challenges the general assumption that the contractor is the equivalent to the assembler. This point of difference is made because the individual owner of a facility has a larger impact on the supply chains than the individual owner of an automobile and, generally, the contractor – it is almost like construction is a 'pull chain' and automobile manufacturing is a 'push chain'. Some may disagree with this observation and it is an interesting point to debate and consider at another time.

By comparison, the *automobile assembler* and the *construction industry client* have a similar stake in effective supply chain management and a similar capacity to impact supply chain organization. There is a greater propensity in the construction supply chain for contractors to abrogate their role in supply chain management and pass on any risks to either the client or the subcontractors. In terms of longevity, financial risk, initiation of the supply chain and potential control, the client in the construction industry is the key stakeholder. Having said that – this is a controversial point I am making – many will not agree and hold firm that the *construction contractor* is the equivalent to the *automobile assembler*.

The commitment to the 'product' by the client is typically over a greater time-span, as buildings and infrastructure can have a lifecycle of at least 50 years. The contractor and various other construction supply chain participants typically move on to the next project once they have completed their contractual obligations – unless we are talking about public–private partnerships or some other similar situation whereby the

Box 5.2 Tier by tier

A focal organization could be a combination of financial institutions, contractors, clients and facility owners, as they join together to form consortiums and joint ventures. In many instances, there are contractual relationships across this tier, and this could impact upon the construction supply chain. For example, project financing strategies such as build-own-operate-transfer (BOOT), build-own-operate (BOO), public–private partnerships, private financing initiatives and alliance schemes, are altering the location of previous downstream supplier firms; as they take equity in the project and alter their positional relationship to the primary project contract because of their involvement in the total life-cycle of a project.

The first tier of project suppliers is typically of two types: primary consultants and prime contractors. Second tier suppliers are contracted to the prime contractors or primary consultants and are grouped as subcontractors or subconsultants. The second tier is where there is potentially a large number of different specialist types of subcontractors and subconsultants as a result of the different trade packages and specialist design services required. On even the simplest and smallest projects, up to 15 trade packages is quite normal. However, on multimillion dollar projects the number of suppliers at tier 2 can rise to well over 50.

Second tier consultants contracted to the project management or architectural consultant include: specialist engineering, specialist design and/or management services. In some cases the client engages project management, construction and design management firms which form three chain nodes – clusters of firms from which various chains are formed.

Third tier suppliers typically involve product suppliers who contract to subcontractors. The number of different types of commodities is at least of a similar number to the number of subcontractors but, conservatively, there can be two, three or four times as many firms. In many cases, the third tier commodity suppliers are supplying a component and they in turn source a number of products or materials from suppliers in order to construct that component.

Tier 4 suppliers may include materials suppliers who are agents/distributors who typically co-ordinate logistics of supply for tier 5 manufacturers. Alternatively, materials suppliers may take a product from a manufacturer and process it further for the subcontractor. At times, materials and component suppliers may supply direct to contractors. This has a significant impact upon total chain costs, information flows and commodity flows.

builder/developer/contractor take up the facility management and ownership of the facility during occupation – but then effectively they become a client. The client is often also located in any number of supply chains.

Generally, throughout the project procurement process, little control or management of the entire supply chain is taken up by the focal organization, since most attention is paid to the immediate contracts. As mentioned previously, each firm may act as a node with a number of contracts with firms at their successive tier. The focal organization can co-ordinate or manage through direct procurement relationships or even indirectly through incentives, rewards and other mechanisms useful for successive tier management. In this situation, firms act as co-ordinators and often as procurement gatekeepers. Globally, each tier acts as a *procurement gatekeeper*. The first tier contractors, typically, act as gatekeepers to the remainder of the subcontracting tier. Each trade subcontractor subsequently acts as a gatekeeper to the materials suppliers of the third tier. Architectural and project management consultants often act as procurement gatekeepers to other consultants. The role of procurement gatekeeper may be quite critical if the firm is a node and at the centre of a firm cluster that is large. Problems of integration may escalate if the number of firms is beyond a manageable level.

The supply chain organization can become quite complex as there are multiple commodity types with many firms supplying each commodity. It is suspected that there are cases when firms are located in more than one tier and/or in different tiers. Because of this diversity each project supply chain for a particular primary trade commodity type – such as glazing, concrete, aluminium, etc. – can be unique and we need to be thinking more *loosely* about the industry to allow for such market flexibility. The overall mapping of the structural organization of firms in the chain for various commodity types allows greater understanding of the structural and behavioural characteristics of the various supply chains.

5.3.3 Horizontal chain organization

The horizontal structure of a tier is the arrangement of firms in tiers away from the focal organization. However, it reflects the degree of specialization within the one commodity type and at successive tiers. It also indicates the level of project complexity, commodity fragmentation and vertical integration. The greater the number of tiers, then the less vertical integration in the chain.

The horizontal chain organization takes one particular commodity type and examines the organization of the firms in the chain for the supply of that particular commodity. There can be many different commodity types in a particular tier. Therefore, it is important, when analyzing the horizontal market structure of a supply chain, that only those firms that supply a similar commodity are considered. Figure 5.13 represents firms that are not competing with each other; for example, architectural consultants and civil engineering consultants and bricklayers and façade subcontractors. These specializations reflect unique markets. Within each market there are typically many firms who are able to supply this commodity. It is within the specialist commodity group that the market structure and individual firm characteristics and capabilities will be reflected in the procurement relationships that are finally formed. The way in which firm–firm procurement relationships are formed and a method for describing the underlying structural and behavioural characteristics is discussed in Section 5.3.4.

Each particular primary commodity type may have different degrees of vertical integration down the productive chain; that is, be fully, partially or non-fragmented. The level of fragmentation may be impacted by the type of commodity and its degree of complexity; that is, whether or not it is customized or standardized. Typically, the less unique a commodity that is required, then the greater the production efficiencies that are able to be obtained. If a commodity is standardized, it typically reflects a higher level of demand and, therefore, firms are able to command a greater position of power in the supply chain. Greater standardization has often been associated with the residential sector. Demand ultimately relies upon the number of projects and, therefore, the impact of multiple projects on structural organization is now considered.

5.3.4   *Multiple projects and supply chain structural organization*

The model has represented the construction supply chain for individual projects. This is a narrow view of the construction supply chain and the next stage of development introduces multiple projects to more accurately reflect the real world of construction (refer to Figure 5.13). The concept of multiple projects is introduced through introducing two more clients in the diagram (refer to white box on diagram). It also proposes the scenario where firms supply to multiple projects (refer to grey shaded box). The potential inter-supply networks that arise at various node firms is evident. Firms may have many procurement relationships with other firms. Projects will have many firms; firms will typically only have one relationship with one project, but can have many procurement relationships through supplying to many firms and then associations with many projects.

5.3.5   *Aggregated project supply chain organization: supply channels*

Comparisons between supply chains of the one commodity type and supply chains of different commodity types – that is, differences or similarities with

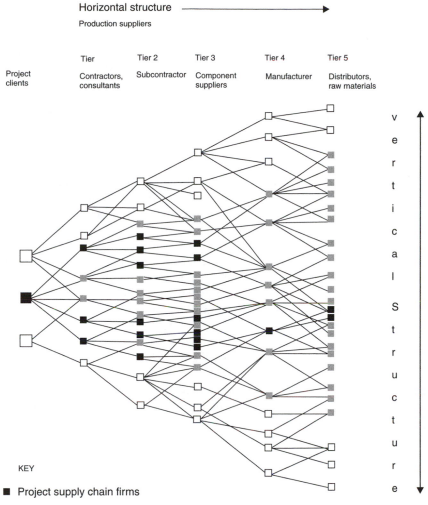

Horizontal structure ⟶

Production suppliers

| | Tier | Tier 2 | Tier 3 | Tier 4 | Tier 5 |
| Project clients | Contractors, consultants | Subcontractor | Component suppliers | Manufacturer | Distributors, raw materials |

KEY

■ Project supply chain firms

□ Project supply chain firms: adjacent projects/clients/suppliers

▨ Project supply chain firms: supplying to multiple projects

■ Focal/client firm

Figure 5.13 Complex construction supply chain organizational structure – multiple projects.

respect to such attributes as distributions of firm size, degree of horizontal and vertical integration and number of relationships – could well be indicators for such characteristics as competitiveness, power distribution, innovation, effectiveness and efficiency – concepts related to market performance.

There are a range of comparative analyses that would be available to the construction research community and policymakers alike once a deeper understanding of the underlying structural and behavioural characteristics are mapped.

Within the context of this model it would be possible to locate such types of relationships and the likelihood of such relationships occurring between particular types of firms on projects with certain characteristics. The value of this model is to understand change in the structuring of relationships and, in time, model the residual impact of change on the industrial market and the industry as a whole. The model also creates a mechanism to understand and locate any differentiation that occurs between firms and create new relationships accordingly.

This discussion has developed the argument for reporting on simple descriptions of horizontal and vertical structure; who supplies to whom along the supply chain. A key concept that underpins the supply chain literature is the nature of the relationships between firms. In the construction industry these relationships are constantly forming and re-forming for each project. An industrial organization economic model for the construction supply chain is not complete without a discussion on the behavioural characteristics of supply chain formation. Figure 5.14 is only possible after we have developed an understanding of multiple projects. Construction is typically regional – although I do acknowledge that exporting and internationalization has experienced a growth in recent years – and so we can begin to develop regional supply channel maps. Once this is done companies are able to make comparisons and policymakers are also able to make comparisons – both groups can then identify policy, process and practice impacts on various supply chains.

5.3.6   Firm–firm procurement events

The model described is static in nature and has focused upon describing the attributes of each entity that are potentially of significance to a construction supply chain model. The real world is not so static and this is a dilemma of this representation thus far. Firm behaviour has been partially addressed by the procurement relationship; however, this also has been represented as a static entity. The sequence of events that describe the manner in which the supply chain is formed – that is, the way firms approach projects and other firms – needs consideration. This will allow for modelling of project supply chain formation. The discussion in this

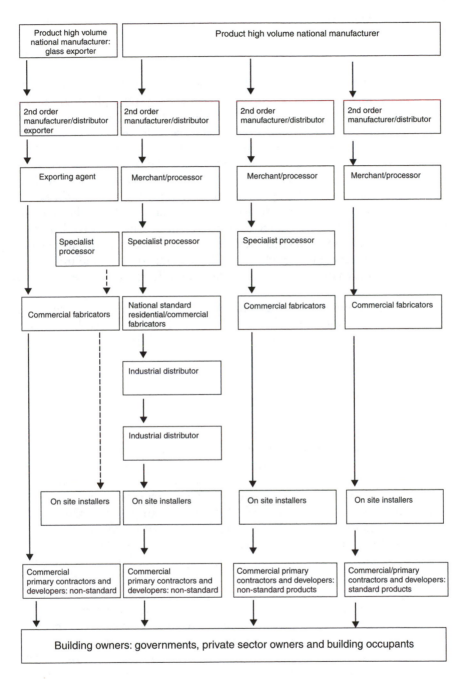

Figure 5.14 Aggregated project market supply chains: industrial market chains.

section considers the formation of the construction supply chain, a little-discussed topic in the supply chain literature and yet this is such a critical part of supply chain management for construction and for our industry as a whole.

A project, when first considered, represents the catalyst for an infinite range of options of actual supply chains that may take place. As noted previously, there is the *project market*, which includes the firms that are within the tendering environment and the *industrial market sector*, which represents all possible firms that could supply the commodity. This suggests that there is the *project chain* which is the final crystallized state of firms that are contracted, the *project market chain*, which includes those firms that are tendering on the project and the *industrial market chain*, which represents the wide pool of firms that could have been asked to tender and could be contracted on the project. The industrial market chain includes all the competitor firms before contracts are established and the project market chain includes only those competing for the individual project. In some cases there may be little or no difference between the two markets.

There are various *events* and *interactions* between customer firms and supplier firms leading to the formation of the chain, and it is these events and the sequence of events that are now considered. Box 5.3 Procurement

Box 5.3 Procurement events

1   an upstream firm considers the project and assesses the commodities required to fulfil the contract and assesses their own commodities and ability to fulfil the requirements; the project description informs the firm of the type of commodities required

2   an assessment of the industrial market by the upstream firm is then undertaken, which may then be followed by an approach to a group of these suppliers; which forms the project market

3   each potential supplier who has been approached by the upstream firm then assesses the project and then their own capabilities;

4   this sets off another chain of events where the original downstream firms approach their potential suppliers for tenders; and

5   downstream firms provide tenders upstream; there may be a period of negotiation and then finally an offer and acceptance is made to an individual firm and a firm–firm procurement relationship is formed; thus setting off a chain of firm–firm procurement relationships forming sequentially down the chain.

events describes a possible scenario for the formation of a single generic construction supply firm–firm procurement relationship. In this scenario there are five events:

There are different types of commodities being transacted in each firm–firm procurement relationship which may impact upon the events that occur. The commodities may differ by their physical characteristics; that is, they could be steel, glass, concrete, etc. Commodities may also differ by their significance to the upstream customer and may be core to the fulfilment of the contract or non-core; for example, it could be structural steel or fixings for the steel. Further to this, the commodity could be in a high-spend or low-spend category for the upstream customer. The commodity could be something that is unique to a project or standard for many projects. There could also be only a small number of firms who can supply the commodity or, alternatively, the market could be quite large and competitive. The firms that supply the commodity could all be located internationally or they could be located in the same region as the upstream customer; thus making the interaction during negotiation and the duration of the contract quite different. The risk in contracting with an international supplier may be deemed higher than a local firm. All these factors could impact upon the events and, particularly, the negotiation phase. Figure 5.15 graphically summarizes the key concepts discussed in this chapter.

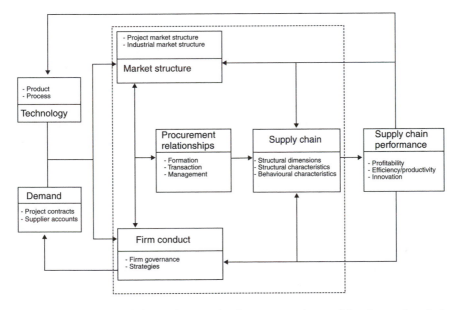

Figure 5.15 Project industrial organization economic model of supply chain procurement.

5.4 A final word

Chapter summary

1 This chapter has sought to consider the core elements of chain organization, market structure and firm conduct from the industrial organization economics literature to develop a project-based industrial organization economic model of supply chain procurement. The primary elements of demand, technology, market structure, firm conduct, procurement relationships, supply chains and supply chain performance are now discussed and brought together in Figure 5.15.

2 Markets typically include an industrial market and an individual project market; which may have different structural characteristics. One of the key considerations of firm conduct in the construction sector is that firms organize themselves in response to the project nature of the industry; that is, their governance structures are typically developed as a strategic response to those commodities for particular markets and often for particular projects or at least project types. Market structure and firm conduct are both determined in part by underlying demand conditions and technology. Demand in the construction sector ultimately relies upon the number of projects. Firms respond to the number of projects as either individual project contracts, or as aggregates of individual contracts (supplier accounts). As firms move further away from the project environment downstream from the client, they tend to view demand as a series of supply contracts and less as project contracts. Many firms supply to the one project. It is also apparent that when a sector level perspective is taken, there are many projects and firms supply to more than one project.

3 Even though there are many project suppliers, they have some common characteristics which assists in understanding the economic organization of the sector and the generic manner in which firms relate to each other in the sector. Some of these characteristics are now considered. Firms supply commodities which are located within an industrial market. Firms may supply more than one commodity. On a project, firms typically supply one commodity. It is suspected that there is a project market operating as well as the industrial sector market; the project market is a unique characteristic of the firm–firm procurement relationship. Each firm is linked to a project by virtue of its association with an upstream customer which has been described as the firm–firm procurement relationship. Procurement relationships may differ from firm to firm, but they all have similar characteristics.

4 Market structure and firm conduct interact. Structure affects firm conduct, but conduct (strategic behaviour) also affects structure. Structure and conduct interact to form firm–firm procurement relationships. Procurement relationships have three primary characteristics: formation, transaction and management. Firm, project, commodity and market are all considerations of the procurement relationship in formation, transaction and management and have some association to the procurement relationship. Formation refers to the manner in which customers and suppliers approach and respond to each other prior to a contract being formed. The market and the commodity details are important. Transaction refers specifically to the individual characteristics of the unique procurement relationships. Projects vary and the nature of the actual commodity that is being transferred can vary with each project, in terms of: complexity, time-frame, volume, service, etc. Finally, the nature of how the relationship is managed is part of the procurement relationship. Management refers to project co-ordination and also the management of suppliers across projects; similar to the supplier development concepts discussed in the previous chapter (Hines, 1994).

5 A key component of the industrial organization economic model for the construction industry is the role of the project; in terms of both demand/sales efforts and its potential to create new market structures. Structure and conduct are both determined by underlying demand conditions and technology. Market structure in the construction industry can be considered on two levels: it is both the market structure of the sector for the commodity and the market structure for the commodity related to the project, which can often be segmented from the main sectoral market. Conduct is also the manner in which firms organize themselves and determine their boundaries. Although this has not really been considered for the construction sector, it is equally applicable; it is the products and/or services that a firm deems necessary to procure to enable it to fulfil a contract and, therefore, those firms that they need to enter into a firm–firm procurement relationship to procure to enable them to bid for the project. Therefore, downstream procurement relationships are significant. Demand conditions in the construction sector relate to the number of projects that firms could potentially *bid* for. The number of projects that firms are invited to bid for may be less for various reasons relating to their firm capabilities/expertise/ innovation/product differentiation, or the relationship that the firm has with the upstream client, which could include past performance/ previous agreements, etc.

6 An aggregation of procurement relationships forms supply chains. There are three main considerations for the supply chain: structural

chain organization, structural characteristics and behavioural characteristics. The horizontal and vertical governance structure of the construction supply chain can be considered in terms of tiers, vertical integration, vertical fragmentation and specialization – terminology and concepts borrowed from industrial organization economic supply chain literature is useful for describing the economic organization firms in supply chains at an individual project level (Williamson, 1975; Ellram, 1991; Lambert *et al.*, 1998). It is suspected that there are common patterns of organization of chains, similar to the channel structure descriptions described graphically in the previous chapter, which aggregate to a description of various types of chains and supply paths at the sector level. However, until further empirical work is conducted it is still uncharted territory. Each of these entities assists in explaining the structure and behaviour of the construction supply chain through their individual properties and their individual behaviour. Each of these entities described is associated in some manner with another entity. Although they all have individual properties and behaviour, the actual properties and behaviour are the same. For example, each firm differs but it differs by one of its properties.

7 Market structure and firm conduct interact to determine performance. Supply chain organization, structure and behaviour determine the performance of the chain in terms of profitability, efficiency/productivity and degrees of innovation; which also feed back to both market structure and firm conduct. Supply chain performance ultimately impacts upon how progressive an industry is and feeds back to product and process technology advances. Sales efforts (that is, how much a firm produces and then manages to sell) is an element of conduct which impacts upon demand. Sales efforts are typically tied to project contracts. Performance ultimately feeds back to technology and structure. Progressiveness relates to the available technology and the rate of technological progress. Finally, profitability determines how attractive it is to enter the market which has an effect on market structure over time. Many of these concepts have a high degree of explanatory power and it is a useful framework. It is also clear, however, that markets are quite complex and their characteristics can be ever-changing.

8 The components of this model – including: *market structure, firm conduct, procurements relationship*s and *supply chain* – are of primary concern for this thesis (those elements included in the dotted square in Figure 5.15).

9 There are still difficulties encountered when using the industrial organization economic methodology as a framework to represent

a model of supply chain procurement. These difficulties have been defined as the concepts of *multiplicity*, *interaction* and *types*: multiplicity of associations between entities; interaction between structural and behavioural characteristics of entities; and types of entities which have common characteristics.

10 First, multiplicity refers to the manner in which there are multiple associations between firms, between firms and projects, between firms and commodities; arising due to the constant forming and re-forming of firm–firm procurement relationships for each unique project and the sheer volume of transactions on each project. The sheer volume of associations between entities is immense. Although there are numerous associations that arise in the real world, the firm's characteristics do not change; nor do, for example, the project's characteristics. Therefore, there are numerous associations between entities that can be described once. Entities couple and decouple as projects arise, firms work on these projects and then projects are completed.

11 Second, the interaction between structure and behaviour as each object possesses a duality of structural characteristics and behavioural characteristics at the same time. The industrial organization methodology attempts to overcome this problem by simply separating the two concepts of market structure and firm conduct and market performance. Although a useful abstraction, this problem becomes more acute when we begin to think about the objects that have been described in the model and how they interact. Objects, such as firms, commodities and firm–firm procurement relationships, have attributes that describe underlying structural characteristics; however, when they interact with other objects to varying degrees the underlying structural characteristics are affected and may ultimately change. Each object in itself has structural attributes that describe the object in its static state. However, each object also has a way of behaving or operating in the real world.

12 Finally, types of entities refers to the way that the system has groups of similar objects with similar characteristics. For example, an object known as a firm is unique; however, there are many firms that have similar characteristics and therefore may behave in a similar manner. Even if the industry is project-based and each project provides for unique circumstances, there are patterns to this seemingly highly diverse world. The critical point is that both uniqueness and similarity needs to be accommodated at the one time. The industrial organization methodology accommodates a restrictive and rigid view of the way in which objects behave in the system.

13 The industrial organization economics methodology has limitations when trying to represent the dynamic nature of a project-oriented industry. A methodology for describing and representing the concepts of multiplicity, interaction and types is needed. A methodological framework that allows for this complex real world representation and thus modelling of the supply chain and the many individual project scenarios is now sought.

6 Multiple project environment chain structural organization

I manage individual projects and suppliers but what is the overall perspective of multiplicity of firms and firm–firm relationships in supply chains?

6.0 Orientation

Box 6.1 Chapter orientation

WHY: In a region many players often work to the same client and with the same consultants, contractors and subcontractors. The multiple project environment provides us with an environment of multiplicity of firms and relationships and projects that we may be able to use to our advantage.

WHAT: Chapter 6 provides an overview of the projects, clients and first tier suppliers to clients and sets the scene for Chapters 7, 8 and 9 case studies. It begins to describe the 'supplier' policy environment of clients (either government or corporate policy) and identifies a lack of policy beyond a single tier and even a lack of lateral thinking of the different types of relationships that can be developed at the government level. The material presented begins to dispel the myth that upstream relationships impact upon downstream relationships in a positive way – unless there is a sustained effort to reach **down** the chain and **across** multiple projects and develop explicit systems to deal with supply chain management. The chapter also presents information about the range of supplier types and the range of ways suppliers procure their own suppliers in an implicit manner – from this I have then provided very simple but explicit ways to begin to think about how to categorize suppliers.

HOW: The research method is briefly outlined, including data collection and analysis techniques. The aim is to begin to develop an understanding of the attributes of projects, client firms, 'supplier' firms at tier 1 and the attributes of the procurement relationships; as well as an overall picture of the structural organization of the chains. There are five main ways of presenting the information: matrices summarizing the information collected from the interviews and documentation, graphics portraying key attributes, selected maps describing the organizational structure between firms and also relating firms to types of commodities, and finally matrices that have a more interpretive quality.

6.1 Introduction

This chapter begins the series of chapters which present the results of a study aimed at developing the structural and behavioural model view for an industrial organization economic model of construction supply chain procurement. This chapter specifically provides an overview and sets the scene for the next three chapters, which take the form of a series of case studies. In the original study, eight quite large case studies were undertaken; however, I have elected to report in detail on the following:

*   façade
*   steel
*   mechanical services, formwork and concrete.

In the original reporting of the study a contextual overview was provided, followed by case study reporting describing the structural view and then this was followed by case study and statistical analysis which formed the behavioural view. In this text I am going to integrate and describe the structural and behavioural view of the supply chain within each chapter through the case study material.

Within the structural view, descriptions are developed using attribute data from the *real world*. The primary 'objects' of the structural model include the projects, the firms, commodities, markets and firm–firm procurement relationships. This part introduces the projects and provides overall descriptions at each tier of supplier firm type by commodity type – which gives one perspective of structural organization of chains where firms are grouped by commodity type. Various attributes of firms are mapped against commodity types across projects. This section concludes by describing in detail the structural organization of chains indicating individual firms in the chains and the types of commodities for the various projects – it is by no means complete, but it certainly starts to provide a comprehensive picture.

The results tend to focus upon the supply chains which evolve from the contractors, although there is some discussion on consultants.

This chapter includes descriptions of projects, clients and firm suppliers at each tier. The description of the supplier firms relies upon describing the

Box 6.2a Methodology: summary of nature of research problem

The following table summarizes the characteristics of the research problem which is essential to the establishment of the most appropriate methodology. They are grouped according to Discipline Knowledge and Methodological characteristics. A more detailed discussion on ontology, epistemology and methodology can be found in my PhD dissertation.

Table summary of nature of research problem

Concept	Characteristics
Discipline knowledge characteristics	
Real world procurement	The research problem is to represent a real world situation; procurement in the construction supply chain and specifically the industrial organization economic context of procurement.
Procurement within context Epistemology: inductive empirical	The research in this study describes supply chains through descriptions of the firms and involved, the economic environments that they are located within and the relevance of this to the organization of the supply chain. It is inquiry that is inductive and is empirically based. The underlying premise to the research is that supply chains do exist and that they can be objectified; however, this process needs critical interpretation.
Varied data sources and interpretations	The data to understand the supply chains is in various locations, including: it can be extracted from the people within firms involved and their descriptions of what they do, who they interact with and how they organize themselves; and it is also within the project and individual firm documentation. The information about firm–firm procurement relationships can be sourced from both participants to the transaction; that is, the customers and the suppliers. The perceptions on the nature of the project market can also be sourced from the customer or the supplier – the customer in their role as observer and decision maker across a number of firm suppliers and the supplier as a participant within the market reacting to competitor firms.
Various scenarios: ideographic	The research involves descriptions of the 'things' (objects) in the system of procurement supply chains and grouping those objects that behave in a similar manner together. However, it is also about the ability to be able to

Continued

Concept	Characteristics
	interpret the system of supply chain procurement uniquely – that is, the individual instances and scenarios that occur versus the common instances and scenario patterns. This research is concerned with exploring the individual project supply chain and the procurement relationships between firms. However, it is also about understanding that particular formed project supply chain in relation to the extent of other choices of firms. It is concerned with uncovering the different types of supply chains that could form and in what situations they do form in a particular manner.
Relationship of particular to the universal: ideographic versus nomothetic	There is a dearth of studies concerned with investigating to relationship between the individual project supply chain and the construction supply chain population. There is some research that describes individual supply chain instances in detail, typically by case studies. This raises a methodological tension between the nomothetic versus ideographic debate; that is, the abstract positivist approach versus the particular individual interpretive approach. This debate is taken up later in this chapter. It is research designed to look globally across the industry and yet it is as much about the development of a methodology to describe the industry through the particulars of projects – and to question if such a global view is feasible.
Descriptions of procurement the industry participants would have us believe versus reality	This research challenges the accepted approaches in a field of research, approaches that rely upon us appropriating theories and concepts from other fields but fail to explore the underlying assumptions and relevance to the construction field. That is, how and when are theories developed from observations in automobile, engineering and retail supply chains appropriate to construction industry supply chains? Describing real world scenarios is one attempt towards a more sophisticated approach to construction industry policy development that aims to change industry structure; alter firm conduct; and improve market performance through increased productivity and increased innovation in the industry. It looks to economic theories of industrial organization in markets to explain forces that affect the competitive behaviour of firms in markets.

Continued

Concept	Characteristics
Current understanding reality	This research is also concerned with the suspicion or of notion that the 'client' is largely shielded from the inner secret workings of the industry. The findings of this thesis may be an important starting point to making transparent the workings of a key sector to every nation's economy.
Finer grained understanding the real world supply chains	This research attempts to refine the abstract classifications of the groups of suppliers to the industry that have been accepted as industry norms. The structure of the supply chain organization is too often described through broad groups and thus attributed broad characteristics; for example, subcontractor, supplier, manufacturer, etc. Again, critical interpretation of the norms that we already place on our descriptions.

Methodological characteristics

Methodology development	The research is concerned with the development of a methodology as well and is characterized by observation, description, measurement and modelling. It is as much about the particular information and data collection about the 'real world' as the development of a methodology for observation, description and modelling of the construction industry supply chains and using the data collected to develop the methodology to describe and represent project-based supply chains. This research unravels the deeper structures of the industry as it attempts to develop a methodology to describe and represent the multiplicity and complexity of the industry.
Interdisciplinary	This thesis develops a methodology for describing the economic organizational structural and behavioural characteristics in supply chains and it draws methodologically from the fields of industrial organization economic theory and information sciences, and the object-oriented methodology.
Methodology for integration of concepts of structure and behaviour	The research pursues the line of inquiry that structure and behaviour are encapsulated in 'things'. That these 'things' that are important to the topic of inquiry have relationships and that these relationships help to create structural characteristics that underpin the system under study. Scenarios in the real world can be representative of occurrences and interactions between 'things' and also give an

Continued

Concept	Characteristics
	indication of the behavioural characteristics of the system under study and the topic of inquiry. The topic of inquiry is procurement in the construction supply chain.
Techniques	A methodology that incorporates this perspective is presented that aims to provide a set of representation techniques for capturing, specifying, visualizing, understanding and documenting construction supply chain procurement scenarios within an industrial organization economic construct. The method will capture procurement scenarios, in terms of a structural and behavioural view. Specifically, observations and measurements will be undertaken that will capture and represent information about the firms, markets, commodities, projects and firm–firm procurement relationships. The method should capture and represent information about the way in which customer firms approach commodity markets, negotiate with suppliers, and make decisions about choice of suppliers. The information will be represented by the UML techniques. However, UML is not a process and does not give guidance on how to make observations and measurements – only on specifying and visualizing.
Positive model/ methodologies versus normative model/ methodologies	The research is concerned with contributing to theoretical debate within the construction supply chain literature. The development of theory for the supply chain concept is a critical component of this thesis. There is an emerging realization of the importance of supply chain management to the construction industry; however, there is less realization that there is a need to develop a theoretical and methodological construct to underpin the concept. Much of the research is about what *should be done* in the ideal supply chain and very little attempts to explain the reality of *what is done* in the vast majority of supply chains. There is a need for more positive models of construction supply chains rather than normative models. The development of theory is assisted through the development of a methodology that attempts to move beyond simple descriptions of case studies that represent an ideal type that the industry should aspire to and towards developing methodologies that assist in describing the underlying structure and behaviour.

Box 6.2b Method

A much more complete discussion can be found on the methodology for those wishing to seek out more. In a nutshell, so to speak, the following is a summary of how the data was collected and the source of the data.

Data collection

Data was collected from firms who were contracted on six major building projects in an Australian city. One thousand two hundred and fifty-three procurement relationships were mapped using data collected from forty-seven structured interviews and forty-four questionnaires.

There were three parts to the survey:

Section 1 Organizational background – finding out about their firm, competitive advantage, competitors, what their firm does in general terms, etc.

Section 2 Project characteristics – finding out about what they did on the particular project

Section 3 Customer/supplier relationships – finding out about who their suppliers were, how they sourced, how they were chosen, how they chose their suppliers, how they approach the markets

A complete list of the questions and the list of respondents is included in the appendices of this text.

Data analysis

The data analysis was both qualitative and quantitative. Matrices were used to reduce the data from transcripts. Various data display techniques were used to describe common themes. The main aim of the study was to *describe* procurement practices in construction supply chain. In particular, how firms approached the markets of their *supplier groups and the way in which they* suppliers, how they categorized their supplier groups and the way in which they negotiated with them during procurement. The analysis was *interpretive* when comparisons were being made between firms and the nature of their decisions in different supplier markets.

It is noted that pilot interviews were conducted – these were extremely useful. There are euphemisms used in the industry to explain behaviour that is perhaps somewhat unethical and may even be illegal. It is borderline. People are nervous when you begin to ask questions about tendering, suppliers, long-term agreements...hints of collusive behaviour creep in.... In time, I was able to build trust with many of the respondents – this was extremely lengthy and exhaustive, but a worthwhile process. I personally found I started to get under the skin of this wonderful industry but, of course, there is so much more you can always learn.

A whole lot of other statistical analysis was conducted on the data as well – this book does not report on this except to show a summary in the Conclusion chapter of patterns of behavioural characteristics. Again, if you wish to know more, check out the dissertation reference. Suffice to say, it was a categorical analysis where I compared observed procurement frequencies versus likelihood of expected frequencies. Those statistics whizzes will recognize this as one-way and two-way tests of categorical data. I think this has really only just opened up this area and I have only just really scratched the surface in a range of areas.

Unit of analysis

A final note – what is the unit of analysis in this study?

The unit of analysis has always been the supply chain. As you will see when you begin to unravel the industry, it does tend to unravel. At one point I was overwhelmed by data but, to be honest, that was the whole intention (and, besides, that feeling only lasted a day!). The big picture was always the aim. Whatever study you do, there is a time when you get overwhelmed by the data...well, you should if you are doing your job. If you don't become immersed in the data and get lost for a while, then very little new and exciting knowledge is forthcoming.

The unit of analysis – all firms were contracted to these major public sector projects which were being built at the same time. What binds the chains together? I started off only focused on the client. The client was the same for three of the projects. As I progressed, it became clear that it would be worthwhile to pursue another major mixed use – residential high-rise project, as one of the contractors who was on two of the three projects was also the main contractor on the residential high-rise project. It was also not public sector – so I wondered if that was useful. Finally, when I then pursued the supply chains on the residential project, it opened up more residential supply chains and therefore I included two more residential projects of a different scale. The beginning point was the project – the end point is to develop sectoral commodity views of the supply chains which merges all types of projects – residential, civil and building – into the one map. At the end of the day, policy does not discriminate...

commodities which are supplied and grouping suppliers by commodity type. In the first instance commodity type is either product, product and service or service. Following this, specific details of the commodity are provided, such as if it was an architectural design service or the supply and installation of bricks. Then generic characteristics are described, such as transaction characteristics of complexity and frequency. The physical structural organization of these supplier groups at each tier and descriptions of structural organization of supply chains at an individual firm level are then provided. The purpose is to give an overview across the entire project of the type of suppliers, the range and number of product/service and product and service suppliers at various tiers and some general representation of key structural characteristics which underpin the supply chain. The overview of the supply chains and various objects are organized in this section in the following manner:

Projects and firms:
• project, client and contractor attributes.

Types of suppliers, associations and commodity groupings:
• tiers 1 and 2 service suppliers: project 1
• tier 2 suppliers: projects 1–4
• tier 2 suppliers: projects 5 and 6
• tier 3 suppliers: projects 1 and 3
• Firm attributes by project.

Structural organization of supply chains at individual firm level.

6.2 Projects and firms

6.2.1 Project attributes

All participants in the interviews, and the firms that they were employed by, were involved contractually in one or more of the six projects. The projects which the firms were supplying to are now described. Key project attributes are summarized (refer to Figure 6.1). This information was obtained primarily from the interviews with the clients and documents that they provided during the interview.

Project 1 is one of the major projects for the Victorian state government in Australia, is considered an icon for the capital city Melbourne, includes a programme of multiple art and art-related functions and is classed as a building mixed-use arts project. Projects 2 and 4 are similar in building type to each other, both being sporting stadiums and classed as commercial sports stadiums. Project 4 is more complex and much larger than Project 2. Project 4, similar to Project 1, is a major architectural and construction statement for this city. Projects 3, 5 and 6 are all primarily residential projects. Project 3 is a high-rise apartment building, has elements similar to Projects 5 and 6, but has

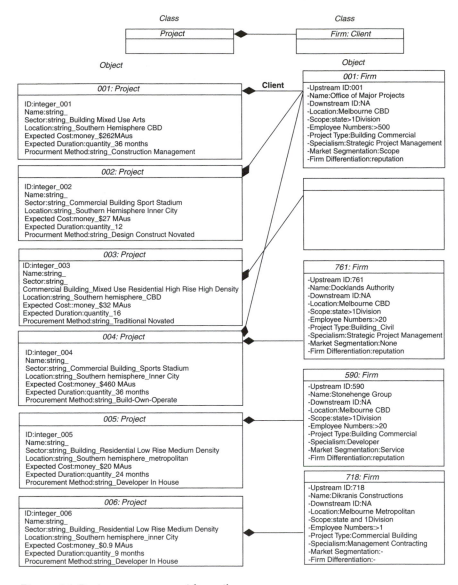

Figure 6.1 Project contractor with attributes.

more complex building systems. Projects 5 and 6 are similar and are classed as residential, low-rise medium-density. Project 5 is much larger in scale and complexity than Project 6. Project 3 is classed as high-rise, high-density residential and has building trade elements similar to Projects 5 and 6.

For further details on function, sector, location, size, budget, client and procurement method, refer to Table 6.1. Figure 6.1 summarizes the project

Table 6.1 Project details

Project	Function and sector	Location, size and budget	Client and project procurement method
1	The project is a mixed-use art centre and includes a multimedia centre, retail and commercial area, an art gallery and car parking. It also includes external paved and landscaped areas for civic square, performance venue and visitors/information centre. It is a major public building. 'The project objective was to develop a new people oriented centre incorporating civic, cultural tourist and complementary commercial uses over the rail yards …to provide a 3 hectare platform for a new civic plaza, a Museum of Australian Art, an Australian Centre for the Moving Image, public atrium spaces, exhibition and performance areas, tourist information services and associated visitor services and amenities' (Briefing document to Premier, October 1999).	Project 1 was approximately 36,500 m^2 of floor space of mixed use and 24,000m^2 of external paved areas. It is located in the CBD in a state capital of Australia. Specifically… '…on a prominent site at the corner of Swanston and Flinders Streets. The project involves deckingover the rail yards between Princes Bridge and Russell Street'.	The project was financed by the three tiers of government: Victorian state government, Australian federal government Melbourne and the City Council. The Principal (client) to the construction and consultant was the Secretary, Department of Infrastructure, Victorian State government, through the Office of Major Projects.
2	The project… 'Provides five indoor netball courts, including an air conditioned show court seating 3,000, four lit outdoor netball	The project is located in a park in an inner city suburb. It is $27m.	The client is the state government and the procurement method was novated detail design and construct. Contractors were

(Table 6.1 continued)

Table 6.1 Continued

Project	Function and sector	Location, size and budget	Client and project procurement method
	courts, two international standard hockey pitches with lights, and catering/function facilities. The centre will house the head office of Hockey Victoria and will be capable of supporting Commonwealth Games competition for both hockey and netball' (Briefing document to Premier, October 1999).		involved in a selected preregistered tender process. *The documents were completed to a stage and then our role was that the consultants will know what they are doing…and that we would complete the design documentation. At the contract signing, the design team was novated to us and they therefore became our responsibility.*
3	Project 3 is a mixed-use building with predominantly a high-rise apartment building. *Effectively, a 13-storey existing office building with a brick façade that we stripped the façade off, gutted the inside and went with a skeleton structure of concrete which we are going to add 12 floors on top of. At the end of the day, the building will be 25 floors of apartments with 6 floors being car parking for all the residents with a restaurant on the ground floor, grand entrance, and apartments ranging from 1-bedroom studio-style apartments up to the penthouses worth over $1m at the top of the building.*	This project is called the Philadelphia Building, a $32 million design and construction contract. The project site is located in the CBD in a state capital of Australia.	The client was a private developer from a neighbouring state capital. The procurement method involved a preregistered selected tender for the construction following a traditional design phase. The designers were then novated to the successful construction firm. It was a form of design and construct with 70% of documentation complete. *In tendering for the construction, we were up against four other construction companies. When we were tendering it, the developer had engaged an architect and when we engaged in a contract to the client the consultants were novated over*

| 4 | Project 4 is a commercial building, a sports stadium. *The best way they determined to do that [i.e. change the perception of this part of the city] was to build a stadium, because that would immediately attract people into the area, they would get used to coming across Spencer Street and into the area, and as a consequence there would be a natural overflow into the stadium area. They would come to the stadium and say, 'well, let's just walk down into the Docks area, come to the boats, make a day of it – whatever'…. Plus the fact that stadium is not just a footy ground, it's an entertainment complex. So the aim is to come, have a meal, see the game, then go to a meal, disco, cinema, whatever. It's not just a straight 'come to the game and disappear'. Or you could come down to the Docks area in the morning, have lunch and go to a footy game in the afternoon. So it's really an entertainment complex.* | The building is located in the inner city adjacent to the river. It is worth $430m. *Traditionally Melbournians see Spencer Street railway line as the end of the city, and one of the things that they had to combat was the fact that it is an area West of Spencer Street. Docklands Authority determined that some areas they went into there were absolutely no people, no life, it was just dead. They came up with the issue that they wanted 30m wide promenade all the way around the water. So that people have very good access.* | The state government created an independent commercial entity which managed the development of a large tract of land; they released the land parcels to consortiums who The bid on proposed developments. consortium would the then be client. The procurement method was build, own, operate and transfer. *The State government actually formed the Docklands Authority, so they report directly to State government. The Docklands Authority did a pretty good job, they went worldwide – they did a bit of a trip right around the world, looking at all of the dock land developments, looking at what actually worked and what did not work. They went to some areas where, for example South Africa, London, Italy, NYC. That's a private stadium.* | *to us. It was a design and construct contract where we take the design when it is 60–70% complete and then we finalized that design. We start building prior to having that design finalized.* |

(Table 6.1 continued)

Table 6.1 Continued

Project	Function and Sector	Location, size and budget	Client and project procurement method
	It's going to be a different level at the finish architecturally than something like MCG, or the Carlton Footy Club ground or one of the traditional footy grounds where you just have the straight complex. So, it's a totally different concept now. (Consortium rep) *So, Colonial Stadium, Sept 1997, Docklands Stadium Consortium $460m stadium no claim, 54,000 seats, closable roof, retractable seating.* (Docklands Marketing Manager)		*So the difference between a private (Docklands) stadium and the MCG. At the MCG run at the weekend. They might run Friday night, Saturday, Sunday, and that will be it for the week. They might have a few corporate boxes going during the week, but traditionally it will be dead. Whereas at the Docklands stadium they're aiming for about 270 events per year. If you have 270 events, you're really running one per day. So, because it's a private stadium, it's (excuse the expression) 'bums on seats' If you haven't got bums on seats in a stadium, you don't have a return. They've got this, the consortium have got this for 25 years. It's a boot operation, build, own, operate and transfer. It transfers to the AFL at the end of 25 years. They have got 25 years to re-coup as much money as they can, which is a fairly long period of time. . . . $430 million.* *But I think, from memory, it's worth $250m in materials and erection, and the remaining*

5 | Project 5 is a low-rise, medium-density residential subdivision.

Metropolitan area, 51 houses and approximately $20m.

The client is a medium-sized residential developer who purchased the land from the government. The state government released the land to developers in a competitive tender with criteria for how the land was to be developed. Individual home purchasers are offered the houses at

money was financing. I've got a bit of information here (Consortium representative). Because we are a marketing and bid development and management authority. So we marketed the project, we called the bids, shortlisted the best bids, got them in, assessed them and awarded the precinct. One of the conditions in bidding for the precinct was that the cost that we'd incurred to date would be at no cost to this professional consortium. As it turned out, that was Docklands Stadium Consortium, and they're in effect the client. They build the stadium to our standards and to the AFL specification, and they provide the finance and when the facility is complete next February, they'll be the owners for 20–25 years. Then the AFL will pay $20m and they will become the owners after that (Docklands Marketing Manager).

When in fact what we're doing is the project that we're building at the moment out at Williams Town is just 50 homes a sub division we're basically what we've done is future proof the houses with technology and we're putting a digital

We are building 51 houses and they are averaging in construction costs about $320,000 per house. The whole project is about $20m and it will probably take us 2 years all up. ... These homes are in the Smart Homes Group. This is at the

(Table 6.1 continued)

Table 6.1 Continued

Project	Function and Sector	Location, size and budget	Client and project procurement method
	home manual in every home (CEO, Developer).	*beginning of that whole concept. I guess what we will be doing Smart house wise in 5 years time and what we are doing now is well we will sort of look back and have a bit of a laugh about it. It is the start though. There is nothing out there that you can take and use. We are on the leading edge of introducing that here. What we have clicked as the Smart House is where it started from (Project Manager, Developer).*	a set price with small variations permitted. The procurement method is design and construct to the homeowner, but the developer organizes in-house the land purchase, building design and construction.
6	Residential low-rise and medium-density *Axelton Street in Cheltenam and this was how many units? 6*	Inner city. 6 units, $0.9m *Okay projects we could probably start talking about the particular project.... Where is it? Axelton Street in Cheltenam.*	Developer. *6 units and this was the type that you did everything yourself you did ...It's a development of our own Of your own yea and its also one that you organized all the subcontractors in Yea everything. ... Yea yea okay and the finance as well okay is it just to talk a little bit about the risks or the factor is there do you feel it's a hard risk those sorts of projects or...?*

descriptions. Each project has an association with a customer firm which is the project's client. The individual clients are now discussed.

6.2.2 Client firm attributes

The client firms, initiators of the supply chain, are now described to provide context for the descriptions of the suppliers. The client for Projects 1, 2 and 4 are essentially the same, namely the state government (refer to Table 6.2 for detailed information). Two strategic project management-type agencies have been set up to steer and manage these projects and the governance structures are quite marketplace-oriented – quite commercial in outlook and 'professional and slick' in much of how they present to the world.

The management of the project is undertaken by different government departments, with Projects 1 and 2 being undertaken by a unit referred to as Office of Major Projects (OMP), who are responsible for the project management of all major projects in the state. The most significant aspect of the government project management unit responsible for Projects 1 and 2 is that they trade as a corporate entity and the legal entity enters into contracts rather than the Portfolio Minister. Project 4 is governed by Docklands Authority (DA).

Box 6.3 Government client: corporate trading entity

'We are quite different to most other government departments who enter into the contracts in the name of the Minister. We actually enter into contracts as a corporate entity' (Director, Office of Major Projects).

Project 2 was being developed and funded by the state Sport and Recreation Department and Project 1 was being developed and funded by the Arts Department, but the conceptual idea for the projects came from the highest level of government office holders; namely, the Premier and his senior cabinet team. The OMP can be classed a strategic project manager and it seems to act in a more flexible and responsive manner than the usual bureaucratic government agency because of the direct relationship to the Premier and cabinet. Decisions about projects which are of strategic importance to the state are made by the Premier. OMP is responsive to the marketplace. However, I suspect not as responsive as they could be.

Table 6.2 Matrix of client details for Projects 1, 2 and 4

Firm type and projects	Structure and scope, size – employees and turnover	Commodities and competitors
Client, project 1 and 2	The client is within a government department which employs 17 staff, 5 of which are executive (1 Director and 5 Project Managers) and 12 support (including project, administrative and technical support). The unit operates independently from the larger department they are located within. *Now we trade here as a corporate entity. Now we have own legislation. We are quite different to most other government departments who enter into the contracts in the name of the Minister. We actually enter into contracts as a corporate entity (Client 1 Director).*	Their commodity is a service. The service is strategic project management. *It is our responsibility to scope up the project. In many times it is that the client has an idea and we need to flesh it out and come to grips with what it could be and how it might work and how it might go and what it might cost. We develop up the projects from an idea that the government has. We will try and structure up the project into something that makes sense and negotiate and get involved with the private sector and get the best deal for the government, not only in terms of land development but also the financial return for the land. We have to spawn the project if you like.* The competitors are other government departments who understand the political issues in structuring and managing major capital works projects for the state government. *You might say we have competitors within the government. Our competitors within the government, well there is the Department of Health and there is a group within there that manages hospital work. There is a group within the Education Department that manages education work and there is a group in the Justice Department and they call themselves Major Projects Office. They manage courts and prisons and things. We have never seen it as our role to gain sides or ground. Our role is to do the projects that the government has asked us to deliver. We don't normally go and find projects that will keep us busy. At the moment there is plenty of projects on. They are not really our competitors. But then now also from time to time … well there is the Urban Land Corporation would see themselves as*

competitors, while we are doing project work we also have these land development projects. Their business is land development and they say well how come you are doing that and nobody has been able to define when a certain project would go to one camp and when to another. So there are no real commercial competitors but there are sibling competitors within the government who might have aspired to undertake some of the projects that we have done because they were rather grand projects. There is not a commercial competitor.

| Client 4 | The client is a state government statutory authority and operates within the capital city of the state. Their mission is to manage land. The size of the authority has fluctuated and is responsive to the uptake of the developments. In the early stages there was little interest – at the time of the interview there were 40–60 employees. Annual reports from 1998 indicated 45 staff and 10 board members. In 1999, eight board members and nine staff, however, with a number of consultants who effectively worked full-time. We are a statutory authority, but probably the most private sector statutory authority you've come across. As public servants go, we're well paid – probably in the 50th% quartile that say in the private sector would be quite high up because the nature of the job, private sector focus. We are a government authority. | The commodity is a strategic project management service for the state for a particular precinct of land in the city. the redevelopment of a large tract of inner city We did, the Docklands Authority too – our main job back then, and it still is particularly if once areas of Docklands which are complete are progressively transferred back to the city of Melbourne – the job is well because we are a marketing and bid development and management authority that is what we do. So we marketed the project, we called the bids, shortlisted the best bids, got them in, assessed them and awarded the precinct. … What we did was an international marketing campaign in 1995/1996. We got 250 expressions of interest. We shortlisted the best of those, in typical Australian fashion, it wasn't until a lot of the international developers came on board that the Australian developers really got interested in it. How we assess all the bids is across these 5 criteria. … We're the planning and referral authority, what that means is that we sign a precinct development agreement, that agreement says that you'll pay the money when there's a chunk infrastructure of the land, or whatever, you'll design to this quality, and so we police them to make sure they do that. We monitor on behalf of government. |

(Table 6.2 continued)

Table 6.2 Continued

Firm type and projects	Structure and scope, size – employees and turnover	Commodities and competitors
	The Authority is more private sector than the private sector if that makes sense. Staff number range somewhere between 40–60. Really has grown, last year we made a significant loss because we employed more staff for managing this big project. This client is responsible for managing the development of 220 hectares of inner city waterfront land for the state government. The large parcel of land includes 7 km of waterfront and has been parcelled into seven major precinct developments. This project for the stadium is one of the precincts. The precincts are packaged up and offered to the private sector for financing. Originally in 1996 our business plan the government signed off, the whole Docklands project was going to be worth about $2billion. In Dec 1999 it was going to be about $900m of construction underway. The reality is that the project's looking more like between $4.5–6billion. Already we've got over $2b unconditional out there. In the last 12–18 months the Authority has grown incredibly. We were originally a marketing and bid assessment authority. Now we're still a marketing authority but now also a project management – we manage contracts, so the skill set of the authority has changed.	The competitors are really not other departments or firms but other developments. Yes, from a basic development point of view we compete with other developments. Same share of wealth that we're going after at the end of the day with regards to people who want to invest in residential/commercial. Even when Docklands is fully developed, do they want to go to St Kilda, South Bank, Docklands, or the city…so from that point, yes we do. From a land based point of view. The other interesting thing, more so in the previous government, more so with regard to statutory authorities – one of the first things which really got me thinking was that I compared ourselves to the city, from the point of view of share of mind, or where we sit with government, major infrastructure projects. So from a marketing share of mind scene we do have competitors.

Box 6.4 Strategic government project managers

'We have to spawn the project if you like. There is not someone out there who is preparing the project brief and comes along and says here is the project brief, here is what I want to build. Will you do it for me? We are another rung up from what I would call Project Managers out in the business world. Out there people get asked to manage something that the client has already conceived and has defined and perhaps has made a funding or investment decision against' (Director, OMP).

Both Projects 1 and 2 have other statutory bodies that have been formed to manage the facility post-construction. Both projects had numerous stakeholder groups and Project Control Groups were developed to manage the interface between client, user stakeholder groups, the primary consultant and the contractors during the design and construction process after the investment decision has been made.

Box 6.5 Project control groups

'After we develop the brief and work out what it should contain, that then goes back to the government for an investment decision if you like at Cabinet and then it might come back out and it will then move into a more defined project and we shall form a Project Control Group' (Director, OMP).

Project 4 was managed by a separate statutory body specifically set up for the management of a suite of projects in a precinct of the city, the Docklands Authority (DA). Project 4 was initially facilitated by OMP through early feasibility work and OMP were the Principal in some contracts.

Box 6.6 Private sector statutory authority

'Docklands is a statutory authority, but probably the most private sector statutory authority you've ever come across. As public servants go, we're well paid – probably in the 50th% quartile; that is say, in the private sector that would be quite high up because the nature of the job, private sector focus. We are a government authority. Docklands Authority is more private sector than the private sector, if that makes sense' (Marketing Manager, DA).

However, unlike the other two government projects where funding is by state coffers, the private sector funds Project 4 through a consortium using a public financed infrastructure (PFI) financial model. These government bodies are autonomous and have the capacity to act in a similar manner to a private sector client. In reality though, they still come under various government legislation; for example, for probity and equity issues.

The agencies are located within a highly political environment. Project 4 client does seem to operate in a much more entrepreneurial manner.

Box 6.7 Entrepreneurial approach

'What we did was an international marketing campaign in 1995/1996. We got 250 expressions of interest. We shortlisted the best of those, in typical Australian fashion. It wasn't until a lot of the international developers came on board that the Australian developers really got interested in it' (Marketing Manager, DA).

It is noted that such changes in how government projects are secured should see an ability to adopt more innovative approaches to supply chain thinking... but perhaps not.

The Docklands Authority is responsible for a large tract of land west of the inner city CBD divided into five development precincts. Each precinct is presented to the private sector for redevelopment along specified guidelines and each consortium bidder was required to fund the development.

Box 6.8 Project 4 location decisions

'What happened with the Colonial Stadium was that in late 1996/early '97, after the Dept of Infrastructure had done their study on where would be the most optimal place to have a multi-purpose stadium, originally for the 2006 Olympics, they were thinking about the Melbourne Olympic Park precinct or Docklands' (Marketing Manager, DA).

The client for Project 3 was a developer, similar to the client for Projects 5 and 6. These clients are completely private sector and organized the project financing. Project 3's client was not, however, involved in the construction industry, as was the clients for 5 and 6, and was primarily a property investor located in another country.

The original client for Project 5 was the government as they released the land with specific criteria for development, in a similar manner to Project 4, but with much less complexity, size and public stakeholder issues. The client for Project 5 is the developer who is involved in the construction industry and was also the primary contractor. Project 5 was one project amongst a suite of projects.

The client for Project 6 is again a developer who is involved in the construction industry and who was the primary contractor (refer to Table 6.3 for client and contractor details for Projects 5 and 6). The contractors for Projects 5 and 6 are primarily concerned with low- to medium-rise residential developments. At times these two firms might compete; however, the contractor for Project 5 typically operates on a larger scale and scope than the contractor for Project 6.

The following discussion begins to move us down the supply chain and summarizes the supplier firms at tier 1. The firms engaged at tier 1 is, of course, tied to project procurement strategies.

6.2.3   Tier 1 supplier firm attributes: contractors

This section provides an overview of the key tier 1 supplier firm attributes, type of suppliers and the structural organization of the suppliers at tier 1. The project procurement strategy impacts upon the type of suppliers located at tier 1. As we know, tier 1 suppliers are commonly referred to as the project team – that is, the primary consultants and the prime contractors. The following discussion provides an overview of consultants and contractors at tier 1. The discussion is focused on the contractor supplier firms.

Table 6.4 summarizes firm attributes for the primary contractor for Projects 1, 2 and 3 and also the primary contractor for Project 4. Occasionally these two firms are competitors on selected projects, when the contractor for Projects 1, 2 and 3 is tendering at the upper limit of size of a project or the contractor for Project 4 is tendering for a smaller project. The contractor for Project 4 is often involved in a portfolio of projects that require complex funding arrangements which are oriented towards civil and process engineering projects; whereas the contractor for Projects 1, 2 and 3 is primarily involved in commercial building projects.

The primary contractors for Projects 5 and 6 are discussed in Section 6.3.3. The following Figure 6.2 maps the firms that supplied commodities to the client (which is termed the focal firm) at the first tier by commodity type attribute (firm–firm object) for each project.

Project 1 had twelve service suppliers at tier 1 who were either design, project or construction management-related firms. Project 1 also had one firm supplying a product and a service – this firm was supplying a major piece of art work. Project 2 had seven service suppliers and again they were either design, project or construction management-related firms. The difference between the two projects in type of firms at this level is that Project 1 had more design consultants (particularly engineering) and more project management-related firms. Within the design services suppliers on Project 1

Table 6.3 Matrix of contractors for Projects 1–4

Firm type and projects	Structure and scope, size – employees and turnover	Commodities and competitors
Client and main contractor 6	This residential development firm concentrates on a particular locality in the city which includes a few large suburbs. They tender for work and they find their own blocks of land for subdivision and development. Their tender clients are local. *We've got one guy 1 or 2 guys that constantly feed us with work.* Their development work is local as well: *I just know a couple of the local developers probably sort of around…. Cheltenham and they're not real competition the only competition for us in those areas is who is going to buy the block.* As such, their scope is state with one division. *There's 3–4 full-time people that work there's myself Mike the estimator a full-time supervisor and a full-time labourer we have a part-time book keeper and a part-time accountant that come in once a week and if we need extra help on site as far as labouring it's just day labour.* This firm would not state their turnover.	The firm is in three residential markets: 50% of own development work and 50% of non-development work. Development work simply means that the firm organizes the finance and acts as a client. Non-development work refers to those projects where they tender for design and construction or simply construction management services. Within the work where the firm does not organize finance, etc, the work is split between design and construction and construction. *So in a sense we can do we do our own development we can do a design structure with a developer or just do a construct…so really there are three types of work…in the residential development market…you develop you organize you purchase the land you organize all the planning. Work with our own development we basically buy the site, I hunt it down and purchase it. And we initiate everything from there. We do everything. Permits, everything? Yea everything and the finance. So 50% of our work is our own development….* With regard to size of projects the individual house units tend to be of a similar value to the markets that Project 5's developer focuses on. However, this developer would not attempt the project development size that the previous developer would: *We do strictly residential and its all multi unit stuff our sort of 2 3 5 10 units that's the size of our jobs so Yea we kind of think that our main sort of area of work is between*

$1.5m and 10m but we do sort of venture out you know a little bit further we try to stay...of the highway the main highway we want to be known as sort of boutique bay side builders. We don't look at small jobs we don't touch renovations and extensions at all so we just do brand new multi unit stuff so what else is there? In Brighton 3 or 4 on the block job value would probably be in excess of $1,000,000 for us value of each unit will probably be sort of $7–800,000 for him to sell.

There are 3 markets: development market, 2–3 unit and then up to 15 unit:

Just cause we go for the larger sites on the larger sites obviously there's less people who can do them so there might 2 or 3 that are your real competitors when you look at a little 2 unit site all of a sudden there's all these other people that want to do them so there might be 10 or 15...the only competition for us in those areas is who is going to buy the block.

There is a range of commodities and associated markets and therefore a range of competitors. However, primarily this feasibility through to site management. They then also provide products through purchasing from various manufacturers for the projects and then as a separate business.

Correct correct we also have the more project management property acquisition business development. ...And we have got the design arm in here and also the construction management arm in here as well. ...The basic divisions of that is that we supply design and construction services... and we supply products that we design those projects for and then we just supply products nothing related to any design work we do....Essentially we...what we do today

Client and main contractor 5	The firm operates in the residential sector only. The firm supplies design and management services from firm is international, operating in this state and one country overseas. The international projects and the export of associated products is a major part of the business along with the local residential market. There are 3 licensed businesses under the firm's banner and 4 operating divisions within the firm. *There are 3 licensed businesses now under the banner of Stonehenge group...interior design group, custom home renovation company and there's a new division that we set up called Stones Development after 6 years now of working in Japan we've done 40 projects we've exported more than $10,000,000 of Australian product*

(Table 6.3 continued)

Table 6.3 Continued

Firm type and project	Structure and scope, size – employees and turnover	Commodities and competitors
	last month we sent 30 containers of Australian product into the market. Licensed business includes: interior design, custom home renovation and property development for project homes. The internal divisions include: real estate division for property development, construction division for boutique renovations, export division and graphic design division. The firms were in a state of growth and were soon to launch a new division: Information Technology, to further develop 'smart housing' market. As a guide, the custom home renovation or boutique market is homes in the $400,000–$1m range and the project home is generally $150,000–400,000. There were 80 employees in total, including 15 designers: architects, interior and landscape designer. The firm employs approximately 50 people across the divisions in the core group and employs approximately 30 people across the 3 licensed businesses. This client would not state their turnover.	*is we're as much a consulting company. People come to us for a range of things I mean our architecture and our construction covers a variety of things we're in renovations we're in custom housing we're in property development we're in land sub divisions we're involved in aid care facilities we're doing an aids care project in Japan at the moment we're doing another retirement facility in Japan and one in New South Wales. We've done apartment projects we're involved in sort of town house development but its generally on where people come to us they come in for a feasibility we do a kinda design service we can also give them financial kinda modelling of how a project might go and then we're also now getting involved in this technology side so when they're doing the development we will bring our technology expertise to bear and also supply that as a total solution.*

Table 6.4 Matrix of client and contractors for Projects 5 and 6

Firm type and project no.	Structure and scope, size – employees and turnover	Commodities and competitors	Competitive advantage
Contractor Projects 1,2 and 3	Multinational. 190 employees in state and 800–900 nationally. $1.6–1.8b turnover. _Effectively, the structure of Multiplex is the National Board of Directors where each state has at least one and in some cases two representatives at the national level. Each state has its own state Board. At state level, these people are the day-to-day managers and executives and control the business on a daily basis in the states. There is some duplication with state representation and national representation. That makes it the key group in each state control the business and then get strategy and feed off the larger parent company at national level through an exchange of information and ideas is through key individuals in the state. More recently, in the 90s, the business has grown where the Victoria Operations now also supports South Australia and New Zealand,_	Contract and asset management of commercial, retail, health, tourism and residential projects. six main competitors across various markets. _Multiplex – probably 60% of our work, we are the largest residential private builder in Australia, when you add up what we do right across the country. We would have at the moment 1,500 apartments under construction at the moment. In Victoria, in terms of our market competitors there are, we tend to come across, it doesn't matter which market, there is a core group of competitors. They tend to participate in most of the market. It would be Baulderstone Hornibrook – it seems like we are both across the same markets. In specific markets we often come across regulars like Walters, Grocon, ProBuild. They mostly do retail, health, commercial, industrial, tourism, hotels. In the retail, it is probably Probuild, in hospital it's probably Walters and Baulderstone,_	Organizational structure, integrated management, subcontractor management, strategic outsourcing. _There are a couple of reasons why we have competitive advantage. A lot of it has to do with the way we structure our setup, very hands-on directors, that get involved one day talking to the labourer to the subcontractors and then he walks in through the door with a client, the government or the Premier or whoever it be and wherever his involvement is required. Our attitude is that we can always find something for an employee – we tend to overwork people rather than underwork them and then we might get to a crisis point where people won't cope – we outsource a lot. We supplement when we need to. We have a core of people and then if we have a peak of 3 to 4 months in estimating or finance we might put some people on for a short term to get us through that. We outsource a lot at the corporate level; we might outsource in estimating, in programming in construction programming and scheduling._

(Table 6.4 continued)

Table 6.4 Continued

Firm type and project no.	Structure and scope, size – employees and turnover	Commodities and competitors	Competitive advantage
	and the WA office supports Singapore and Dubai; and Sydney supports London, which is our more recent new business that we have started. And in more years we have expanded out of pure construction and the construction property market and also created a mining business, which is both contract mining and mining investment in ownership of mines as well as contracting for other mines. So, in Victoria, effectively there is a Managing Director, with 3 key directors, one looking after construction, one looking after new business, the other one looking after the money side. We have teams that work off the three directors… then Delivery of the contract and this is the control part of the business, Financial Effectively then the MD rides across the lot and then there is succession planning in place that is an evolving	in the hotel and commercial office building it's normally Grocon and industrial it's normally people like more medium range builders – not in the same league . We don't play in the industrial market. Vaughans and Cordukes do those – they are the mid-range builders and LU Simon. With apartment constructions we tend to come up against Baulderstone, Walters and LUSimons and more recently Thiess and Leightons, but they are really only emerging. They haven't had a lot of success in building in recent years. They would be the main players. Hollands has fallen away. They are not a big player any more. Baulderstones and Multiplex are across all those big markets. In the residential market it is also Mirvac, although they are different as they finance as well – that is developers and we don't do that a lot.	They are the main areas. On projects we tend to hold the best people; we don't want to have a hire and fire mentality. We don't want to upsize and downsize – we have only had to do this once in 12 years – so we are pretty proud of that. So far we only had to downsize and that was 12 years ago, up to 30 people but in the good time up to 160 people – sustainability. Rather than peaks and troughs we. But with this type of industry it is hard to maintain this. Try to win the work to suit your market.

organization where people are moving up through the organization. The projects are supported by managers, project teams are fairly autonomous businesses in their own right. We give a fair amount of autonomy to the teams managing each project as they report through to the office on a regular basis on time and cost. And this head office is merely a service facility for our projects as well as creating new opportunities for new projects.

The entire business turns round about $1.6–1.8b a year. As a group in about 50–60% of that is in NSW, about 20–30%, 25% is in Victoria and 10% in Queensland and the balance in WA and overseas. Sydney is about half because of the Olympics. But that changes and the markets change they are different. Geographical markets are important because of different highs and lows in the market at different times.

. . . We have directly 190 in Melbourne and nationally about 800, 900 and growing.

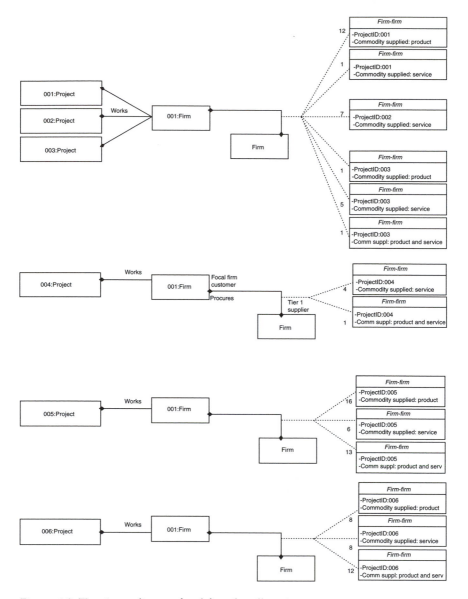

Figure 6.2 Tier 1 suppliers to focal firm for all projects.

there was an internationally renowned landscape/urban designer and a façade consultant.

Within the project management-related cluster of firms on Project 1 there were more direct project management firms and also more firms supplying project management-related services. The project management-related firms have been classed as professional support that was non-construction specific. These types of firms include, for example, contractual lawyers, art works professional advice and marketing management. The following Section 6.3.1 discusses the tier 1 service suppliers for Project 1 in more detail.

Projects 3 and 2 have a similar number of service suppliers. The key difference between these two projects, as mentioned previously, is that Project 3 has a product and service supplier and a product supplier as well. The product and service supplier was an external cladding specialist contractor and the product supplier was an agent for imported kitchen equipment.

Project 4 was the largest project in terms of scale and value and yet the number of firms at the first tier is surprisingly lean. This is solely due to the procurement method. The build-own-operate (BOO) method meant that many of the design, project and construction-related firms moved down to the next tier. The product and service firm in this case was the development consortium. The development consortium involved a major construction and development firm. The development consortium then engaged the construction management firm who in turn engaged engineering design divisions on the project. The four service suppliers were: design, architectural and engineering, project management and art works consultant. These design consultants were required, in the initial stages, for conceptual and developed design. They were also involved during the construction phases, although still directly engaged by the government. The project management firm in this situation is largely an impartial firm engaged to report on the project for the government.

Projects 5 and 6 have a similar profile of suppliers by commodity type; there were proportionally more product and service and product firms in relation to the three previous projects. Projects 5 and 6 have a similar supplier type profile at tier 1, although in different proportions. The developer for Project 5 has in-house design and project management services and so the services are direct construction labour and some land and building surveying consultants. Project 6 client engages directly construction-type firms and also various design consultant firms. Both firms employ direct labour firms for carpentry. Project 5 developer subcontracts direct labour for plastering and Project 6 developer engages direct labourer for bricklaying subcontracting packages. In turn, this means that they purchase directly from the materials supplier; that is, the brick manufacturer and the large national multi-materials wholesaler and/or the large international materials manufacturer.

It is anticipated that the firm–firm procurement relationships at tier 1 would typically involve complex and significant transactions and as such the contractual type would reflect these attributes. An investigation of the commodity attributes frequency, complexity and significance of tier 1 for Project 1 transactions matched against the contract type clearly indicates that, although the transactions may be highly frequent, significant and/or complex, the contract type does not reflect any further development of the relationship beyond the project boundaries, that is, every contract is a project contract. Of the thirteen transactions, seven were highly frequent, complex and significant to the supplier.

In a comparison with tier 1 suppliers for Project 2 and Project 4 (which are both government sector projects) a similar scenario is found; namely, that all transactions are project contracts even though they may be highly complex, require a high level of interaction and/or are highly specialized markets. Within each supplier category the customer is sourcing from either one supplier or a small group (2–5). This is explained by the approach taken to strategic procurement by the OMP unit.

Box 6.9 Strategic procurement and public sector probity and equity demands

Even though the unit is quite progressive in that it is a corporate and commercial entity, the pressure of operating within a government environment and the associated probity and equity issues reflects in the conventional and non-commercial-oriented procurement philosophies. This is a dilemma for government agencies – how to develop strategic procurement relationships balanced with equity and probity demands of public office?

Alternatively, does it matter? Let us see what happens as we progress down the chain.

The results for the other three Projects 3, 5 and 6 indicate a small number of different contract types strategically matched to the characteristics of the transaction. For example, Project 3 has seven suppliers at tier 1 and five are project contracts, one is a purchasing agreement and one is a company-wide strategic alliance. The purchasing agreement is with a product supplier and the product is white-goods and the company-wide strategic alliance is with a design management firm, that manages all the developments by the client/developer. The prevalence of more long-term relationships can be explained through Project 3 being a commercial project, whereas Projects 1, 2 and 4 are bound by state government probity and equity regulations. The decision for a different procurement relationship was based upon the attributes of the transaction only and not wider stakeholder implications. For both suppliers the decision was based upon a combination of frequency of transaction across numerous projects and complexity of commodity (namely, quality of service or quality of product).

The tier 1 suppliers for Projects 5 and 6 are the construction site suppliers as well as design consultants, because the focal firm is the residential client/developer. There are thirty-five suppliers for Project 5 and of these there are three company-wide strategic alliances, four purchasing agreements, one project alliance and twenty-seven project contracts. There are twenty-eight suppliers for Project 6 and, of these, two are engaged through ongoing purchasing agreements and one is through a company-wide strategic alliance and the remainder are project contracts.

The service suppliers for Project 1 represent the typical project procurement method that client 1 would prefer to use; it is the traditional method of project procurement.

Box 6.10 Managing control of complexity

'Every project is a pyramid really and one of our Project Directors sits at the top and occasionally there is a Project Manager inside this office. But from there on in everyone else is contracted to the project. I typically prefer a traditional method of project procurement. I don't think that as a state agency we can go down the path of strategic alliances and I like to keep control of the complexity of developing a brief and control consultants. I don't particularly want to hand over control to contractors as OMP should be the direct liaison between the consultants – that is what is expected from us by our own agency clients' (Director, OMP).

Even though there are alternative methods to the traditional method being trialled in Australia and have been trialled over the last two decades, it is the current preferred method for this state. A published journal paper provided by the director of this unit clearly indicates the current philosophy of this state's major government projects manager. This was also upheld in the interviews. But...sometimes attitudes change. ...Is this really the only way to 'control complexity'?

Box 6.11 Power and project development structure

'There is an interesting article that describes this, it was written by some lawyers here. It was writing about the project development structure. This explains the relationships between the parties and how we get power and how things work and so on' (Director, OMP).

It is worthwhile now to describe the attributes of tier 1 service suppliers to client 1, since client 1 is the major procurer and project manager of public sector projects in the state.

6.3 Types of suppliers, associations and commodity attributes

6.3.1 *Tiers 1 and 2 service suppliers: Project 1*

This section develops the structural model view by describing the types of firms and their relationship to other firms and the associated commodity which links each firm. As such, this section develops the structural organization maps of individual commodity suppliers and groups them according to type. Tier 1 and tier 2 service suppliers are described.

Project 1 was initiated as a selected invitational international design competition. The design firm that provided the winning scheme was then asked to develop the scheme in conjunction with an architectural firm from the city where the project site was located. The design architectural firm had worked with an engineering firm to create the winning scheme: although there was no contract, there was an understanding of a future procurement relationship. Eventually, the two architectural firms created a separate company and contracted with the Principal and the sub-consultants. The strategic alliance between the two architectural firms did not occur as soon as the project began; it took almost 12 months to realize.

The Principal had also engaged thirteen consultants, of which the new project strategic alliance (PSA) architectural firm was one. The PSA architect then engaged ten engineering and specialist design consultants and the project manager engaged three other firms.

Figure 6.3 maps the type of firms by commodity type and indicates the nature of the commodity which they provide at the first tier for Project 1. Clearly, the majority of commodities are services. The commodities supplied were classed by the client in four groups for this project: project management cluster, architectural design management cluster, construction management cluster and project marketing. The architectural design management and construction management clusters are the largest. At the time that the construction contract had been let, the Principal had appointed the three key consultants:

Box 6.12 Design management cluster

- 'The architect on a joint and several basis to design and document the project
- The quantity surveyor to provide cost advice and quantity surveying services
- The building surveyor to provide building certification services
- The architect has subcontracted a number of sub-consultants to provide civil, structural, façade, mechanical, electrical and landscape design services for which the architect is responsible' (Managing Contractor Agreement, V1).

According to the contract between the managing contractor and the Principal, the intention of the contractual arrangements between the client and the contractor was that the design team were to be novated to the contractor.

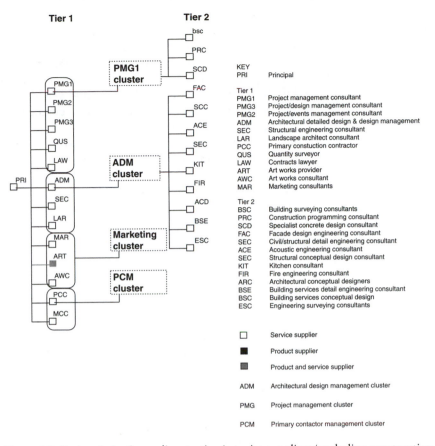

Figure 6.3 Project 1 tier 2 suppliers to tier 1 service suppliers (excluding construction management cluster).

Box 6.13 Sometimes things don't happen as they were intended to happen…

'…within 5 days from the date of the completion and approval of the Project's schematic design by the Principal, the Principal will novate the Agreement it has entered into with the Architect to the Managing Contractor. The Managing Contractor must enter into a deed of novation' (Managing Contractor Agreement, V1).

This means that the contractor assumes responsibility for the management of the consultants and is contractually bound to them. However, this did not happen on this project; the consultant team were not novated across and the Principal managed the tier 1 design team contracts. The reason given in this instance by the Director of OMP was related to his desire to maintain control of the competition-winning design architectural consultants as they were very inexperienced in this scale of a project. In time, we were able to see that this was a good assessment of the situation.

The Design Management Cluster contracted by the Principal at tier 1 consisted of three firms, comprising: the PSA architectural firm (an international and a local architectural firm), an engineering firm and a landscape architectural firm. The international architectural firm used an engineering firm and, in particular, this firm had two divisions – an engineering and a façade engineering firm – and these two firms were used by the local architectural firm during design development.

The local architectural firm were the lead or primary consultant responsible for managing developed and detailed design and to do this they contracted and co-ordinated various sub-consultants. This sub-consultant group formed the conceptual and developed design suppliers and there were ten of these, including a structural engineering firm, four multidisciplinary engineering firms, an acoustic specialist engineering firm, a façade engineering firm, the international design firm, engineering surveying firm, and a specialist kitchen design consultant.

The Project Management Cluster consisted of five firms, including three who provided project management services, one who conducted early feasibility studies and then progressive cost monitoring and one who provided contract law advice. There were five major design and construction elements for this project:

Element	Description	Area m²
Element 1	Completion of the upper section of the crash walls from the early works element and decking over existing railway	
Element 2	Museum	12,500
	Cinemedia centre	9,200
	Atrium	10,500
Element 3	600 space car park	300
	Information centre	7,000
	Civic square	17,000
	other external spaces	
Element 4	Boutique bar and pub, function centre and other commercial uses	3,500
Element 5	Early works, demolition, foundation works for piling and crash walls	

One of the project management firms was involved with the early civil and structural work for the deck, whereas the other two project management firms were brought on to co-ordinate part of the building elements. The quantity surveying consultant was involved from the early stages as well. The project management consultant firm 1 was engaged to co-ordinate and monitor Elements 1 and 2. Project management consultant firm 2 was engaged to co-ordinate and monitor Elements 3 and 4 and project management consultant firm 3 was engaged to co-ordinate and monitor Element 5. Element 5 was completed prior to the start of work on the remaining elements. The same contractor was engaged to construct all five elements, even though the second stage of the development of the site which included Elements 1–4 was conducted as a pre-registration selected tender process, in a similar manner to how the Stage 1 contractor was sourced.

The construction management group involved the contractor who had two contracts with the Principal; the first for early civil works and the structural deck and the second for the building elements. Fifty-five firms supplied to the construction management group for the second contract and are classed as construction site production as they supply direct to the site. Typically, there were four classes of construction site production suppliers identified by the construction site project manager, namely those that:

- Supply products
- Install products
- Supply and install products
- Design, fabricate and install products.

We start to see explicit categorization of suppliers. As the following chapters will indicate, this is prevalent in the industry. It seems to be the way we can process and manage complex environments. At times the categorization takes place, but it is not explicitly 'tagged' as I have done and others have done. It is noted that within the supplier firms there is also a further categorization which includes those that supply products through distribution and those that supply products by first processing. Therefore, within this group there was a combination of two types of firms: those that supply directly to construction sites and those that distribute products and process products to others that supply direct to site, as well as supplying direct to site. The construction management clusters for this project and the other two Projects 2 and 3, which this contractor was engaged to construct, are now considered in detail.

6.3.2   Tier 2 suppliers: Projects 1–4

This section continues to develop the structural model view through descriptions of the structural organization of types of firms by commodity type and numbers of these types at tier 2. This helps us to picture which firms are sourcing and managing how many supplier firms and also the degree of complexity of the commodity being transacted.

Figure 6.4 maps tier 2 suppliers for Projects 1, 2, 3 and 4 and they are grouped by commodity type (product, service or product and service). Each supplier is also grouped with the other suppliers by upstream (service) customer according to the actual commodity supplied. Each cluster is

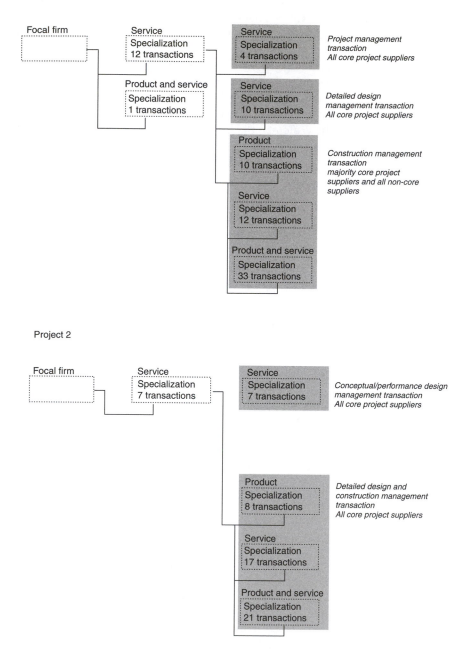

Figure 6.4a Tier 2 suppliers: Projects 1 and 2.

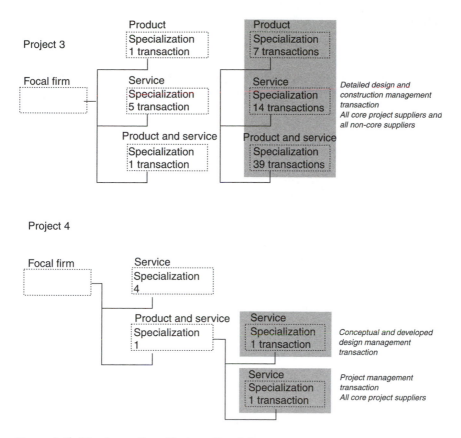

Figure 6.4b Tier 2 suppliers: Projects 3 and 4.

indicated by a shaded box with the description of the commodity beside the cluster of transactions.

For example, on Project 1 there are three tier 2 transaction clusters, which describes the suppliers to a project management firm, the design management firm and the construction management firm. These three supplier groups supply respectively to a service firm at tier 1. The project management transaction involves four services transactions; the detailed design management transaction involves ten services transactions and the construction management transaction involves fifty-five suppliers, including ten product transactions, twelve service transactions and thirty-three product and service transactions. As we know, suppliers to the construction management firm are collectively and commonly referred to as sub-contractors – perhaps a not very useful term really.

It is to be noted that not all firms indicated all their suppliers when responding to the questionnaire or during the interviews, but they did indicate their most significant suppliers. Therefore, beside each cluster the suppliers' significance to the upstream customer has been noted by the following:

> **Box 6.14 Supplier significance to customer**
>
> - core supplier(s): which are the main suppliers central to the transaction (e.g. aluminium, steel, glass and connexions for façade)
> - non-core supplier(s): which are the support suppliers (e.g. logistics, administrative, IT)
> - primary supplier: the main supplier central to the transaction (e.g. clay for brick manufacturer).

On Project 2 the design management firm has seven service suppliers for conceptual/performance design for the early phase of the project and the construction management firm has engaged forty-six suppliers for detailed design and construction management. Of the forty-six suppliers, eight were product suppliers, seventeen were service suppliers and twenty-one were product and service suppliers.

The contractor engaged sixty firms on Project 3 and of these seven were product suppliers, fourteen service suppliers and the remaining thirty-nine were product and service suppliers.

Typically, it seems that a contractor manages equal numbers of product suppliers and service suppliers and more than twice as many product and service suppliers. The greater number of service suppliers is accounted for through a design-and-construct procurement method (Project 3) or novated design contracts (Project 2).

Clearly, the type of financial structuring of projects impacts upon the governance structure and the number of firms at each tier, the relationships between firms and subsequently the eventual length of chains.

6.3.3 *Tier 1 and tier 2 suppliers: Projects 5 and 6*

This section describes and develops maps for the suppliers to Projects 5 and 6 who were located at tier 1 and tier 2. Projects 5 and 6 were both funded by the developer who also managed the on-site construction and design phases. Therefore, tier 1 includes both the consultants as well as the subcontractors. This differs to the previous project suppliers found at tier 1 and tier 2; tier 1 primarily included consultants and tier 2 included subcontractors.

The developer who was the client and contractor for Project 5 had a number of in-house consultants. At tier 1 there were thirty-five suppliers, including: sixteen products, six services and thirteen product and services suppliers. From the sixteen product suppliers there were another thirty-five suppliers mapped and from the thirteen product and service suppliers another fifty-two suppliers mapped.

For the smaller residential Project 6 there were twenty-eight suppliers at tier 1, including: eight product, eight service and twelve product and service suppliers. From the eight product suppliers, ten suppliers were mapped, these being nine product suppliers and one service supplier. From the twelve product and service suppliers, two product suppliers were mapped.

Figures 6.5 and 6.6 map all suppliers for tier 1 and selected suppliers for tier 2 for Projects 5 and 6 respectively and the suppliers are grouped by commodity type.

Box 6.15 Supplier profile: commodity and industrial market characteristics

The supplier profile at tier 1 is similar for both projects, except that the developer for Project 5 is able to purchase more products directly from manufacturers or distributors. They are able to achieve this as they have a higher purchasing volume, which enables them to exert more power in the marketplace. They also work more closely with selected manufacturers or distributors; however, their relationships differ across product suppliers. For example, with a major national plasterboard manufacturer their relationship is based upon purchasing volume and price; however, their relationship with the brick manufacturer is based upon quality of product, service and price. The national plasterboard manufacturer is larger and more powerful in the marketplace than the contractor; whereas the brick manufacturer is a boutique or 'niche' market supplier and the power in the relationship is with the contractor.

This indicates that each market is considered on its own merits and procurement relationships are developed based upon commodity characteristics and countervailing market power.

It is more difficult to make comparisons at tier 2 as less information was available from the tier 2 Project 6 suppliers. However, tier 2 suppliers for Project 5 provide interesting situations to consider.

Box 6.16 Supplier co-ordination categories

Four categories of supplier co-ordination:

- suppliers who co-ordinate a large number of suppliers
- suppliers who co-ordinate a small number of suppliers
- suppliers who co-ordinate different types of suppliers; and
- suppliers who co-ordinate a different level of complexity.

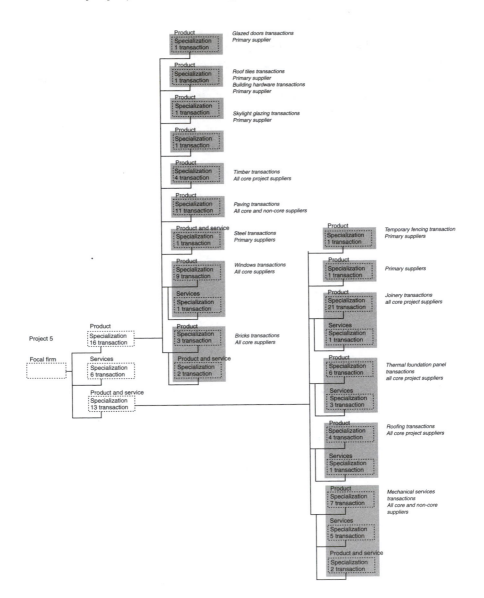

Figure 6.5 Tier 1 and tier 2 suppliers: Project 5.

At tier 2 of the six suppliers to the product and service supplier group there is one supplier, the mechanical services subcontractor, who manages all three different types of commodity suppliers (product, product and service and service). The mechanical services subcontractor co-ordinates seven product suppliers, five service suppliers and two product and service suppliers. Although not of the same variety and volume, the thermal

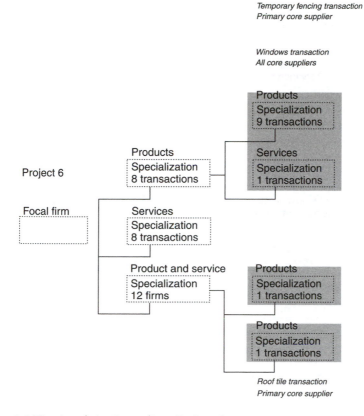

Figure 6.6 Tier 1 and tier 2 suppliers: Project 6.

foundation panel subcontractor co-ordinates six product suppliers and three service suppliers. The joinery subcontractor manages a large number of suppliers; however, they are primarily product suppliers; twenty-one product suppliers and one service supplier.

The following Figure 6.7 gives an indication of the extent of supplier co-ordination for Project 5 across two dimensions: supplier total numbers and supplier commodity type numbers. It also indicates selected details for Projects 1 and 6 to allow comparisons. Such a matrix can be used to map internally a project to evaluate the suppliers or to compare across projects the same supplier type.

The matrix indicates that on Projects 5 and 6 a tier 2 supplier is managing well over 10 suppliers and is co-ordinating 3 different types of suppliers; it happens to be the mechanical services supplier for both projects. For Project 5 at tier 1 the contractor manages nearly thirty-five suppliers and for Project 1 the contractor manages well over thirty-five suppliers and again is classed a category 3 level of co-ordination. Project 1 contractor is mapped adjacent

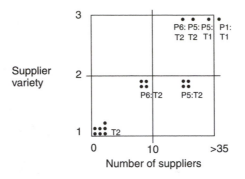

Figure 6.7 Project 5 matrix of tiers 1 and 2 suppliers by supplier co-ordination dimensions: supplier number by supplier variety (selected details for Projects 6 and 1 for comparisons).

Box 6.17 Supplier variety

Type of supplier is defined as:

Product supplier

Product and service supplier

Service supplier

Supplier variety is defined as:

Category 3: Co-ordinating the 3 different types of suppliers

Category 2: Co-ordinating 2 different types of suppliers

Category 1: Co-ordinating 1 type of supplier

---

to the right for tier 1. For both Projects 5 and 6 at tier 2, four subcontractors manage two different types of suppliers; for Project 6 these subcontractors manage just under ten suppliers and for Project 5 it is well over ten suppliers. Finally, there are seven suppliers for Project 5 that are co-ordinating just a few suppliers of one type.

The supplier co-ordination matrix gives an indication of the volume of co-ordination required by suppliers in terms of the number of commodity types – it could be a proxy indicator for complexity. However, is it more complex to manage a service firm as opposed to a product firm? Firms were asked about level of complexity of the firms that were supplying to them – that is, how complex did they think the commodity was that was being supplied to them? The following Table 6.5 maps the subcontractors to the contractor for Projects 5 and 6 in terms of transaction complexity and by

Table 6.5 Transaction complexity by commodity type across projects

Transaction complexity	High	Medium	Low
Project 5 (20/35)			
Products	9		
Services	2	1	1
Products and services	4	2	1
Project 6 (21/28)			
Products	8	1	
Services	2	1	1
Products and services	6	2	

commodity type. The contractors could only provide information about complexity on selected subcontractors (for example, twenty transactions from thirty-five for Project 5). It is anticipated that there is a high level of complexity for a number of subcontractors who install on-site and co-ordinate other suppliers or those that provide a service. Perhaps what is more surprising is the high number of product suppliers who have been categorized as providing a high level of complexity. This is accounted for primarily through off-site fabrication.

6.3.4   *Tier 3 suppliers: Projects 1 and 3*

This section continues to develop the structural model view by exploring the next tier of suppliers for Projects 1 and 3. Maps of commodity type and the matrix for supplier co-ordination are developed for the tier 3 suppliers for each of these projects; they provide useful information to develop supplier commodity abstractions and comparisons because Projects 1 and 3 have the same primary contractor. However, they differ in other respects; for example, Project 1 client is public sector and Project 3 involves a private sector client. Project 3 is similar to Project 1 in that the client out-sources these services. The project procurement method for Project 1 is construction management with a form of traditional procurement of design services (albeit an international design competition) and Project 3 is traditional and then novated.

Figures 6.7 and 6.8 map the various suppliers to the tier 2 suppliers at tier 3 for selected suppliers. More data from the product and service suppliers (subcontractors) at tier 2 was available, collected and mapped than for tier 3. The commodity maps give an indication of the overall structural characteristics of the supply chains, particularly with respect to the commodity type and number of suppliers at each tier. It clearly indicates that there are a high number of transactions at each tier on each project and that each supplier is co-ordinating a number of transactions.

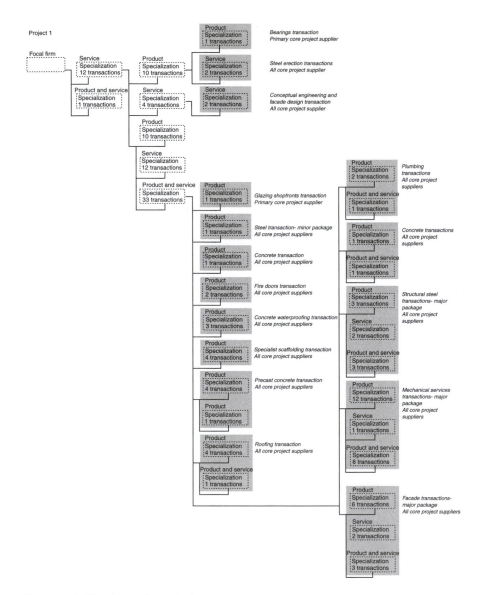

Figure 6.8 Tier 3 suppliers: Project 1.

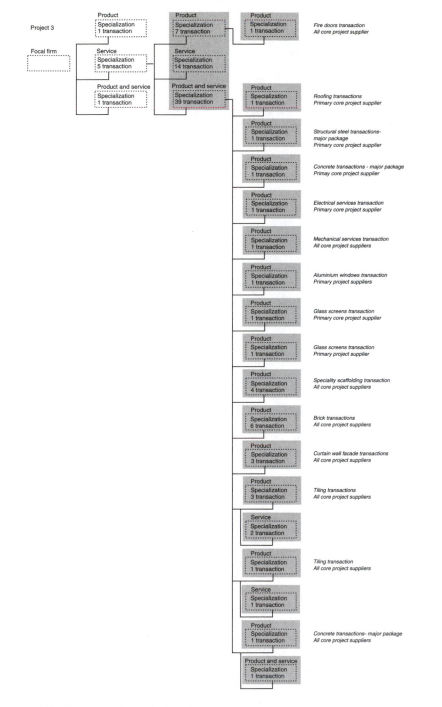

Figure 6.9 Tier 3 suppliers: Project 3.

Box 6.18   Specialization, integration and supplier variety co-ordination

What does this really tell us though?

Clearly, firms do not just purchase products – they have a range of commodity types with a range of complexity levels. There are diverse levels of variability in supply and thus varying levels of capability in supplier co-ordination.

If we wish to reach down another tier and co-ordinate, manage, integrate or develop at the next level – do we have the capabilities to do so? How can we connect to the next tier and enhance rather than meddle? What are our own systems for managing the different types of suppliers and then how does our system mesh with our key suppliers? The words 'integrate business systems in supply chain management' really are powerful but they have serious implications.

We need to look at our own vulnerabilities. Which of our own supplier is strategically critical to us and in what way? In the following chapters we shall see examples of firms who do categorize their suppliers by a mix of the traditional way – by commodity type and then also by how that supplier relates to risk and expenditure. Although seemingly simple, it is surprising how many firms may not do this explicitly. It is quite interesting – our construction industry is highly flexible and yet we see this as a negative where other industries are looking to flexibility in their manufacturing systems. Our industry looks at suppliers from a commodity and a process perspective – that is, suppliers don't just push products on to us, they provide a suite of attributes. How much do we know about those attributes? The procurement relationship object described in the previous chapter encapsulates those attributes.

The matrix in Figure 6.10 again maps tier 3 suppliers for Projects 1 and 3 by number of suppliers that they co-ordinate by number of different types of suppliers that they co-ordinate. There are nine suppliers on Project 1 that manage suppliers of one type.

The differences between the supplier co-ordination profiles can be explained by a combination of tier 2 supplier characteristics, commodity characteristics and tier 3 market characteristics. There are a range of explanations for the more firms who have a higher level of supplier variety. For example, the façade subcontractor for Project 1 out-sources more than for Project 3. Each façade subcontractor develops their unique method for supplying their commodity. Also, the mechanical services commodity was more complex for Project 1 than Project 3; therefore, the mechanical services package was split into two major packages. The same supplier won both packages at tier 2 and determined that they would project manage one package and construction manage the other; accounting for a large number of suppliers again accounting for a great variety

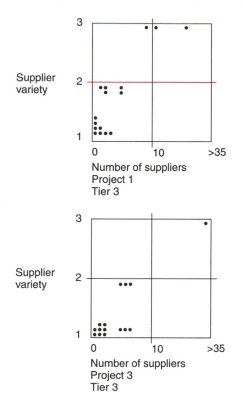

Figure 6.10 Projects 1 and 3: matrix of tier 3 suppliers by supplier co-ordination dimensions: supplier number by supplier variety.

of suppliers for Project 1. A similar situation occurs for the steel subcontracting package in Project 1, thus accounting for the other high-level supplier variety firm and medium-sized number of suppliers being co-ordinated. The project procurement method for Project 3 enabled more design and construct packages at tier 2 and therefore at tier 3 for Project 3 there were more single product transactions of primary core suppliers. This discussion on variability provides some background to the more detailed structural organization maps and discussion on case studies in the following section.

6.3.5   Firm attributes by project

This section provides descriptions of the firms related to commodity type. Table 6.6 summarizes the size of firms by commodity type across the projects for the data collected. Not all firms supplying to the projects are reflected in the results.

Of the thirty-seven firms supplying to Project 1, fifteen of these firms employed +5000 staff and seventeen firms employed +100 or less. This result of similar proportions of very large firms to small- to medium-sized firms occurs to some degree across each project; although, in some cases, not as marked.

Table 6.6 Firm size by commodity type across projects

Firm size	>500			>20			1–20	
Project 1 (37)								
Products	9			1		2		
Services	2	1	1	2				1
Products and services	4	2	1	7	2		1	1
Project 2 (35)								
Products	8	1		1			1	
Services	2	1	1		2			
Products and services	2	1		1	5	2	2	5
Project 3 (31)								
Products	5						1	
Services	2	1						
Products and services	4	1	2	4	7	2	1	
Project 4 (37)								
Products	6			1				
Services	2		1		1			
Products and services	3	1		3	4			
Project 5 (26)								
Products (8/14)	5	1				1	1	
Services (2)	2							
Products and services (8/10)	1			2		1	3	1
Project 6 (11)								
Products (3/5)			1	2				
Services (1/5)	1							
Products and services (1)		1						

The most significant result arising here is that, contrary to common reporting about the size of firms in the construction industry, there are many more larger firms than anticipated. However, this is partially explained by the fact that firms that are of this size tend to be the manufacturers or distributors of products and they tend to supply to all of the six projects repeatedly in various supply chains. Regardless of this, it indicates that each project has suppliers of a significant size and that they tend to supply products or products and services. The smaller to medium (SME) sized firms tend to supply products and services; perhaps reflecting the craft or task orientation of the SMEs' role on the construction site or fabrication off-site.

Firm size is one indicator of the nature of a firm. The assumption is that large firms tend to supply standardized products and smaller firms tend to be more responsive and flexible and will supply those commodities requiring the capabilities for supply and installation. The results tend to support this and the ideas of flexible specialization are relevant. However, the results also indicate that there are very large subcontractor firms and very small product suppliers.

The large subcontractors can be explained by these firms being multinational or national in scope and, when pressed in the interviews, were often operating as individual units and divisions within countries or states. Discussions with these types of suppliers indicated that generally they operate as a regional firm in operational matters with a much smaller capacity than the multinational status; however, at strategic times for strategic projects to fulfil strategic contracts or to procure strategic suppliers, the full resourcing capacity and purchasing leverage of the national or multinational firm is used. How useful!

The small product suppliers can be explained by these firms being national or international in scope and operating as distributors of specialized products. For this reason it is useful to consider the scope of a firm in relation to the commodity it supplies to further understand the underlying structure of the supply chain.

The following Table 6.7 summarizes the firm scope by commodity types for the Projects 1 to 6. Not all the data was supplied for all firms supplying to the projects; only those that were surveyed or interviewed. In summary, product

Table 6.7 Firm scope by commodity type

Firm scope	Multi-national	Inter-national	National	State with > division	State with 1 division
Project 1 (37)					
Products	7	3	2		
Services		3	3		1
Products and services	4	5	3	2	4
Project 2 (35)					
Products	5	1	1		
Services		1	2	3	
Products and services	1	1	4	3	9
Project 3 (31)					
Products	4	2			
Services		1	2		
Products and services	3	1	2	12	3
Project 4 (37)					
Products	3	3		1	
Services			2	1	1
Products and services	3	1			
Project 5 (26)					
Products	3	2		1	2
Services			2		
Products and services		1		1	6
Project 6					
Products (5)	1	1	1		
Services (5)	1				3
Products and services (1)			1		

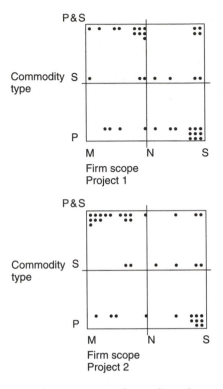

Figure 6.11 Projects 1 and 2: matrix of suppliers by supplier co-ordination dimensions: commodity type by firm scope.

suppliers tend to be national, international and multinational in scope; service suppliers tend to be national, international and state with more than one division in scope and product and service suppliers tend to be state and/or national based. The larger the project, as in Project 1, then the more likely the product and service suppliers are national, international and multinational in scope. This is explored in more detail in the following chapters. Figure 6.11 maps graphically firm scope for Projects 1 and 2 against commodity type for tier 2 suppliers to the contractor. Perhaps what is most surprising is that the industry is not as regionally based as we once thought – of course, construction is largely regionally oriented – we build on a particular site at a particular time. This theme is picked up again in later chapters.

6.4 Structural organization of supply chains at individual firm level

As indicated in Chapter 4, in diagram 4.10 the channel structure can be graphically represented through grouping of types of suppliers at each tier.

The approach taken in this study is that to develop those aggregated descriptions of supply chain structure a more detailed consideration of individual firms and specific chain structure is required. This section maps the structural organization of the chains of individual firm–firm procurement relationships for Projects 1, 2 and 5 (refer to Appendices for Projects 3, 4 and 6). The structural organization of project supply chains is developed for the following:

- Project 1 from the focal firm (the client) through to tier 6
- Project 2 from focal firm through to tier 3
- Project 5 from the focal firm to tier 1.

The following Table 6.8 summarizes all the structural organization maps at the firm level which have been developed based upon the interviews. These can be found in my PhD in Appendix K – it would be too exhaustive to repeat all the diagrams in this book; however, this does provide in some way a sense of how the comparisons across supply chains can be made and the validity to the aggregated channels which are developed in the following chapters.

Table 6.8 Structural organization maps developed

Supplier type	Project 1	Project 2	Project 3	Project 4	Project 5	Project 6
Tier 1 consultants		•	•	•		
Tier 2 subconsultants		•				
Tier 2 subcontractors		•	•			•
Suppliers to subcontractors						
Woodwork/ carpenters		•		•	•	
Structural steel	•					
Windows	•	•	•	•		
Roofing	•	•	•	••	•	
Partitions	•	•				
Masonry	•		•			
Hydraulics	•					
Mechanical	•	•	•		•	
Concrete	•	•	•	•		
External rendering					•	
Resilient finishes		•		•		
Furnishings				•		
External elements				•		
Tiles	•	•				
Fire door products	•					
Access floors	•					

Figure 6.12 describes the overall structural organization of many of the supply chains involved for the procurement of the design and construction for Project 1. It indicates the relationship between firms. It is the structural organization of firms for the project supply chains. At the time of the fieldwork, not all the internal fitout trade packages had been let and therefore not all the tier 2 firms were known. At that time, the project was approximately two-thirds complete. The maps clearly indicate the volume of transactions across various supply chains. It also indicates the complexity in terms of the variety of firms, commodities, markets and procurement relationships that evolve on construction projects and across construction supply chains. Coupled with this partial view of the project, only a selection of suppliers were mapped, including: fourteen of the tier 1 customers' suppliers; three of the tier 2 customers' suppliers; three of the tier 4 customers' suppliers and two of the tier 5 customers' suppliers.

The Principal engaged the contractor firm who then proceeded to let a number of trade packages. The tier 2 supplier firms are grouped according to the type of transaction that the firm provides:

- Service: professional design and engineering services
- Product/service hire: contract labour hire firms, scaffolding, equipment hire, crane hire, etc.
- Design, fabrication and installation: typically, firms who fabricate entire building systems and subsequently integrate a number of other specialist suppliers
- Supply and installation: firms who purchase products and install and who purchase primarily products from other suppliers
- Product supply: firms who include delivery to site.

This is the manner in which the contractor described and grouped the suppliers. The map in Figure 6.12 provides a graphical representation of the firms and detail on the actual commodity that is being transacted in the relationships.

The maps are static and simply list the firm supplier by commodity type. However, what lies beneath the static graphic is emergence of a greater level of complexity of the organization and governance of the supply chain that is dependent upon commodity, firm and market characteristics. For example, across the project there are three structural steel fabricators at tier 1, including the following: the first providing design, supply and installation; the second supplying a product and installing and the third is simply fabricating off-site and supplying the structural steel. Why did this occur? Is it because of the project – size, design or construction technology? Is it because of the characteristics of the commodity? Is it because of the supplier industrial market(s)? Or is it because of the customer (contractor)?

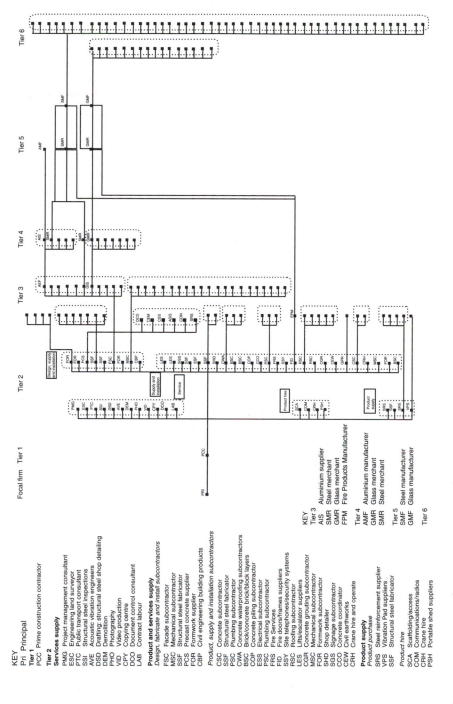

Figure 6.12 Structural organization map of Project 1 tiers 1–6.

There are numerous other scenarios which indicate interesting market, commodity and supplier characteristics. For example, the formwork firms were involved with designing a system to span the decking over the railway tunnels. The package was split into two because there were a number of physical building locations, not because of complexity of design. Whereas, there are two precast concrete suppliers and clearly one is supplying a commodity that involves a high degree of customization and the other is simply supplying a product that can be ordered and installed by another firm – the contract labour firm.

Firms often compete in the same market; however, some firms are more suited to a particular transaction than another. The formworkers' commodities are substitutable commodities; however, the precast concrete commodities are not. One of the precast concrete suppliers is more suited to a particular market than the other and yet they both competed for both tenders on this project. For the high degree of customization precast concrete commodity, a specialist market, the contractor particularly wanted to choose one firm over another. Therefore, the project tendering market sourcing strategy was a form of dual sourcing with the intention of one supplier enabling the contractor to bargain and renegotiate the tender price and lead time if required. This was a common sourcing *strategy* discovered throughout the fieldwork and examples are discussed in Chapters 7–9.

The structural organization map describes in more detail the nature of the commodity rather than simply the generic commodity type; it allows more detailed understanding on the supplier and the market.

When asked, the project manager for the contractor firm for Project 1 grouped the suppliers according to high, medium and low complexity levels and we can now match the generic groupings with the structural organization map detail. For example, in services the engineering land surveyor and project management consultant were considered to be of a high level of complexity, the copying centre was considered to be of a low level of complexity and a medium complex service was the structural drafting. The contractor grouped the service suppliers in terms of task complexity as follows:

- high: 5 suppliers
- medium: 3 suppliers
- low: 4 suppliers.

It is noted that within the low level task complexity category was the contract labour group and even though the task was often considered low level the contractor's project manager stated that the sourcing, co-ordination, management and payment of this firm was a high priority as they were critical to workflow on the site. Therefore, they were considered to be a highly significant supplier with a low level of task complexity.

Further to this categorization process, the contractor also grouped the product and service suppliers in terms of complexity. The suppliers were typically grouped as either high or medium task level complexity categories, which corresponded to either a design, fabricate and install subcontractor or a supply and install subcontractor and were grouped as follows:

- high: 6 suppliers
- medium: 21 suppliers.

Finally, even the product suppliers were classed into two groups: product purchase and product hire and again product purchase was considered to be low level task complexity and product hire to be within the medium level task complexity category. Their suppliers were grouped as follows:

- medium: 4 suppliers
- low: 3 suppliers.

Similar to product and service suppliers, even though there were three suppliers that were considered to be providing a product and it was a low level of task complexity, this does not necessarily equate to low significance. One of the firms was supplying the steel reinforcement which is highly significant as it was a large package in terms of contract value.

At the next tier of suppliers, tier 3, fourteen groups of suppliers were mapped, including those firms supplying to the following tier 2 suppliers:

- design, supply and install product and service suppliers: formwork supplier, precast concrete supplier, structural steel fabricator, façade subcontractor and mechanical subcontractor
- supply and install product suppliers: structural steel fabricator, concrete waterproofing subcontractors, precast concrete supplier, fire doors and frames suppliers, roofing subcontractor, concrete subcontractor and concrete grouting subcontractor
- product supplier: steel reinforcement supplier and vibration pad suppliers.

The number of suppliers at tier 3 who are supplying to each of these at tier 2 varies from one through to twenty-one suppliers for the mechanical services subcontractor. The more complex product and service suppliers, then the more likelihood that there are more suppliers being sourced. From the façade subcontractor the aluminium, glazing and steel supply chains were mapped and from the fire door frames supplier the glazing and steel supply chains were also mapped. The steel supply chain forms part of the structural steel supply chain as well as the fire door frame and the aluminium supply chain. The steel supply chain also forms part of the mechanical services supply chain.

These chains are discussed in more detail in Chapters 7–9, as well as the mechanical services, formwork and concrete supply chains. Steel supply and glazing supply is characterized by tier 5 merchants and/or distributors that are divisions of the manufacturers. However, in both situations the distributors compete with other merchants. There are fifty suppliers to the steel manufacturer at tier 6 and sixteen suppliers to the glass manufacturer.

At the manufacturing level tier 5, individual projects related to commodity supply chains are less identifiable, whereas at the merchant and distributor level projects can be identified, even though customers are considered as accounts over a twelve-month period.

A similar comparison between the three projects at tier 2 of task complexity results in Table 6.9.

The three contractors did not consider any product and service suppliers as providing a task of low-level complexity and they did not consider any product suppliers to be providing a high level of task complexity. The high level of task suppliers are skewed for Project 2 because of the procurement strategy. At tier 2 the contractor is managing quite similar profiles of suppliers in terms of variability of commodity type and complexity level.

If we take away the impact of the design and construct procurement strategy for Project 2 creating a number of more engineering and architectural consultants and design and construct trade packages, then it seems that the likelihood of major supply chains to be structurally organized in a certain manner is dependent upon the structural characteristics of the supplier markets and the task complexity. At no stage were the specific characteristics of the products a criterion for selection of suppliers. The customer's (contractor's) understanding of the supplier firms' ability to perform and the knowledge of dynamic nature of the market are contractor

Table 6.9 Projects 1, 2 and 3 tier 2: task complexity and commodity type

Task complexity	High	Medium	Low
Project 1			
Products	0	4	3
Services	5	3	4
Products and services	6	21	0
Project 2			
Products	0	6	4
Services	14	6	5
Products and services	12	20	0
Project 3			
Products	0	3	7
Services	5	4	4
Products and services	7	15	0

capabilities. The ability of the contractor to 'carve up' the works into packages unique to their knowledge of the capacity of the market to perform and their capacity to manage the project typically creates different supply chains from one project to the next. At the end of the day that is one of the key services you are paying for when you select one particular contractor over another. We would do well to remember this synthesizing skill – that is, their know-how and how they see and respond to the management of their suppliers. This is important when we choose the lowest price for a bid! Nowhere (yet) have I seen a tender criteria which allows for supplier co-ordination and or management to be rewarded. There are pseudo-criteria; for example, past performance, time and staff curriculum vitaes – nothing explicit, though.

A structural organization map for Project 2, 5 and Project 3 has been developed and is provided in Figures 6.13, 6.14 and 6.15 respectively.

It is interesting to compare Projects 1 and 2 as they have the same client and the same contractor; however, a different project procurement strategy. There are numerous more service suppliers for Project 2 since the project procurement strategy is design and construct. Similarly, there are more design, fabricate and install packages for Project 2 than Project 1. Regardless of the project procurement strategy, some suppliers provide a design, fabricate and installation package; for example, façade, mechanical, precast concrete and structural steel. Also, regardless of the procurement strategy, the contractor splits various packages in a similar manner, indicating a highly specialized market. For example, concrete is still split into various packages, including: concrete waterproofing subcontractor, concrete product supplier, formworker, steel reinforcement supplier, precast supplier, subcontract labour hire and concrete piling subcontractors.

A similar process is occurring for the steel on both projects. This specialization within markets and then integration by the contractor is occurring at tier 2; however, it only occurs for selected markets. At tier 3 this supplier integration is achieved by other subcontractors; for example, particularly the façade subcontractor and the mechanical services subcontractor. It might be anticipated that at tier 2 either the concrete and/or the steel markets could be aggregated for projects. There does not seem to be a reason for either of these markets to be differentiated.

However, when we look closely at both of these markets they are highly differentiated. The structural steel fabricators who provide a design, supply and install product and service on both projects primarily focus on the extremely complex and difficult projects. There are more structural steel fabricators competing in the structural steel design and supply markets. Both the firms who won the tenders on Projects 1 and 2 compete against each other regularly and are of a similar size, scope and capability – and the same can be said of the concrete market.

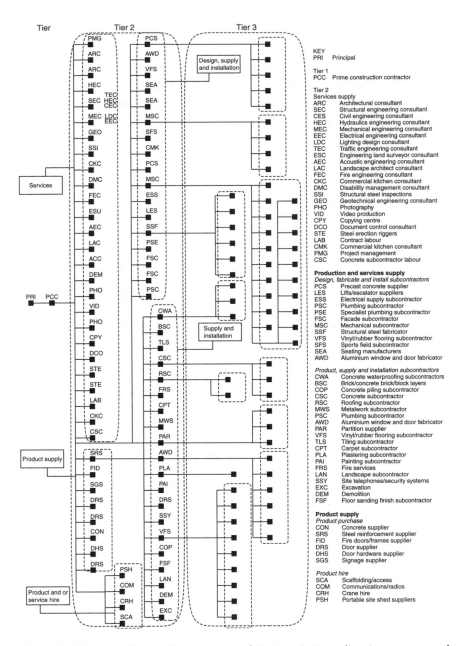

Figure 6.13 Structural organization map of Project 2 tiers client/contractor and subcontractor.

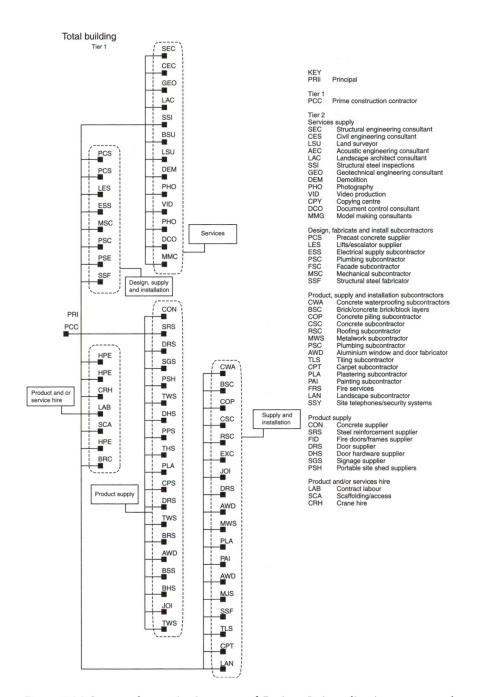

Figure 6.14 Structural organization map of Project 5 tiers client/contractor and subcontractor.

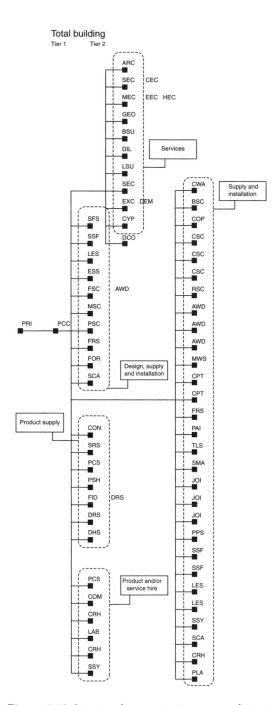

Figure 6.15 Structural organization map of Project 3 tiers client/contractor and subcontractor.

6.5 A final word

Chapter summary

1 Government clients have a major impact upon our projects. The typical government client is not as entrepreneurial as the two examples in this particular region.

2 Strategic Government Project Managers are able to develop a position on project feasibility in relation to the current political environment and the investment decision in relation to broad governmental and societal objectives. There is opportunity for those objectives to encompass improvements to the performance of the industry as well as project objectives.

3 A government client is typically not a single entity. To manage this diversity and to clarify 'so who is the real client?', Project Control Groups are typically formed. The key issue is that there is continuity between early government decision makers and this Project Control Group.

4 Entrepreneurial approaches can assist in the adoption of more innovative approaches to supply chain thinking; however, it appears that this opportunity has been missed in this region. There was little evidence of awareness, strategies, tools and techniques in relation to supply chain economics.

5 Strategic procurement and public sector probity and equity demands: Even though the units were quite progressive in that it is a corporate and commercial entity, the pressure of operating within a government environment and the associated probity and equity issues reflects in the conventional and non-commercial-oriented procurement philosophies. This is a dilemma for government agencies – how to develop strategic procurement relationships balanced with equity and probity demands of public office?

6 Managing control of complexity is identified as an important objective but is limited to the first tier of suppliers and the immediate contractual relationships with consultants and contractors and the complexity in relation to the project. Control does not have to be through one mechanism and can be multidimensional – a wider and more sophisticated way of managing at each tier may be possible with more innovative approaches to project procurement. There is a convergence of thinking to the individual project contract which does not acknowledge the historical place that these clients can play in industry and economic regional development. This wider perspective on 'control of complexity' indicated by the clients can be transferred to business systems and supply chain economics and across multiple projects and the supplier clusters; however, there needs to be strategies, tools and techniques in place to achieve this.

7 There are a range of clusters identified at tier 1, including project management, architectural design, construction management and project marketing. The project clusters are embedded in regions and multiplicity could be more explicit.

8 Explicit categorization of suppliers is possible and was evident by clients and contractors in relation to their suppliers. There are times when the categorization is implicit. Categorization is not one-dimensional – it relies on a range of attributes related to the supplier significance to the customer based upon quality (risk, complexity), volume (expenditure). However, there was little evidence of supply chain management.

9 There are various ways to categorize suppliers; for example, by way of their core business and its relationship to the project; core and non-core; product, service and product and service. Alternatively, we can categorize suppliers by the way they conduct their core business; for example, supplier co-ordination and variety. Supplier co-ordination is based upon the number of suppliers, supplier variety (i.e. types of suppliers they are co-ordinating) and the level of complexity of those suppliers. This is a first step to managing supply chains – gathering information about the environments of our own suppliers which can then inform the way in which we interact with them.

10 There is a propensity to think only about the relationship of customer–supplier. It is perceived as difficult to implement supply chain management if a customer does not have a direct relationship with suppliers' suppliers. The situation does not have to be so black and white – knowing how your suppliers interact with the next level is a good start. To reiterate previous comments: specialization, integration and supplier variety co-ordination – clearly firms do not just purchase products – they have a range of commodity types with a range of complexity levels. There are diverse levels of variability in supply and thus varying levels of capability in supplier co-ordination. If we wish to reach down another tier and co-ordinate, manage, integrate or develop at the next level – do we have the capabilities to do so? How can we connect to the next tier and enhance rather than meddle? What are our own systems for managing the different types of suppliers and then how does our system mesh with our key suppliers? The words 'integrate business systems in supply chain management' really are powerful, but they have serious implications. We need to look at our own vulnerabilities. Which of our own suppliers are strategically critical to us and in what way? In the following chapters we shall see examples of firms who do categorize their suppliers by a mix of the traditional way – by commodity type and then also by how that supplier relates to risk and expenditure. Although seemingly simple – it is surprising how many firms may not do this explicitly, it is quite interesting – our construction industry is

highly flexible and yet we see this as a negative where other industries are looking to flexibility in their manufacturing systems. Our industry looks at suppliers from a commodity and a process perspective – that is, suppliers don't just push products on to us, they provide a suite of attributes. How much do we know about those attributes? The procurement relationship object described in the previous chapter encapsulates those attributes. This chapter has provided the context of a multiplicity of not only projects, but firms and firm–firm relationships in the supply chain system.

7 Case study

Complex core commodity supply chain – façade chain cluster

What are the structural and behavioural characteristics of a complex core commodity supply chain?

7.0 Orientation

Box 7.1 Chapter orientation

WHAT: Chapter 7 provides a detailed investigation into the nature of the procurement relationships that are formed in relation to the façade supply chain, including discussion on firms, projects and market attributes – which underpin the nature of the sourcing strategies and approach and negotiation interactions between customer and supplier firms. Underlying questions which are considered include:

What are the sourcing strategies at various tiers to deliver a complex core commodity?

Are the chain paths different across the sector?

What are the connections between markets, firm types, commodity types and procurement relationships? What is the sequence of events which takes place during procurement?

WHY: We tend to make assumptions about the nature of the industry without a detailed understanding. Although we suspect that there is variety and complexity in the chains that make up the industry, we lack any real data and detailed descriptions.

WHO: A complex core commodity supply chain is a commodity which is core to the project contract. A complex commodity chain is one where the nexus of contracts to the project contract is complex in either technology or managerial complexity; that is, requiring unique, specialist and innovative design and/or construction solutions or a high level of integrative managerial capacity. These types of supply

---

chains can be characterized by innovative design, new materials, juxtapositions of new materials, numerous different types of suppliers, and a requirement to source and integrate suppliers not typically managed previously.

Core	Core and simple	Core and complex façade
Non-core	Non-core and simple	Non-core and complex
	Simple	Complex

Chapter orientation categorization of chains.

7.1 Introduction

This chapter presents the results from the interviews with various project managers, firm executives, production and procurement managers involved in the supply chains for commodities that are clustered around the façade specialist subcontractor on Project 1. The chapter is organized in five main sections:

- Firm attributes
- Industrial markets and commodities
- Supplier types
- Procurement relationships
- Supply channels.

After the clients and/or contractors were interviewed, then specific subcontractors were interviewed and various chains were followed in detail related to a commodity product. For the façade subcontractor this also involved interviewing the structural steel fabricator, merchants/distributors and manufacturer, the glazing supplier, the glazing distributor and manufacturer, the aluminium fabricator and the steel painter. The suppliers related to the steel supply chain are discussed in Chapter 8. During the interviews the range of projects that were involved in was also uncovered and these are listed:

- Façade for Project 1
- Structural steelwork fabrication for Projects 1, 2 and 4
- Glazing supplier for Project 1
- Glazing manufacturer for Projects 1–6
- Aluminium for Projects 1–5
- Steel distributor, merchants, manufacturers for Projects 1–6.

As was discovered, many of the commodities are found in all six projects; feeding into numerous other chains. The firms were all allocated a number and will largely be referred to by that number. A large database was developed for all the firms which was used for the statistical component to the study.

One of the most significant factors of Project 1 was the complex building form: angular in three dimensions. The outer skin of the building was a curtain wall façade with triangular infill panels that were either stone, zinc or glazing. The façade was not only complex in its 3-dimensional geometry, but also in detailing. As a result of this complexity, the façade was a significant element in relation to design, construction and project management and many firms were involved in some manner. As well as this, the façade subcontractors were required to source and integrate suppliers that they typically did not co-ordinate and manage on projects.

The structural organizational map in Figure 7.1 indicates four clusters of firms that the client procured: design, project, construction and marketing. In many ways, all clusters relate to the façade, even marketing in this case, since the façade was controversial. However, the design, project management and construction clusters are more directly related to the construction of the façade than the marketing cluster. Within these clusters, the architectural consultant (Arc), the quantity surveyor (QS), the building surveyor (BS) and the contracts lawyer (Law) are significant to the façade and, as the façade was particularly complex, a specialist façade consultant (Fac) firm supplied a service of specialist advice to the architectural consultant as well. The detailed design, construction and installation were provided by the façade subcontractor who was procured by the primary contractor.

Figure 7.1 indicates the chain of contracts from the façade subcontractor at tier 2 through to tier 4 suppliers. The façade subcontractor is firm 191. To deliver the façade to the client, the façade subcontractor (firm 191) contracted twelve suppliers.

These suppliers to the façade subcontractor are organized in three groups: common products, common services and unique products. These categorizations were offered by the façade subcontractor in the following manner:

- common products: aluminium extrusions, specialized glazing (seraphic), gaskets, silicon, pop rivets
- common services: façade cleaning, aluminium welding, façade site installation
- unique products: steel, stone, zinc.

This section describes selected firm attributes, their firm suppliers and the market environment for the following key firms in the chain: the façade subcontractor, aluminium extrusions fabricator, specialist glazing manufacturer and the glazing manufacturer. The steel fabricator was also part of this project supply chain; however, as noted previously, they are discussed in the next Chapter 8 as part of the case study on the steel supply chain.

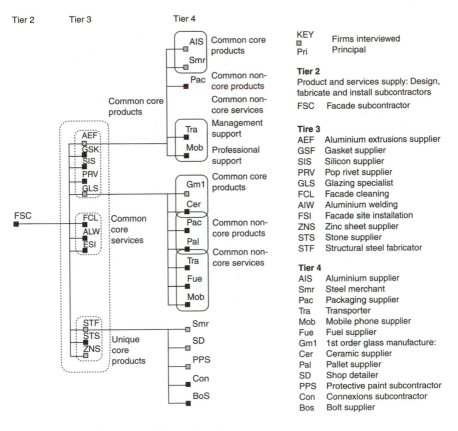

Figure 7.1 Façade structural organization map.

7.2 Firm details

The following matrix in Table 7.1 summarizes attributes for the following firms: the façade subcontractor, the aluminium extrusions fabricator, the specialist glazing supplier and the glazing manufacturers. The discussion focuses on describing the characteristics of the façade subcontractor, aluminium fabricator and the glazing suppliers firms and then the industrial market and commodity types to provide a context for the firm–firm procurement relationships developed. The procurement relationship begins to develop the behavioural model of the supply chain as I describe the events and interactions which took place.

The façade subcontractor firm 191 employs over 1,000 people world-wide with headquarters in Europe and a PacificAsian regional operation. It is a multinational and the particular division in Australia employed approximately 50–60 people. There were two main divisions in Australia,

Table 7.1 Matrix of supplier details for Project 1 facade

Firm type and project	Structure and scope, size – employees and turnover	Commodities and competitors	Competitive advantage
191 Specialist façade subcontractor Project 1	12 manufacturing plants located in Asia, Australia, Europe and United States. Within Australia there are two divisions, both in capital cities. Multinational, $6–25m t/o. >1000 employees and approximately 60 employees in this division.	Designs, manufactures and tests prototypes of curtain walls and then manufactures offsite and installs onsite. two types of markets: 1 standard curtain wall: five competitors and highly competitive 2 complex customized façades: one competitor and is a niche market Competitive advantage: international scope. Integrated firm. Innovative.	
532 Aluminium extrusion supplier Project 1	2 divisions in capital cities with joint ventures in Pacific Asia; international, 4,500 employees and >$250m t/o.	Designs and fabricates metal and plastic extrusions three main divisions: copper, brass and aluminium extrusions. Plastics was an acquisition and diversification. Two types of markets for the construction industry: 1 project specials 2 standard extrusions: high volume work supplied to large national manufacturers (e.g. window mfrs) four main competitors including a general import competitor – Refer to table on aluminium extrusions fabricator for specific discussion on the commodities and competitors.	The competitive threat is high. *It is a hugely competitive market that we operate in. To give you an idea from the dollars and cents just quoting some figures. An average price in Melbourne for an extrusion at the moment form the mill might be say $4.80/kilo. Chinese can get it in for say $3.80/kilo. No one – firm 146, Capral, Boral or US can do this at the moment.* The importers have to have an agent in Australia. These companies tend to be Chinese based. However, the

majority of these extrusions go towards the standard high volume and the automobile industry markets and not in the project work. For Australian firms in this market the competitive advantage is lead time and service. The project is the same generally across the board. The ability to respond to the unique project, customer requirements and/or customer changes is particularly important for this market.

In the project type work Firm 191 are competing against Firm 146. When 191 win the project we are competing against Capral or imports. Imports for project work is very risky. I don't think Rocco would be too comfortable having half his load stuck on a ship. It would be a huge risk. In project work there is the ongoing 'oh and we need this'. And they have forgotten about it. It is the responsiveness that is important. I think physical

(Table 7.1 continued)

Table 7.1 Continued

Firm type and project	Structure and scope, size – employees and turnover	Commodities and competitors	Competitive advantage
			closeness is important, for example, the project is in Melbourne and we are in Melbourne, also if the project was in Sydney or vice versa that sort of closeness is ok. Because you always have an architect in inverted commas. You know no architect ever wants the same shape even though it is exactly the same in our opinion, they will want a little tweak of it. Something different ... to the human eye there is no difference but to us as manufacturers there is a huge difference. And you need to be able to respond to that quite quickly or they will change their mind half way through and they do that as well.
472 (170/640) Specialist glazing firm	Division operating independently within a multinational. 3,000 employees nationally; 125 in division. $26–75m t/o.	Processing of raw glass for production of seraphic glass. Four markets: 1 appliance glass, that is, kitchen appliances, cooking appliances, which is the Australian market 2 export market appliance glass	

3 architectural products (export and local)
4 other products, primarily for Japanese market: automobile industry,
   solar panels, street lighting – various products
In the architectural market, there are three primary products: cladding,
internal partitions and internal/external signage
Highly specialized product; aesthetic quality.

473 (170/64)
Raw glass
processor and
merchant

Firm 472 have a large share of the market, although it is noticeable that they
do not produce high-performance glazing products and that firm 146 is the
only supplier of these types of products in Australia.
*We are in a commercial market to a certain extent but not the level that we
used to be in. We don't have a glass which is suitable for the high-rise
building, for example, the Rialto or any of those in Collins St. We used to
produce a glass for that type of building that is, it would be a coated glass
to get that semi-reflective surface. Not very high but semi which of course
need now to keep your air conditioning loads down. The glasses that we
would produce these days and this is as a result of the building bust that
came about in the late 80s where we sold our coater. The coater puts the
reflective surface on. We sold it to firm 146 up in Qld our customer up
there. We used to sell it under the Suncor name that is the brand name for
it. It went on numerous buildings around Australia. We sold that and
concentrated more upon the low-rise buildings be it in suburban area or
otherwise, apartment blocks with what I would call a medium performance
glass using tinted glasses and those sorts of things. As distinct from what
I would call high-performance glass. In summary we are in the commercial
market for low-rise buildings, factories to a certain extent, apartment
blocks to quite a big extent but not in the 30/40 high-rise building. That
tends to now to be serviced by either firm 146 but mainly by the US suppliers.
By that I mean people like Virocon is a typical example where they have
large production runs of the appropriate product and therefore can get
economies of scale.*

(Table 7.1 continued)

Table 7.1 Continued

Firm type and project	Structure and scope, size – employees and turnover	Commodities and competitors	Competitive advantage
		When firm 767 (external to firm 170) and firm 473 (division of firm 170 at the state level) act in competition at the state level they are treated the same as customers. *From a national viewpoint we overcome that by ensuring that our Sales Centres buy at no better but they buy at the best available price within that particular state. In other words, the prices that I sell to my major independent, for example, in Victoria that is firm 767, our firm 473 Sales Centre would buy at that same price. I am not treating one more favourably than the other. This seems to work out well.*	
170 (640) Raw glass manufacturer	Multinational, 3,000 employees nationally and >30,000 overall. >$250m t/o. *To manufacture glass this firm has three building product plants, two interstate (Sydney) and one located in this state. (Melbourne)Another plant does laminating in Geelong (sw of Melbourne) for the automotive*	Manufacturer of raw glass. Firm 170 has a large share of the market but they do have competitors depending upon the product. If you take raw float glass, which is the glass that generally makes up most of the residential windows, they have about 75 to 80% market share. However, that product could go down to be cut to size for windows or go through another process, that is, laminated, etc., or it could go to another path. At the level of laminated glass firm 170 tend to have sales of laminated glass they tend to hover around a bit less than 50% of market share. In terms of mirror glass they tend to hover around a bit less than 45% of market share. In terms of toughened product it is probably about 55 or 60%.	

There is quite a difference in market share for firm 170 between the products. Apart from the firms that are their competitors in Australia, the other major threat is from overseas supply and this tends to concern the firm more than the other local processors as they tend to supply the majority of the raw glass at the beginning of the chain. The main concern with imported glass is the low price. However, firm 170 claim that their competitive advantage is a principle that they have described as 'mother warehousing', which is simply that they have large warehouses which means that the customer can buy from them without having to wait for long lead times. Lead times are at least 4–6 weeks shipping times. In reality the lead time is actually a lot longer than that because there is the lead time for the manufacturer as well. Firm 170's major competitor is from Asian imports and to a lesser extent European glass. The glazing market in Australia is characterized by firms competing and also supplying to each other, referred to as the customer/competitor dilemma.

Firm 146 and the others are purchasing Pilks glass and then processing it to form another product that Pilks actually produce as well. We have this customer competitor dilemma that people like myself would have to deal with. However, somebody at Mike Gleeson's level don't have to. Mike in Victoria sees 767 as a competitor to him, whereas to me it is a customer. We have to walk this very fine line of selling to and having a very close partnership with them. But at the same time they are competing with the people that buy from them. It is a tricky diplomatic game to say the least.

industry and one of the plants does laminating; one does toughening and one does patterned glass. There is an encapsulation plant (which reprocesses) and a small plant interstate (South Australia) for automotive. The patterned glass plant in Sydney does a number of distribution centres in each state.

one in Melbourne and one in Sydney. Firm 191 designs a curtain wall, manufactures off-site, tests the prototype and installs on-site. They have two markets they operate in: standard curtain wall and niche market of complex curtain walls. This project falls into the latter category. The firm's competitive advantage is the capacity to deliver creative solutions and they have an international reputation for their innovative design and construction capability.

The aluminium extrusions fabricator, firm 532, is international in scope; however, it primarily operates within the state. The firm employs over 4,500 people. This division is concerned with aluminium metal products and the organization is primarily involved with aluminium and copper products.

Firm 472 is a specialist glazing supplier and is within the organizational structure of firm 473 and firm 170. However, the three firms operate as separate divisions. Firm 472 was acquired by firm 170 approximately 15 years ago. Firm 473 is the state sales division and firm 170 is the glass manufacturing national sales division. They are part of a multinational that employs over 30,000 people worldwide with headquarters in Europe and has PacificAsian regional operations.

7.3   Commodities and industrial market details

Firm 191's market for Project 1 was the niche market for complex curtain walls. The subcontractor competes in two markets: standard and complex curtain walls. The division is a medium-sized enterprise embedded within a large multinational. There is one other major competitor in Australia for the complex curtain wall. The façade subcontractor sources from twelve suppliers and the strategies and manner in which the firm groups the suppliers is discussed later in this section. They were indicated graphically in Figure 7.1.

Box 7.2 Degree of predictability

It appears that there are strong historical ties between various firms in supply chains which relate the procurement relationships that are formed to the nature of the industrial market to the project market to the commodity type. Once a decision is made at the subcontractor level of a particular supplier, then there is a small number of typical supplier procurement paths that will follow, based upon the interdependency between these four objects; namely, commodity type, industrial market, project market and procurement relationship. This does not mean that there is only one path; more that the choice of supplier is not entirely random and that there will be a likelihood of certain suppliers being chosen; that is, a degree of predictability.

The strategic direction of firms, firm governance and decisions regarding firm boundaries impact upon the nature of the market and eventually the types of procurement relationships that form. Such long-term characteristics of markets and movements may evolve over many years and be structurally embedded within the sector and the supply chains. The historical context assists in explaining the eventual structural organization of the supply chain.

The aluminium extrusions fabricator, firm 532, has four other major competitors, including 'imports' as a general category. The Table 7.2 indicates the firms and the product types in the Australian market and was developed by the General Manager of the aluminium extrusions fabricator during the interview. Similar to how the façade subcontractor competes in two types of markets based upon level of complexity, there are five aluminium extrusions markets based upon degree of complexity. The façade subcontractor competes in two markets: standard and specialized. The aluminium extrusion supplier competes in the construction industry in two markets: project specials and standard extrusions. The five aluminium extrusion markets include the following: standard high volume work (typically residential window sections and standard commercial sections for shopfronts/light industrial/

Table 7.2 Key suppliers and customers market leaders by commodity type allied to the aluminium extrusions

Firm	Type by production phase		Type by customer/sector		
	Commodity type 1 Distribution	Commodity type 2 Extrusions	Commodity type 1 General market	Commodity type 2 Standard high volume work	Commodity type 3 Project work
Firm 146 (customer and supplier)	•	•	•	•	•
Firm 191 (customer)			•	•	•
Firm 765 (customer)			•	•	•
Firm 766 (customer and supplier)	•	•	•	•	•
Firm 141 (customer)			•	•	•
Firm 194 (supplier and customer)	•	•	•	•	•
Firm 532 (supplier)		•		•	•

offices/showers, etc.), project work (customized sections), general market (supply directly to contractors), distribution (standard sections mass production to distribution centres) and extrusions (fabricate special extrusions to various customer types). This table does not include all other firms that purchase extrusions; however, it includes the major players in the market or what is commonly referred to as market leaders.

The table includes two of Firm 532's major customers, including a customer in project work and the other in standard high-volume work – the previous upstream façade/curtain wall subcontractor supplying to Project 1 (Firm 191) and another national window manufacturer (Firm 765). Firm 765 also competes against Firm 194 and they are both national manufacturers and suppliers of standard windows. They are market leaders in the supply of residential windows. Firm 194 is also a market leader in the supply of commercial standardized aluminium window sections and also a market leader in the supply of numerous other construction materials and components as well, such as concrete, quarry and timber products. Firm 146 and Firm 141 are the two firms that compete in the specialist curtain wall glazing market (Firm 191 is the façade subcontractor for Project 1). Firm 141 is closely allied to Firm 146. Firm 532 does not supply to Firm 146 and, in fact, Firm 146 and Firm 766 (another commercial window section manufacturer and supplier) are their major competitors. The table was developed based upon descriptions of the market by Firm 532, Firm 141, Firm 191 and Firm 003 (the primary contractor for Projects 1, 2 and 3). The descriptions were provided by company documents and interviews. Each interviewee confirmed the data in the table.

As previously discussed, the commodity type and market characteristics impact upon the procurement relationships that are formed in the supply chain. The previous discussion gave examples for the aluminium fabrication market in Australia. The customer and supplier firms were all easily identifiable as separate firms and yet there is an interdependent nature between firms based upon procurement relationships; that is, some firms always align themselves with another firm. Therefore, if a choice of one firm is taken then there is another firm that will most likely be sourced. As noted, Firms 146 and 141 were closely allied. Firm 141 was historically a division of Firm 146 until approximately a decade earlier. Such alliances existed in the glazing supply chain and the steel supply chain as well.

The competitive advantage for the aluminium fabricator is price and lead times; followed by the service they offer – particularly on complex project work. Price for all metal products is based upon the London Metals Exchange and fluctuates daily; therefore, the ability to hold prices for a period of time on behalf of their customers is a significant advantage. This is essentially risk management.

The fabricator, Firm 532, was historically a part of the national aluminium manufacturer and so they too have close ties with the only Australian aluminium manufacturer. The aluminium market discussion would be enhanced by an understanding of the aluminium manufacturer and its supply channels; however, this firm was undergoing major upheavals in terms of mergers and acquisitions at the time of the study and no-one was unavailable for interviews. The following discussion describes the nature of the glazing market in Australia and the various commodities supplied in a similar manner to the previous aluminium extrusions market discussion.

7.3.1 Glazing commodities and industrial markets

Firm 472, specialist glazing supplier, sources from the glazing distributor, Firm 473 (state sales division) of the glazing manufacturer, Firm 170. Each of these are considered as separate firms as the interviews clearly indicated that, with regard to procurement and pricing, they are treated no differently to any other firm purchasing in the marketplace.

Firm 170, the glazing manufacturer, produces four glass products: annealed, laminated, toughened and mirror. They are the only suppliers of the raw glass material (annealled glass) in Australia. However, three other firms laminate glass, including: Firm 146 (previously discussed in the section on aluminium extrusion suppliers and a competitor to the façade subcontractor), and two others – let us call them Firm 767 and Firm 777. There are also a large number of importing agents who supply laminated glass. The following Table 7.3 summarizes the types of products that are supplied by each player in the glazing market and their role; that is, whether or not they are a window fabricator, a merchant or involved in 2nd order processing as well.

Firm 473 has a large share of the market, although it is noticeable that they do not produce high-performance glazing products. Firm 146 is the only supplier of these types of products in Australia.

Standard high-volume work typically includes both the residential market and the commercial market. In the commercial market it may include products such as glazing for retail shopfronts, standardized curtain walls, glazed balustrades, internal/external screens and shower screens. In the residential market it includes primarily standard aluminium window sections and shower screens. The project work typically includes atriums and curtain walls.

Glazing product supply on projects in Australia generally takes one of three major procurement chain routes. The first is the most direct route, which takes its roots from Firm 473 or the National Sales Division of the only glass manufacturer (Firm 170) to the major nationally operating window residential fabricators/glaziers (of which there are three). The second is more circuitous and involves smaller firms that generally operate at the state level; that is, the State Sales Division (473) purchases unprocessed glass from the National Sales Division of the glazing manufacturer (170) and either simply repackages the product in smaller volumes or alternatively provides secondary processing and then supplies to glaziers/fabricators.

Table 7.3 Key glazier suppliers market by commodity/production

Firm	Commodity					Production phase		Sector (direct supply)	
	Anneal. glass	Lam. glass	Tough. glass	Mirror glass	High perf. glass	2nd order processor	Merchant	Stand. high volume work	Project work
170	•								
473		•	•	•					
146		•	•		•	•	•	•	•
767		•	•			•	•		
777		•				•	•		
779			•						
780			•						
Imports China	•	•	•	•	•	•	•		
781						•	•	•	
782								•	
778								•	
191								•	•
765								•	•

Finally, the third route is where glaziers purchase glass from an importing agent who is acting for distributors in another country, and currently this is most typically China. Having now described the major paths, it must be acknowledged that there were many permutations in this chain, as noted by the Australasian Supply Manager for the glass manufacturer, Firm 170.

Box 7.3 Intricate supply chains

It is very competitive in the marketplace. It is also very intricate as well (Australasian Supply Manager, Glazing Manufacturer).

For example, the glazing for Project 1 is somewhat different again – the specialist glazing supplier, supplying seraphic glass (472), purchases from the National Sales Division (170). However, there are only three customers that the seraphic glazing supplier (472) will allow to purchase their product direct from them, including Firm 191, the façade subcontractor. All other curtain wall or window manufacturers must purchase through the States Sales Centres (for example, 473) – which adds another link in the chain and in reality adds to the price.

Box 7.4 Countervailing market power

They would do it two ways. Usually, Firm 191 comes here direct. There are about two or three companies that we deal with direct and one is Firm 191, Firm 141 glazing and a couple of others. The reason we do that is because they can not be bothered going through Firm 473 and they would rather deal direct. If they go through Firm 473 they will get priced out anyway. It is just adding another commission on it. We have picked two or three or four of those types of people that we will deal with direct (Australasian Supply Manager, Glazing Manufacturer).

Similarly, the National Sales Division, Firm 170, supplies glass to the major residential window fabricators of which there are three and some state-based fabricators. Coupled with this is the residential 'hack and glaze' market which is window replacement. They also deal with many small window fabricators or processors as well. The other major category is the customer that is both a residential and commercial window fabricator. The common element amongst these customers is not the size of the firm but whether or not they have the capability to process the glass as it comes to them in a raw state. There are processed products that this firm supplies as well, which in turn means that they supply to their competitors at times.

Contrary to expectations, the highly vertically integrated firm in this situation does not necessarily equate to a lack of external competition. As noted previously in the matrix that described the attributes of the firms and their commodities and competitors, at the next tier the division Firm 473, within Firm 170, competes on equal terms with the firms outside; therefore, each tier operates purely on a transactional basis from the supplier side. This appears to work out well for the supplier – however, the 'firm' within the firm, 473, who is the customer upstream, does not seem to have the same market independence.

Box 7.5 Thou shalt not stray

A major independent customer is one that is external to Firm 170 and they can purchase from us or one of four competitors, whereas our Sales Centres being part of Firm 473 must purchase from us; they do not have that choice.

'Thou shalt not stray!'

(State Manager, Glazing distributor)

This customer–supplier relationship between the 1st order (glazing manufacturer Firm 170) and 2nd order manufacturers (glazing distributor Firm 473 and Firm 472, processor) is discussed in more detail now, followed by a description of the downstream suppliers to the 1st order manufacturer.

Firm 170 have three major types of construction industry customers: the residential window fabricators, the 'hack and glaze people' and the commercial sector. The 'hack and glaze' are window replacement.

Box 7.6 Market categories

Although seemingly straightforward, the market is further categorized. In terms of building glass, we either have the window fabricators on the residential side, the hack and glaze people, and the commercial side is the Firm 767, Firm 146 and Chevron. This is in terms of the building area, but you need to look a bit further to split the market up in terms of what is the main function of that customer:

Is he a merchant?
Is he a processor of glass? (i.e. making toughened, laminated, etc.)
Is he simply there supplying cut-to-size glass for smaller people to glaze with?

(Australasian Supply Manager, National Sales Division)

Within these three categories the traditional merchant would buy glass from a manufacturer (either this firm or from an importing agency). They would typically purchase 20 tonne loads and then would on-sell that in block form. The next extension to that role would be to cut that product to size to whatever someone would require and sell it as a cut piece of glass for a product. The merchant has largely evolved into value adding glass, that is, he became a processor – either furnacing the glass, printing the glass, putting in a laminating line to supply laminated glass. That is, adding value to the glass he buys from this firm so that he can in turn get a better margin when the firm on-sells to their customer.

Box 7.7 Market competition – changing structure of supply chain

That merchant-processor is tending to meld one into the other now. In other words, they are not making much money on merchandizing. They find that they have to get into value added to stay alive principally. There are still merchants hanging around as distinct from processors. They buy from us in 20 tonne loads. A typical example is Davis Glass, who is a merchant.

(Australasian Supply Manager,
National Sales Division)

For the commercial sector projects there tends to be two main supply chain routes, one through a processor and then the glazier or direct to the glazier. If it went to the glazier, then the chain would include the State sales division. If it went through the processor, then it would not include the state sales division. He continued as follows:

Box 7.8 Three main supply channels – diverse supply chains

Now with our commercial projects. The processor, bearing in mind that for most projects in the commercial project they need to have some sort of value added glass. So be it either laminated or toughened, we would tend to be supplying either the middle man, that is, the processor, or we would supply direct to the glazier. Just depends upon the contract. We would quote for a contract to supply glass for a project XYZ Hotel and it would be quoting the glazier direct via the Sales Centre if the sale went via that route; if it was a Firm 146 or a Firm 767 supplying, they would be buying the glass from us and they would then be processing it. So there could be two paths.

If it went to a subcontractor glazier it would go through our Sales Centre and in turn the customer of the Sales Centre may well be a merchant who in turn may well supply into that same building or compete with our own Sales Centre, because they would be supplying another glazier or another glazier would be competing for that job.

You have all these subsets that are occurring in the marketplace. It is very competitive in the marketplace. It is also very intricate as well.

(Australasian Supply Manager,
National Sales Division)

Another product market that has emerged for the merchant is the fabricating of double-glazed units. A merchant customer of Firm 170 has recently entered this market. This firm merchandizes and also processes. Then there are other firms that have developed various stages of production.

Box 7.9 Alternate supply chain paths – value adding

Then you might have, of course, a pure processor like Mowen Glass here in Victoria who set themselves up in the window game... making windows and then they value added further by putting in a double-glazing lite. So they now produce double-glazed units and in turn put them into windows and out into the marketplace. Of course, they sell double-glazed units to Boral, Stegbar or whatever. Ever so recently they have put in a toughening furnace and again because toughened products are required in their windows and it is this continual value adding process.

(Australasian Supply Manager,
Glazing Manufacturer)

The following diagram as in Figure 7.2 is for Firm 170, the glass manufacturer, and it summarizes the primary types of project, firm–firm procurement relationships and industrial markets based upon customer types. It is presented using representation graphics from object oriented modellings, in particular using the unified modelling language.

It is worthwhile also describing Firm 146. This firm is a major player in the glazing market in terms of processing, fabricating and installation. It is interesting that when the State Manager explained Firm 146, the description melded descriptions of competitor firms and markets.

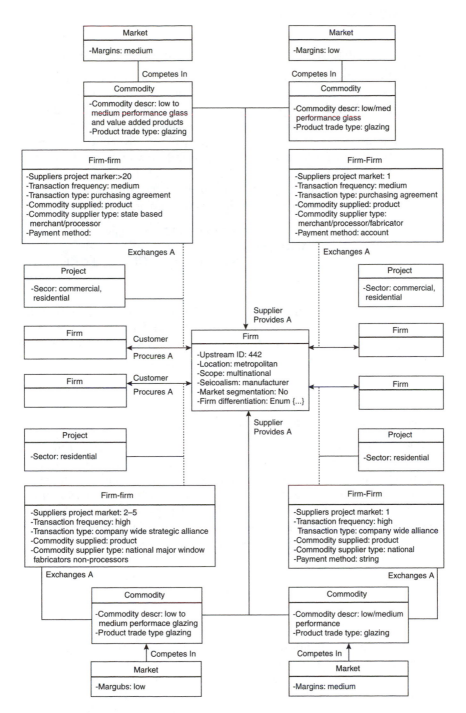

Figure 7.2 Glass manufacturer Firm 170 markets by customer.

> ### Box 7.10 Integrated aluminium and glazing supply chains
>
> Firm 146 are a many many faceted company. They are a huge company. They are a merchant, they are a processor, they make toughened, printed, laminated, double-glazed commercial and residential units glass, they also are into glazing both low-rise and high-rise, they are also into aluminium extrusions. They have grown and diversified tremendously.But they are a bit unique in the glass industry. Firm 767 is going down that track too in as much as they do most of that, they merchandize, they process, laminate, print, toughen, glaze; the only thing that they haven't gotten into which Firm 146 has is the aluminium side of things.
>
> In other words, you could get Firm 146 to do the complete outside of your building, the complete façade.
>
> (State Manager, Glazing distributor)

Firm 146 competes with Firm 191 for the highly complex façades for projects and although, for Project 1, Firm 191 purchased from Firm 170, this is not the usual case as the following quote from the Chief Executive Officer of the specialist glass supplier, Firm 472, indicates:

> ### Box 7.11 Hidden connected supply chain paths
>
> Now we used to deal with Firm 191 who do the same sort of things; we used to deal with them when we produced the Suncor type product, the high performance type product. But we don't any more. They would tend to import and perhaps buy a bit from Firm 146 and Firm 767 in the way of laminate products. The great majority of stuff they import. If they wanted glass in Australia they would not tend to come to us, they would go to merchants or processors.
>
> (Chief Executive Officer, specialist
> glazing, manufacturer)

There are four major national customer accounts for Firm 170, three of whom are residential window fabricators. They require processing and are serviced by the State-based processing/distribution centres; however, the account is managed by the National Division. Likewise the national glass replacement firm is dealt with by the National Sales Division; however, since they have their own processing facilities they are supplied with the product as well – that is, the raw glass product.

Box 7.12 National to national customer to supplier 'arrangements' based upon purchasing volume

We deal with the major window fabricators, people like Boral, Stegbar, James Hardie, plus various state-based fabricators. So we would sell them the glass and they in turn would cut it up or buy it from us in cut size and put a bit of aluminium around it and there is your window.

It would be expected that national agreements would be in place for these accounts based upon volume and pricing arrangements; however, there are no national agreements between this manufacturer and the four major national residential window fabricators.

Box 7.13 Non formal agreements

There is no point in having a tied-in contract with the likes of Stegbar, etc., because a contract is only as good as what either party wants. If either party is dissatisfied whether they will let the other one know and request a change and come to another agreement. They will only buy from you if they are happy and so frankly I don't see any point in national agreements. When they do purchase they purchase from all our Sales Centres all around Australia and they put in daily orders and they would be delivered on a daily basis to their appropriate sites. I personally deal on the national level and deal with customers who principally buy truck loads of glass from the glass plants; in other words, they buy 20 tonne loads of glass. In other words, they are the big guys. I then deal with the four major accounts that I mentioned before; that is, Boral, Hardies, Stegbar and O'Briens. We deal with them on a national basis. But they are actually serviced on a day-to-day basis by the State Sales Centres.

Pricing and margins are an interesting issue in this market and these tend to rely upon world supply trends, even though it appears that Firm 170 has a monopoly. The monopoly is only as good as the prices that the firm provides. The existing upstream customer relationships were somewhat surprising given the multinational scope of the firm and for this reason the entire quote (albeit somewhat long) has been extracted from the transcripts and provided in the following text.

Box 7.14 Gentleman (or gentlewoman) agreements

We don't tend to work on 12 month agreements. We tend to work more on gentleman agreements because we have dealt with them for so long. At times our relationships are rather rocky, like they are at present; but that is because we are trying to put price increases through.

No, we don't have official contracts with them in terms of 'though shalt purchase X tonnes at Y dollars'. They obviously have price schedules from us and they to date well...we have been very very lucky they have purchased around about 95 to 100% of their volume from us. We are very lucky in that we are in that unique situation that we have some very loyal customers.

Prior to November last year we tended to work on decreases; we could not even spell the word increase, mainly because of the world glass glut, particularly in Asia. This came about because of the Asian crisis of some years back. Glass has been freely available; it tends to slush down to Australia and reasonably good prices down here and therefore we were always fighting uphill to get increases.

Since about the beginning of this year glass has become short on the world scene. Asian economies have been improving. America and Europe have been very very buoyant and have tended to use glass in their own confines. As a consequence, glass has become shorter; we then saw an opportunity to follow world trends by putting prices up. From November last year we started to have increases on our customers. So we had a very small increase in November; we then had the next increase on 1st April and we are about to put through another increase on 1st August. That is very very unusual to have those sorts of increases and of course that is the reason why some of our customers are upset with us. We are basically following world trends.

Firm 170 supplies through to all six projects. Firm 170 either supplies a product or a service in the relationship. The product is a common core product and the service is a common core service. The service is a management service that is provided to their four major national accounts (Firms 781, 778, 765 and 194). As stated earlier, these major window fabricators do not process and order daily from each state division.

The following Figure 7.3 describes the suppliers to the glass manufacturer. Table 7.4 describes these suppliers in a little more detail. The sequence of events for creating procurement relationships is considered in detail in Section 7.5. The procurement relationships are mapped in Figure 7.4.

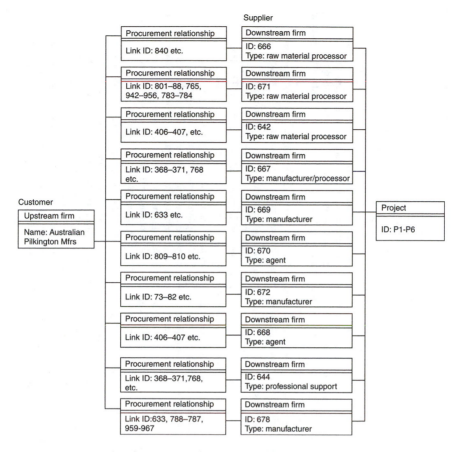

Figure 7.3 Firm 170 glazing manufacturer suppliers.

7.4   Supplier types

The façade subcontractor, the aluminium extrusion fabricator, the specialist glazing supplier and the glass manufacturer all require suppliers to enable them to fulfil their contracts with their customers. When questioned about who their suppliers were, each of the respondents described their suppliers in groups.

Table 7.4 summarizes all the firms interviewed and discussed in this section. The sourcing strategies are discussed in detail in the Procurement Relationships Section 7.5 of this chapter.

The suppliers to firms have been described and in many cases it is evident that the customer firm explicitly or implicitly categorizes the suppliers. The categorization often relies on a number of considerations. The considerations typically rely upon the *volume* of the commodity which is purchased and

Table 7.4 Supplier types categorizations

Customer firm	Supplier types
Supplier types to facade subcontractor: Firm 191	To deliver the product to their client the subcontractor tenders for the supply of various products and services. There were 12 suppliers to the facade subcontractor of which there was a variety of commodities and firm types. Firm 191 conceptually grouped their suppliers by products and services suppliers by the following categorization: common products, common services and unique products. The term 'common' referred to the fact that even though each project may have different specific procurement requirements, that commodity is commonly used by the subcontractor and therefore the firm suppliers were largely well known to the subcontractor. For example, common products in relation to volume and specification would be glazing, aluminium and silicon. Even though the glazing for this project was a specialized type the subcontractor had often used this type of glazing and therefore was not classed by the subcontractor as a unique or specialized product. Firm 191 identified that they needed suppliers for the following products/services:
	'*Common products: aluminium extrusions, specialized glazing: seraphic, gaskets, silicon, pop rivets; Common services: facade cleaning, aluminium welding, facade site installation; Unique products: zinc sheet, stone, steel*'
Supplier types to aluminium extrusion fabricator: Firm 532	To produce aluminium extrusions Firm 532 has two main suppliers of products: steel and aluminium. Firm 532 maintains that aluminium can be purchased from anywhere in the world. Firm 532 purchases their aluminium direct from an Australian manufacturer and the steel from a local steel merchant. The firm also has has three other major suppliers: packaging, logistics and mobile phones.
Supplier types to specialist glazing manufacturer: Firm 472	There are two main types of commodity suppliers; the core products and the non-core products. The core products include glass and ceramic paint. The non-core product suppliers include such items as specialist machinery, packaging and fuel. There is also a third type of commodity supplier of which there were numerous suppliers and this category is services which similar to most other firms includes suppliers of logistics, mobile phones, stationery etc. Logistics (distribution of finished product) is a critical service particularly for export of products and is considered a managerial service and is non-core. The mobile phones and stationery are considered professional support services. These are classed as non-core as they are provided across many commodities.
Supplier types to glass manufacturer: Firm 170	There are 16 primary supplier commodity categories to Firm 170. Of this 16 there are products, product and services and services. There are 11 core products including: sand, dolomite, limestone, soda ash, sodium sulphate, silver nitrate, sodium carbonate, sodium nitrate, cobalt oxide, selenium and plastic. There is also specialist equipment suppliers. Of these there are 6 common ingredients and 5 additives – or rather 6 common and 5 special or unique products. The supply of two products and services are significant to Firm 170 and they include electricity and telecommunications. The primary services are considered to be logistics and shipping. There are many more other products that are supplied for example, gloves, stationery etc. The following figure indicates 10 of the major suppliers, identifies a number of procurement relationships associated with the various projects and also the type of supplier.

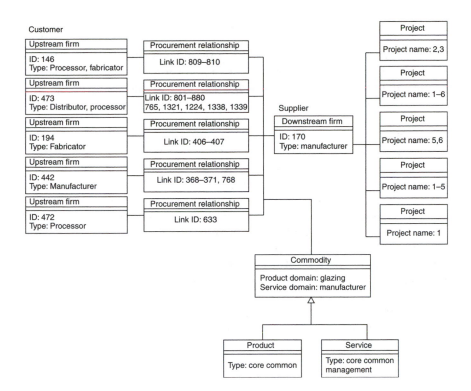

Figure 7.4 Glazing manufacturer procurement relationships.

whether or not it is a commodity that is *commonly* purchased by the firm. The other major consideration is the *complexity* of the commodity. These then seem to create the following supplier groups which rely upon the *significance* of the *commodity*:

* common core commodity
* common non-core commodity
* unique core commodity
* unique non-core commodity.

Within each of these categories firms often describe their suppliers in terms of whether they are supplying a service, a product or a product and service, but this is secondary to the consideration of commodity significance.

The results indicate that the significance of the commodity to the supplier is not the only consideration that impacts upon how firms source, what they source and how they categorize their suppliers. The other major consideration is their own bargaining position in relation to their supplier.

The bargaining position is dependent upon where they are located within their own market and their relationship to their competitors. Indicators of their position are the firms' scope and their turnover in relation to their competitors and to their suppliers. The bargaining position reflects whether or not the customer or the supplier has the power in the relationship. This countervailing power is an underlying structural characteristic of the supply chain and impacts upon the way a firm categorizes and approaches their suppliers.

The significance of the commodity and the countervailing power impacts upon the firms' governance structure and strategies related to outsourcing, which then governs how the firms approach the procurement of suppliers. The structural organization of the chains is also ultimately affected. These decisions are also not static – as the industrial structure of markets changes over time. The firms within the markets alter their policies on whether to outsource or whether to supply commodities in-house. This section has provided a detailed description of structural characteristics of the supply chains; in particular, the nature of the industrial market structure and firm conduct in terms of governance structure and strategies towards firm boundaries in relation to what suppliers are sourced at each tier. This provides a background to understanding the next section, which is concerned with how procurement relationships develop on a particular project.

7.5   Procurement relationships

A behavioural view of the construction supply chain is developed in this section by describing instances of the behaviour of the firms as they procure the commodities required to participate in the project supply chain. The interactions between the customer firm, the supplier markets and the supplier firms are described using a combination of matrices and sequence diagrams. Sequence diagrams are borrowed from object-oriented modelling and are a succinct way to describe the messages between the firm and markets. Matrices and sequence diagrams are developed for procurement of suppliers for the various projects for commodities required by the following customer firms:

- façade subcontractor
- aluminium extrusions fabricator
- glass: specialist glazier and glass manufacturer.

7.5.1   *Procurement of façade subcontractor's suppliers*

This section describes the procurement of the suppliers to the façade subcontractor for Project 1. First though, the following quote summarizes how the façade subcontractor was procured by the main contractor:

> ### Box 7.15 Contractor to façade subcontractor project by project negotiation
>
> We tender for 95% of the projects. There may be the odd occasions when we negotiate for a project. Most of them are tendered because it creates a competitive market for the client and that is what they are chasing.

The subcontractor provided a tender to the main contractor. The main contractor assessed the market and created a smaller project market from five firms; three Australian and two international façade subcontractors.

The façade subcontractor then categorized their suppliers in three groups: common products, unique products and common services, as highlighted earlier in this chapter. Even though the subcontractor grouped the suppliers in this manner, their approach to each supplier and the market was more complex. The interactions between firms is varied and complex across the different commodity markets (refer to the matrix in Table 7.5). The manner in which the subcontractor approaches each of the markets does not necessarily align strictly with this grouping. For example, within common products there are two different sourcing strategies, even though there is a high level of expenditure.

The first is where there is a high level of expenditure and a high risk (the aluminium and glazing suppliers) and the second is where there is a high level of expenditure but a low level of risk of supply (pop rivets, gaskets and silicon).

The subcontractor created a smaller and unique project market for the supply of aluminium extrusions and the glazing products. Four firms were approached for the supply of aluminium extrusions which included two from within Australia, one from Malaysia and one from China. The Australian and Malaysian suppliers were the fabricators, whereas the Chinese supplier was a distributor for a manufacturer in China. The tender price was governed by the prices set by the London Metal Exchange and fluctuated daily. Suppliers submit a rate for a specified time or a specific number of projects and after that the rate will alter according to the prices on the world market – subsequently there is a high level of risk in this transaction. Four tenders were received and the contract was awarded to an Australian fabricator, Firm 532, whose factory was located in metropolitan Melbourne. The other Australian firm and the Malaysian firm produced extrusions and also produced standardized aluminium window frames; however, they would not normally tender for a complex façade project such as this one. The Chinese firm also supplies sheet aluminium. The firm chosen was a specialist fabricator.

The choice of supplier was based upon the rate and the delivery schedule. At the same time a similar process was begun for the supply of seraphic glass; a specialist type of glazing. A smaller market was created of three

Table 7.5 Matrix sourcing strategies by façade subcontractor

Supplier types	Sourcing behaviour
Strategic critical unique products and common products high risk/ high expenditure	The façade subcontractor assessed the market of aluminium extrusion fabricators and glazing suppliers and then created a smaller and unique project market for the supply of the aluminium extrusions and the glazing products. The subcontractor requested tenders from these selected product suppliers. Four firms were approached for the supply of the aluminium extrusions which included 2 from within Australia, 1 from Malaysia and 1 from China. The Australian and Malaysian suppliers were the fabricators, whereas the Chinese supplier was a distributor for a manufacturer in China.
	The aluminium extrusions are from a company called Firm 532. We have been dealing with 532 for a little while over the last couple of years. Again we negotiated our price, our price is not negotiated on a handful of aluminium it is negotiated on 2 or 3 or 4 hundred tonnes of aluminium at a time and so. … We know what their service is like we know that they can produce good extrusions. We also know that they are a lot more competitive than some of the other people that are around. Now there is really only one other supplier in Australia which is YYY and then you need to go to Asia for some of the other Asian suppliers. Or you need to go to Europe for some of the European suppliers, or Europe but in Europe it is more expensive for aluminium. We go to 766, we go to 532, we go to China. We go to 198 in Malaysia. It does not matter. It depends where our projects are. If our projects are in Asia we then predominantly source from Asia and keep it in Asia. If our projects are Australian and we are not getting a good competitive price out of the Australian suppliers we then go to Asia for the supply of aluminium as well. So we do go offshore for the supply of aluminium. The tender price is governed by the prices set by the London Metal Price Exchange and fluctuate daily. Suppliers submit a rate for a specified period of time or a specific number of projects and after that the rate will alter according to the prices on the world market – subsequently there is a large amount of risk involved. Four tenders were received and the contract was awarded to an Australian fabricator, Firm 532, whose factory was located in metropolitan Melbourne. The other Australian firm and the Malaysian firm produce extrusions and also produce standardized aluminium window frames, however would not normally tender for a complex façade project such as this one. The Chinese firm also supplies sheet aluminium. The firm chosen is a specialist fabricator. The choice of supplier was based upon the rate and the delivery schedule. …At the same time a similar process was begun for the supply of seraphic glass. A smaller market was created of three suppliers, two Australian suppliers and one Asian distributor. One Australian firm was a first and second order glass manufacturer and the other a second order glass manufacturer.
	After the tenders were received the subcontractor then assessed the international market for glass supply and focused upon distributors. Initially, the subcontractor had intended on procuring from an overseas supplier, through a Chinese distributor based in Sydney. This set up rather a long chain as the Chinese distributor sourced from an agent in China who sourced either from a Chinese second order glass manufacturer (processor) or direct from a manufacturer. It is well known that this international façade subcontractor sources glazing from all over the world and according to the eventual supplier was 'caught short' and therefore turned to the Australian first order manufacturer for supply. Strictly speaking, there are a number of steps along the way as the supply of seraphic glass is from a smaller division of the larger manufacturer which operates essentially as a distinct firm. At times

the decision choice for a supplier is based upon different criteria, even if the market size is the same. Although both the aluminium and the steel supply market are the same size, the decision choice of the supplier for the supply of aluminium is more concerned with price, whereas the supply of the steel is based upon the quality of the service and the product. Steel is a unique product.

The steel subcontracting company is Firm 28, which is in Clayton and Wangarratta. They are currently doing the steel, they are doing the triangular frames for us, the support structure. They are doing the steelwork in the atrium as well. Again they have a back to back contract because they are a major supplier. Our package with them is not only to manufacture but it is also for them to shopdraw all the steel work, to purchase the materials, to manufacture it to galvanize and then to deliver it to us. We don't go in and say 'here I want you to make this piece of steel here are all the drawings', we have that built into their whole package as well. They take care of the whole package. Also they install it onsite as well. Some of it is ours that we have to assemble here but much of it they have to install onsite. When we tendered we tendered to three suppliers, 687, 28 and 533. Our project manager looked after it but I know that is who he went to. We have taken on a lot more steelwork than what we normally do. We put the steelwork out to tender.

There is a great deal of interaction with the steel people, because there is the supply, there is the design involved, there is the delivery, there is the storage component, there is the installation. So there is a lot more involved in that than the supply of the aluminium, for example. When the director of the subcontractor firm discusses the relationship with the aluminium supplier he talks of volume and cashflow and is more concerned with these matters than the quality of the extrusions. Cashflow and inventory management is much more significant for the aluminium extrusions supplier as the commodity price is much more volatile in this market.

This is the same for our aluminium suppliers, we give them the list of the bulk order, it is 200 tonne in total, it is 10 tonne of this type and 10 tonne of this type etc. we want you to deliver it in three lots and one third of it each lot each month. We plan that as well. Now the reason for that is that we don't want to carry a lot of that stock here number 2 is that we are not going to process it all in one month it will take us 2, 3 months and by doing that we spread the project and we spread supply. We also have the continuity of the work but at the same time we also spread our costs as well. We only pay for that month of work, because of well what would happen…we need to manage our cash flow. And also we can't produce things in one month so it is a bit ludicrous to have it all sitting here it is better to keep it in someone else's yard rather than ours.

Although the façade cleaning and the aluminium welding and the façade installation are all grouped as Common services, they are not treated the same. The quality of the service for welding and installation are more critical and therefore these suppliers are largely treated as an extension to the subcontracting firm. They are both small suppliers and this is rarely tendered. The subcontractor knows their workload and their operation and works closely with them, whereas…

… Whereas the façade cleaning is tendered openly.

Strategic security
common services
low exp/high risk

Tactical common
services low risk/low
expenditure

(Table 7.5 continued)

Table 7.5 Continued

Supplier types	Sourcing behaviour
Leverage common products low risk/high expenditure	The façade subcontractor is an international firm and annually requests tenders for the supply of silicon. The contract is then for the supply of silicon to all branches of the firm worldwide; therefore it is an international supply agreement. Similarly, there are national suppliers for pop rivets and gaskets for a 6-month period at an agreed rate for a set volume. However, when there is a larger project that increases the volume the subcontractor at times then approaches the market again to achieve an even more competitive price. This depends upon the competitive demands being placed upon them from their upstream customer.
	We have a few of these (annual agreements). For example, we have one for silicon, one for aluminium and we would have one for the pop rivets and the screws and fixings from a major supplier. Where we would say well we are buying 2 million pop rivets per year, 'what is the rate?' and the rate is the same for Melbourne and Sydney. We have a corporate rate, it is more on a corporate level, a country level than a divisional level. ... We put most of the packages out to tender, the steelwork, sandstone we didn't because we didn't need to. Zinc packages we put out to tender. Aluminium extrusions we put out to tender. They are the main ones. Everything else from there on in, for example, the supply of gaskets, we put out to tender to 2–3 suppliers. We normally find that the one that we are with now and the one who mostly supplies us is the one that comes up with the best price, unless you have a good competitive supplier out there who wants to get into the marketplace. Or who wants to get into bed with us somehow or other. We basically tender a lot of our subcontract work anyway because we are no different to the managing contractor. Once it comes to us we need to get the best price as possible and the best deliveries. Also the best product that we think is going to be the most suitable for future use on that project. We are no different to the big boys.
	However, earlier the subcontractor had stated that they had annual agreements for large volume standardized commodities.
	The gaskets well we normally design a lot of those. They will cut an extrusion die for us and we check it if it fits in where we want it to fit and it is all ok. It is basically then just an order. But with those people it is I want 5,000m this month and 5,000m next month, 5,000 metres in three months. They are quite amicable with that set up. We don't buy 15,000 metres in one hit.

suppliers, two Australian suppliers and one Asian distributor. After the tenders were received, the subcontractor then assessed the international market for glass supply and focused upon distributors. Initially, the subcontractor had intended on procuring from an overseas supplier, through a Chinese distributor based in Sydney. This set up rather a long chain as the Chinese distributor sourced from an agent in China who sourced either from a Chinese second order glass manufacturer (processor) or direct from a manufacturer. It is well known that this international façade subcontractor sources glazing from all over the world and, according to the eventual supplier, was 'caught short' and therefore turned to the Australian first order manufacturer for supply. Strictly speaking, the façade subcontractor did not source directly from the manufacturer as there are a number of other steps along the way as the supply of seraphic glass is from a smaller division of the larger manufacturer which operates essentially as a distinct firm as discussed previously. Due to the international scope of the façade subcontractor they often source glass internationally because of the higher purchasing power that this provides; this tends to be conducted on a project-by-project basis.

The façade subcontractor is an international firm and annually requests tenders for the supply of silicon. The contract is then for the supply of silicon to all branches of the firm worldwide; therefore, it is an international supply agreement. Similarly, there are national suppliers for pop rivets and gaskets for a 6-month period at an agreed rate for a set volume.

Box 7.16 Annual purchasing agreements: leverage common products; low risk/high expenditure

We have a few of these (annual agreements). For example, we have one for silicon, one for aluminium and we would have one for the pop rivets and the screws and fixings from a major supplier. Where we would say 'well, we are buying 2 million pop rivets per year, what is the rate?' and the rate is the same for Melbourne and Sydney. We have a corporate rate, it is more on a corporate level, a country level than a divisional level.

However, when there is a larger project that increases the volume, the subcontractor at times then approaches the market again to achieve an even more competitive price. This depends upon the competitive demands being placed upon them from their upstream customer firm.

Figure 7.5 summarizes the sequences for procurement of the common core commodities: aluminium extrusions, specialist glazing, pop rivets, gaskets and silicon. For pop rivets, gaskets and silicon supply various forms of long-term agreements are in place. Again, the façade subcontractor is an

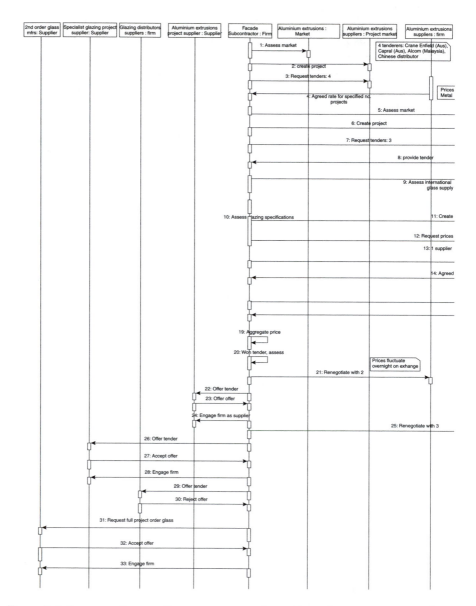

Figure 7.5 Sequence diagram for procurement of suppliers of common core products by façade subcontractor.

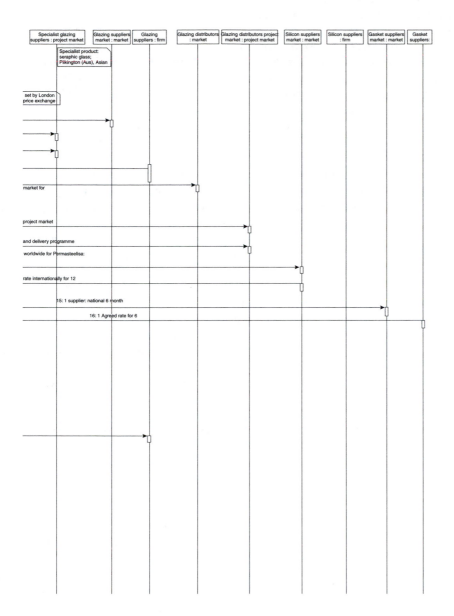

Specialist glazing suppliers : project market

Glazing suppliers market : market

Glazing suppliers : firm

Glazing distributors : market

Glazing distributors project market : project market

Silicon suppliers market : market

Silicon suppliers : firm

Gasket suppliers market : market

Gasket suppliers:

Specialist product: seraphic glass; Pilkington (Aus), Asian

set by London price exchange

market for

project market

and delivery programme

worldwide for Permasteelisa:

rate internationally for 12

15: 1 supplier: national 6 month

16: 1 Agreed rate for 6

international firm and uses this purchasing power to negotiate various agreed rates for more than one project. The procurement of firms downstream of the glazing supplier is also included in this sequence diagram and is discussed in detail later in this section.

A sequence diagram was developed for the unique products that are procured by the façade subcontractor. For the supply of stone, steel and zinc, project markets were specifically created from the broader industrial market.

At times the decision choice for a supplier is based upon different criteria, even if the project market size is the same. Although both the aluminium and the steel supply market is the same size, the decision choice of the supplier for the supply of aluminium is more concerned with price, whereas the supply of the steel is based upon the quality of the service and the product. Prices are more volatile in the aluminium sector than in the steel sector in Australia. This criteria impacts upon choice of tenderers and the subsequent negotiation of prices and eventual decision of which firm shall supply the commodity.

The subcontractor renegotiated prices with selected steel, zinc and glazing suppliers, whereas they couldn't with the stone supplier as there was only one supplier in Australia who could supply this commodity. The architects defined the project market for the stone supplier and therefore the bargaining power lay with the supplier.

Finally, the façade cleaning, the aluminium welding and the façade installation are all grouped as common services; however, they are not all treated in the same manner by the subcontractor. The quality of the service for the welding and the installation are more critical and therefore these suppliers are largely treated as an extension to the subcontracting firm. They are both small suppliers and this commodity is rarely tendered. The subcontractor knows their workload and their operation and works closely with them, whereas the façade cleaning is tendered openly, given that the subcontractor considered that this service had both a low expenditure and a low level of risk. The following Figure 7.6 is the sequence diagram for the common core services.

7.5.2   *Procurement for aluminium supply*

The aluminium extrusions fabricator has two main commodities including aluminium and steel, which are considered to be common core commodities, and then they have a number of non-core services and products (refer to Table 7.6).

Aluminium can be procured from anywhere in the world. However, it is considered to be a high risk and high expenditure-type commodity. The relationship between the fabricator and the aluminium supplier is strategic and long term and price is only one part of the relationship. This firm considers the arrangement with the aluminium supplier a strategic decision. The quality of the product (metal) is critical to the fabricator; however, price is always a critical factor and the market is competitive.

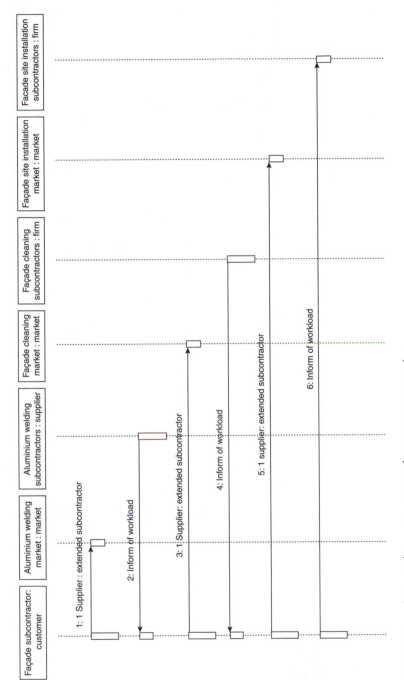

Figure 7.6 Sequence diagram for procurement of common core service.

Table 7.6 Matrix sourcing strategies for aluminium extrusions fabricator

Supplier types	Sourcing behaviour
Strategic critical core common high risk/high expenditure	The aluminium extrusions fabricator has two main commodities, including aluminium and steel. Aluminium can be procured from anywhere in the world. The relationship between the fabricator and the aluminium supplier is strategic and long term; however, price is only one part of the relationship. *Basically you could get aluminium from anywhere in the world …you can get it worldwide. Basically in Australia you have a remelt facility where you take the scrap and remelt it and produce your own billet again, or you buy it from Pescano or Comalco. We buy it from Comalco. But you could get South African billet but well if you asked me what percentage out of 100, 98% would be supplied by those three areas in Australia at the moment. Either remelt, Pescano or Comalco. We used to remelt but we don't any more because of the cost of recooking is prohibitive. We can buy it cheaper now than what we could take the scrap, remelt it produce our own and get better quality out of it.* There are many strategic decisions like that and also this firm considers the arrangement with the aluminium supplier a strategic decision. The quality of the product (metal) is critical to the fabricator. However, price is always a critical factor and the market is competitive. *We get plenty of people offering a lot more billet from overseas. But we don't go back to Comalco and say 'can you tweak the price a bit'. We make the commitment and say that is the way we are going to go. We expect to negotiate. If you give us the best price and if you don't. We don't want to know that someone else is getting a better price, they can be getting the same but not cheaper.* They have regular contact with the aluminium supplier at all levels in the company. *I would have Comalco's person in here at least six times per year and he would be seeing my boss at least that many times per year as well. Our people in our despatch area who look after our inward goods, they would be talking to Comalco roughly once a fortnight. We have interaction at the operational and strategic level. It is a big business when you consider 30,000 tonnes at $100,000 project. They look after us, they want to know if it arrives on time, how it is arriving, any quality issues. How the presses are running. We get Comalco to come in because they also have a background in extrusions, they got out of it but we utilize their people to come and have a look at us. They do an audit on us. If there is anything that they pick up we want to know about it. So that we can improve. It is a good process actually. A good working relationship. With regard to price setting*

the purchase of aluminium is high risk and volatile and is calculated daily. However, for special projects and the Project 1 was in this category because of the volume the relationship between this firm and the aluminium supplier was such that a special pricing deal was arranged. It was termed 'backing into your supplier'.

The other thing that might be of interest to you is that you can choose to back into your supplier. We might say but we don't do it too regularly. But we did it recently. Someone has come to us and said 'we want to buy 300 tonnes every week for the next 3 weeks at a set price irrespective of what is happening around the world' So we will go to Comalco and say 'on top of what we are doing give us today a set price for the next 3 weeks for an additional 300 tonnes per week.

Tactical
common services
low risk/
low expenditure

The aluminium extrusions fabricator (Firm 532) has national supply agreements with the majority of its other major suppliers for non-core products For example, they have national supply agreements with a logistics firm and a mobile telephone company.

Last week we looked at transport. We bought our Copper Division and our Brass Division onto this site, their sales offices. This was to get economies of scale with them came two trucks each. We don't actually need 6 trucks now. We are going to rationalize that. Crane is very much about making the most of our synergies and getting cost down and from a purchasing point of view if you are buying steel and you are in Queensland, you are in Tassie and you are in Melbourne. You have four different companies, get together and buy from one person and get the best price and then give them 100% royalty. Make it worth their while too to deal with Crane. Expenditure in comparison to steel and aluminium is comparatively lower.]

Leverage
core common
products
low risk/high
expenditure

The other major supplier is the steel supplier, which is required for the manufacture of the dies. The steel supplier could be one of a number of firms. At the time of the interview it was Firm 637, a second order manufacturer who then purchase their steel from Firm 458, the distributors for Firm 135, the steel manufacturing firm in Australia. 532 only deal with one of about three suppliers; however, there might be about 30 little firms that supply steel in this market. The steel requirements are for the tool shop in Melbourne for the aluminium extrusions and they also buy steel for their brass and copper toolshops. So, as a group they go to 637 and say 'we need 3,000 blanks delivered over the next 12 months, 250 per year; give us a price.' They use their purchasing power and have a national annual supply agreement. So we can say to you, 'Kerry, your price is $3 and if metal goes through the roof over-night or the Australian dollar plummets it does not affect you, because we have made a commitment to you.'

> **Box 7.17 Aluminium purchase: strategic critical common product; high risk/high expenditure**
>
> We get plenty of people offering a lot more billet from overseas. But we don't go back to Comalco and say 'can you tweak the price a bit?'. We make the commitment and say that is the way we are going to go. We expect to negotiate. If you give us the best price and if you don't. We don't want to know that someone else is getting a better price; they can be getting the same but not cheaper.

With regard to price setting the purchase of aluminium is high risk and volatile and is calculated daily. However, for special projects, and Project 1 was in this category because of the volume, the relationship between this firm and the aluminium supplier was such that a special pricing deal was arranged. It was termed 'backing into your supplier'.

> **Box 7.18 Backing into your supplier**
>
> The other thing that might be of interest to you is that you can choose to back into your supplier. We might say but we don't do it too regularly. But we did it recently. Someone has come to us and said 'we want to buy 300 tonnes every week for the next 3 weeks at a set price irrespective of what is happening around the world'. So we will go to XXX and say 'on top of what we are doing give us today a set price for the next 3 weeks for an additional 300 tonnes per week'. So we can say to you, 'Kerry, your price is $3 and if metal goes through the roof overnight or the Australian dollar plummets it does not affect you, because we have made a commitment to you'.

The other major supplier is the steel supplier, which is required for the manufacture of the dies. The steel supplier could be one of a number of firms. At the time of the interview, it was Firm 637, a second order manufacturer who then purchased their steel from Firm 458, the distributors for Firm 135, the steel manufacturing firm in Australia. Firm 532 only deal with one of three suppliers; however, there might be about thirty smaller firms which supply steel in this market. The steel requirements are for the tool shop in Melbourne for the aluminium extrusions and they also buy steel for their brass and copper toolshops. So, as a group they approach Firm 637 and say 'we need 3,000 blanks delivered over the next 12 months, 250 per year; give us a *price*'. They use their purchasing power and have a national annual supply agreement. The bargaining power of the customer is strengthened through an increase in expenditure and leveraged through the combined business unit's higher purchasing volume. The risk inherent in this commodity is somewhat lower than for the aluminium commodity.

The aluminium extrusions fabricator has national supply agreements with the majority of its other major suppliers for non-core services and non-core products. For example, they have national supply agreements with a logistics firm and a mobile telephone company and again the expenditure is increased as for the purchase of steel. However, the expenditure is not quite as high and therefore the actual purchasing tends to be devolved out to the business units who then devolve it out to managerial staff through credit cards or purchasing orders.

7.5.3 Procurement for glass supply

This section describes the procurement strategies used by the specialist glazing supplier and the national raw glass manufacturer.

Firm 472, the specialist glazing fabricator, sources three commodity supplier types: core products, non-core products and non-core services. The sourcing strategy for the non-core products and services involve annual national agreements which define contract negotiation and terms related to price, quality and delivery. The firm uses their purchasing power to bargain various agreements with the suppliers and at times uses the combined purchasing power within the multinational glass manufacturer.

The core products are glazing and ceramics and the sourcing strategies for these commodities differs from the annual purchasing agreement contract type; refer to Figure 7.7 for the sequences of interactions between firms. The glazing is procured primarily from their own parent firm in another state, Firm 473, which is the state sales division (merchant-distributor/processor) in the state where the manufacturer is also located. Firm 170 is the national manufacturer and Firm 473 sources from 170. There is no state sales division in the state where 472 is located.

The matrix in Table 7.7 Sourcing strategies by glass suppliers: specialist glass manufacturer, notes key quotes which suggest categorizations of sourcing strategies developed by the manufacturer related to risk and expenditure, including:

- common services: low risk/low expenditure and tactical purchasing
- glazing: high risk/high expenditure and strategic critical purchasing
- ceramic: low risk/high expenditure and leverage purchasing.

Procurement is a highly regarded activity of Firm 170, the national manufacturer for glazing. All procurement relationships are considered critical and the company-wide supply policy incorporates methods for purchasing; which is somewhat different to the upstream customer relationships discussed in the previous section.

These are typically specified and tendered and either national or state agreements are arranged for 12 months; except for sand, which is 3 years. There

Specialist facade subcontractor : customer | Specialist glazing supplier : firm | Specialist glazing distributor for 2nd order mfr : Supplier | Glazing primary manufacturer : supplier | Ceramic suppliers : market | Ceramic supplier : project market | Ceramic : firm | Packaging suppliers : supplier | Pallets : supplier | Specialist machinery : supplier | Transportation : supplier

TNT

Pilkington Building Products Division

2 Australian suppliers, one in Japan and 2 in Europe

2 Australian suppliers, ferro and johnson cookson mathy

Note : 20 major suppliers to specialist glazing mfr: major suppliers noted here: strategic relationships entered into on the basis of the appliances, automotive glazing products and not the architectural products. Same suppliers but different products and therefore different pricing and delivery structures.
Export distribution company for exporting of architectural products is the same as for the appliances as is the fuel and transportation supplier. Fuel is company wide but transportation is project based.

1: Notify of order and request price and confirm

2: Confirm stock and

3: 1 specialist supplier: request delivery programme and confirm annual price

4: Accept offer

5: Offer tender for additional

7: Accept offer

8: Fill order

9: Assess

10: Create project marker

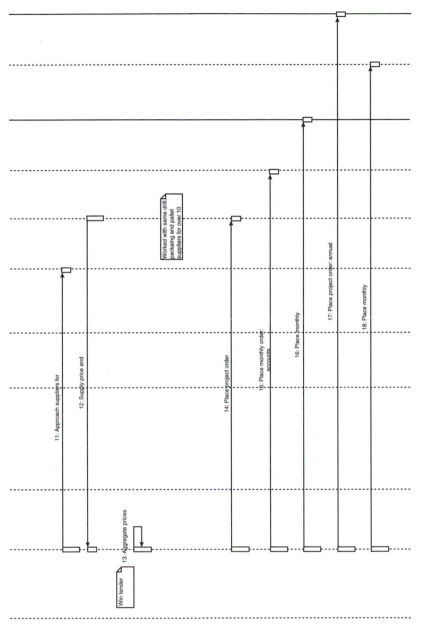

11: Approach suppliers for

12: Supply price and

13: Aggregate prices

Win tender

14: Place project order

15: Place monthly order: accounts

16: Place monthly

17: Place project order: annual

18: Place monthly

Worked with same drill packaing and pallet suppliers for over 10

Figure 7.7 Sequence diagram for procurement of suppliers to specialist glazing fabricator.

Table 7.7 Sourcing strategies by glass suppliers: specialist glass manufacturer

Supplier Types	Sourcing behaviour
Tactical common services and non-core products low risk/low expenditure	**Specialist glazing supplier's method for procurement of suppliers** Firm 472, a specialist glazing fabricator, sources three commodity supplier types: core products, non-core products and non-core services. Annual national agreements are set up for the services and the non-core products and procurement and associated contract negotiation on price, quality and delivery occurs at that time.
Strategic critical core common products high risk/high expenditure	The core products are glazing and ceramics and the nature of procurement for these commodities differs from the annual purchasing agreement contract type. The glazing is procured primarily from their own parent firm in another state, Firm 473, which is the state sales division (merchant-distributor/processor) in the state that the manufacturer is located. Firm 170 is the national manufacturer and Firm 473 sources from 170. There is no state sales division in the state that 472 is located within. *We have two suppliers, but primarily from Firm 473 in Melbourne. If they don't make the particular product that is some specialized products they don't make. You have to remember they have to make for markets and some specialized glasses they don't make. This is when we source from somewhere else. This will then be imported from outside of Australia. There are only two plants in Australia that manufacture glass and they are both Firm 473; that is, one manufacturer. We don't source outside often.* Typically, Firm 472 source from the parent firm in Germany, which supplies a high quality glass.
Leverage core common products low risk/high expenditure	**Specialist glazing supplier's method for procurement of suppliers** The other material is the ceramic paint of which there are two suppliers interstate that the firm primarily sources from. Alternatively, they may purchase from Japan or Europe. *There is no set contract with these guys. This company is built on trust. We don't enter into any contractual arrangements with anyone. It is purely built on trust and performance and if we don't get any satisfaction from one of the paint manufacturers we will go somewhere else.* The manner in which they interact with the ceramic supplier is informal and routine. *We simply order a certain amount each month and then it comes and we pay a monthly account. But sometimes we get a special job and we might get them to mix a special colour for us, if it is a sufficiently large job. For example, 20,000 metres squared of glass, rather than us mix it here we will get them to mix it there and supply it to us. Mainly we do our own mixing here. You take a base colour and add tints to it…it is just like mixing paints for your home really. We do a lot of mixing here. But if it is standard colours we buy them in as a standard colour.*

are a variety of market types and agreements that range from the global to the national to the state and the more critical of these is now considered.

> ### Box 7.19 Annual agreements
>
> We have a national agreement with these companies for at least 12 months. Actually, the sand is for 3 years. Pretty well with every supplier we have this arrangement. We negotiate on the price. It remains relatively stable for the 12 months.

Sand is sourced on quality of product and there is a great deal of interaction between customer and supplier. It is also a state-based negotiation. The agreement is for 3 years. Dolomite is also provided on a national basis by this supplier and there is only one supply location in Australia – dolomite is only used in small amounts. There is a long history between the sand supplier and the glass manufacturer and a high degree of interaction between the two firms. The glass manufacturer was once part of the quarrying group. One of the major considerations in sourcing for quarried products is the cost of transport. The location of the sand relative to the plant is critical to the manufacturer and is the major factor that affects costs. However, at times the physical distance is not the only factor considered; the quality of the sand is also critical and is the first criterion and therefore if the qualities are not appropriate then location is irrelevant.

> ### Box 7.20 Supplier location: quality and distance factors affect sourcing strategy
>
> Sourcing is a separate strategy for the separate regions. Sourcing is based upon where the quarry and the material are located.

Soda ash is a significant product and is sourced from one location in Australia; it is a national agreement. There is only one supplier in Australia. The same supplier is used for limestone.

> ### Box 7.21 Strategic critical core common products; high risk/high expenditure
>
> Actually, soda ash is the highest cost product going into glass making. There is only one soda ash manufacturer in Australia and it is called Firm 667.

A number of other suppliers are sourced from all over the world. These products are not as critical or as risky a purchase as sand is, but nationally or internationally the purchasing power enables long-term agreements. Even though these are 1–3 year contracts at the time of renegotiation, for the markets that are competitive, there is a tender.

Finally, supply of some products and/or services is considered purely on a transactional basis and other relationships are closer and considered to be an alliance. The major Australian construction product/material suppliers (like glass, steel and concrete) typically have joint working parties to solve production problems and Firm 170 considers these as a strategic alliance.

Shipping, logistics and specialist equipment suppliers were also considered to be strategic alliances, where the relationship was not treated on a transactional basis.

Box 7.22 Strategic alliances for leverage/core products; low risk/high expenditure

We have other areas where we are not out to tender all the time. We really have suppliers who understand our business; we understand them and we have good understanding of their costings and therefore we won't retender and will just negotiate and just agree where we are up to.

We do this with trucking...we do this with distribution quite a lot. We certainly have a strategic alliance there; a number of our raw material transport providers also. ...Also specialist equipment providers are an important supplier. ...Also we need a lot of open-top containers for shipping to New Zealand, so one shipping line is making sure that they have in their world system lots of open-top containers. ...'I will always have these available for Firm 170.' And when we renegotiate this is important and another guy does not have them available at the time.

Suppliers and their interactions can be classed according to the following:

- high risk/high expenditure
- low risk/high expenditure
- low risk/low expenditure (comparatively).

In high expenditure groups the manufacturer is able to exert power in the tendering and negotiating stage.

What a long way we have travelled in the construction supply chain.

7.6 Aggregated project supply chain organization: supply channels

Chain structures map the transfer of ownership of commodities on individual projects and this section draws together the project supply chains into supply channel structural maps. This section summarizes eight channel structure maps for the construction industry, including: aluminium, steel, concrete, glass, fire products, mechanical services, tiles and masonry. The maps are of the primary material – for example, glass, steel, aluminium – and do not include all the subsidiary product and raw material chains required at the manufacturing tier, nor the gathering together that occurs at the subcontractor level for site installation. For example, aluminium windows require numerous suppliers, including rubber gaskets, silicon, fastenings, framing, fixings, as well as the aluminium window component. The previous chapter has served to highlight the complexities of chains for individual firms and the immediate competitors. This has assisted in being able to develop less abstract channel structures than previously used when describing the construction industry. The commonly used chain structure that is used is contractor to subcontractor to materials supplier/manufacturer. The following maps clearly develop a much greater picture of the diversity of interactions in channel structure than previously known. This has allowed for a much richer description of industrial market structure. Each channel structure map is now discussed in more detail.

7.6.1 *Aluminium chains*

Figure 7.8 maps the transfer of ownership of aluminium products in the construction industry from the manufacturer who extracts raw materials from the ground to the building owners. The channel maps indicate the structural organization of primary commodity suppliers and the types of firms that are involved. There are currently eight tiers available for the transfer of ownership of the aluminium commodity to flow through, disregarding the importation of aluminium billets; however, there was no channel identified in the study that included all eight tiers. If the aluminium extrusions fabricator imports aluminium then the channel extends dramatically to twelve tiers. Currently, this is the location where imports are introduced into the channel. Thirteen chain options were identified, including one standard import channel.

The organization of the channels extends from three main branches: the extrusions fabricators who manufacture standard as well as special extrusions; the extrusions fabricators who manufacture standard extrusions only; and the large-scale production of residential window extrusions by national manufacturers. Each of these branches has a variety of paths, depending primarily upon purchasing power. For example, if the purchasing power of the curtain wall/window fabricators is high, then an industrial distributor will not be involved; however, if the purchasing power is low, then the fabricator procures extrusions from a distributor. Similarly, if the purchasing power of a primary contractor or developer is high, then they can procure directly from the national window fabricator; or, if is lower, then the purchase route will be through the industrial distributors.

7.6.2 *Glass chains*

Figure 7.9 maps the transfer of ownership of glazing products in the construction industry from the manufacturer who extracts raw materials from

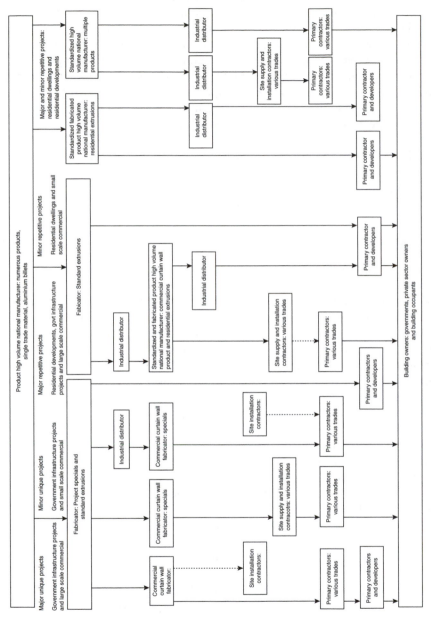

Figure 7.8 Structural organization channel map for primary commodity aluminium.

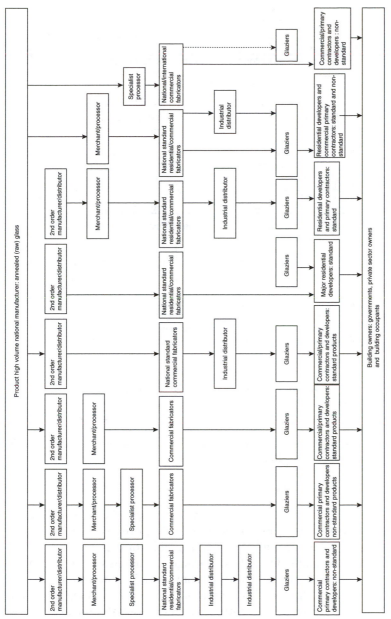

Figure 7.9a Structural organization channel map for primary commodity glazing.

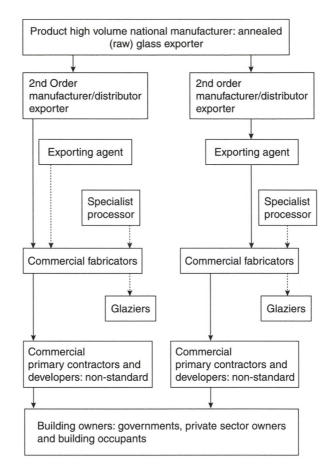

Figure 7.9b Structural organization channel map for primary commodity glazing including exporting.

the ground to the building owners. The channel map indicates the structural organization of primary commodity suppliers and the types of firms that are involved. There are currently ten tiers available for the transfer of ownership of the glazing commodity to flow through, including an import channel. Importation currently occurs at the commercial window/curtain wall fabricator tier in the channel. The length of the tier does not extend unduly (6 levels) as the commercial fabricators are multinationals and large enough that tier purchasing power by virtue of purchasing volume ensures that they procure glass more directly. Eleven channel options were identified. We can begin to develop interesting questions for further research in relation to cash and information flow and also delivery times. For example, does the commodity flow at the same rate in different chains? How and when can we simulate and optimize time performance?

The organization of the channels extends from three main branches: the extrusions fabricators who manufacture standards as well as special extrusions; the extrusions fabricators who manufacture standard extrusions only; and the large-scale production of residential window extrusions by national manufacturers. If the purchasing power of the curtain wall/window fabricators is high, then an industrial distributor will not be involved; however, if the purchasing power is low, then the fabricator procures extrusions from a distributor.

7.7 A final word

Chapter summary

1 It appears that there are strong couplings between various firms in supply chains; and as such there is a certain degree of predictability. Once a decision is made at the subcontractor level of a particular supplier, then there is a small number of typical supplier procurement paths that will follow, based upon the interdependency between four entities; namely, commodity type, industrial market, project market and procurement relationship. This does not mean that there is only one path; more that the choice of supplier is not entirely random and that there will be a likelihood of certain suppliers being chosen; that is, a degree of predictability. The strategic direction of firms, firm governance and decisions regarding firm boundaries impact upon the nature of the market and eventually the types of procurement relationships that form. In some cases the couplings are based on historical corporate ties between the two firms that have arisen because one of the firms was originally part of the other firm and typically as a division. Such long-term characteristics of markets and movements may evolve over many years and be structurally embedded within the sector and the supply chains. The historical context assists in explaining the eventual structural organization of the supply chain. Other coupling ties that were identified were related to purchasing volume, that is, countervailing power, which was typically related to size of firm, and then commodity complexity, that is, standardized versus unique commodities. Participants within the industry acknowledge that there are many different paths and described the marketplace as intricate.

2 Within large firms there can be quite constrained models of market competition for internal functional markets which create artificial markets and difficult relationships in inter-organizational functional supply chains; giving rise to independent customers versus internal dependent customers.

3   There are a wide variety of functional categories of suppliers within the glazing market; that is, merchant, processor and raw material supplier, and then various combinations of these functional categories; that is, merchant/processor, merchant/raw material supplier, etc., thus giving rise to a high level of complexity in the various supply chains. The structural organization is dynamic. Market competition changes the structure of supply chain in glazing supply as the various different functional types of firms take on different functions along the chain to improve their competitive advantage and value add.

4   Supplier types categorizations rely upon volume of commodity which is purchased and whether or not the commodity is commonly purchased by the firm as well as the complexity of the commodity. This then creates the following supplier groups: common core commodity, common non-core commodity, unique core commodity and unique non-core commodity.

5   Procurement relationships and categorization of sourcing strategies which were identified: common services: low risk/low expenditure and tactical purchasing; glazing: high risk/high expenditure and strategic critical purchasing; ceramic: low risk/high expenditure and leverage purchasing.

6   Suppliers and their interactions can be classed according to the following: high risk/high expenditure; low risk/high expenditure; low risk/low expenditure (comparatively).

7   The categorization of suppliers encapsulates how the firms interact with the market and the sequence of events and interactions between upstream firms and downstream firms to create firm–firm procurement relationships. Selected detailed scenarios were developed, described and summarized by both matrices and sequence diagrams. The sequence diagrams were developed for various suppliers to the façade subcontractor and suppliers to the specialist glazing fabricator. Categorizations depended upon an assessment of risk and an assessment of expenditure related to the supplier type. Risk involved both the market risk and the commodity risk, which is a combination of the internal significance of the commodity and the external supply communities.

8   These scenarios can be abstracted to the following five stages:

   •   assess what suppliers are needed by commodity type and create a description of the commodity (including commodity type, transaction complexity, frequency and financial value)
   •   create a project market for the commodity (typically, if firm is supplying directly to site or one tier from site suppliers)

- assess their own firm attributes (typically, scope and turnover) related to the commodity supplier project market; group by common core/non-core and unique core/non-core
- develop sourcing strategies related to market power relationships dependent upon commodity grouping; categorize commodity by risk and expenditure assessments
- and negotiate with various supplier firms; these negotiations are dependent upon countervailing power, which is based upon perceptions of risk, and purchasing power related to the commodity.

9 Eight tiers for the transfer of ownership of the aluminium commodity were identified and twelve if importation chains were included, with thirteen chain options identified. There were three main branches identified, including: the extrusions fabricators who manufacture standard as well as special extrusions; the extrusions fabricators who manufacture standard extrusions only; and the large-scale production of residential window extrusions by national manufacturers. Each of these branches has a variety of paths depending primarily upon purchasing power.

10 There are currently ten tiers available for the transfer of ownership of the glazing commodity to flow through, including an import channel. Importation currently occurs at the commercial window/curtain wall fabricator tier in the channel. The length of the tier does not extend unduly (six levels) as the commercial fabricators are multinationals and large enough that tier purchasing power by virtue of purchasing volume ensures that they procure glass more directly. Eleven channel options were identified.

8   Case study
Simple and complex core and non-core supply chain – steel chain cluster

What are the structural and behavioural characteristics of a supply chain which is both a core and non-core commodity?

8.0   Orientation

Box 8.1 Chapter orientation

WHAT: In a similar fashion, Chapter 8 provides a detailed investigation into the nature of the procurement relationships that are formed in relation to the steel supply chain, including discussion on firms, projects and market attributes – which underpin the nature of the sourcing strategies and approach and negotiation interactions between customer and supplier firms. Underlying questions which are considered include:

- What sourcing strategies are used at various tiers to deliver a complex commodity which is both core and non-core and simple and complex?
- Are the chain paths different across the sector?
- What are the connections between markets, firm types, commodity types and procurement relationships? What is the sequence of events which takes place during procurement?

WHY: Again, this chapter provides information about the industrial organization economics of the markets at various tiers in the chain and the way in which this influences sourcing strategies and composition of the chain.

WHO: The steel supply chain clusters described in this chapter are both core and non-core because at times they support other supply chains providing commodities to other subcontractors who deal directly to the project contract; for example, mechanical services, aluminium fabricators, façade subcontractors, concrete subcontractors.

A complex core commodity supply chain is a commodity which is core to the project contract. A complex commodity chain is one where the nexus of contracts to the project contract is complex in either technology or managerial complexity; that is, requiring unique, specialist and innovative design and/or construction solutions or a high level of integrative managerial capacity. These types of supply chains can be characterized by innovative design, new materials, juxtapositions of new materials, numerous different types of suppliers, and a requirement to source and integrate suppliers not typically managed previously.

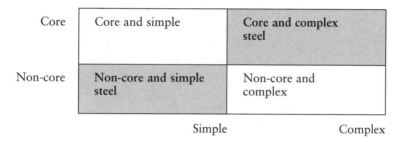

	Simple	Complex
Core	Core and simple	**Core and complex steel**
Non-core	**Non-core and simple steel**	Non-core and complex

Chapter orientation categorization of chains.

8.1 Introduction

This chapter presents the results from the interviews with various project managers, firm executives, production and procurement managers involved in the supply chains for commodities that are clustered around the supply of steel to the construction site. Steel, of course, is involved in numerous products and various supply chains – which has already been discussion in Chapter 7.

The structural steel fabricators/subcontractors were interviewed in relation to:

• Structural steelwork for Projects 1, 2 and 4.

After the subcontractor interviews, subsequent interviews with processors, merchants and manufacturers eventually led to a tracing of more general industry chains for the supply of the following products with the associated projects:

• Steel for Projects 1–6.

The chapter is organized in a similar manner to Chapter 7 in four main sections, including:

• Firm attributes
• Markets, commodities and competitors, including competitive advantage

- Supplier types
- Procurement relationships, including sourcing strategies.

8.2   Firm details

Steel supply is integral to a number of firms as both core products and non-core products in the manufacturing or fabricating process and is now considered through a number of the chains on the projects. Key firms involved in the chain include: steel fabricators (Firms 28 and 157); protective painter (Firm 196); steel distributor (Firm 458) and steel manufacturer (Firm 135).

Firm 28 supplies and erects steel to Projects 1 and 4 and Firm 157 supplies and erects steel for Project 2. Firm 28 is located at both tiers 2 and 3, supplying to both the contractor directly and the façade subcontractor. Firm 157 is located at tier 2, supplying directly to the contractor. Firm 196 supplies and applies protective coating paint for the steel fabricator Firm 28 and is located at tier 3. The steel distributor and steel manufacturer, Firms 458 and 135, supply to all the projects. The distributor is a division of the national steel manufacturer with a similar governance structure as the glass manufacturer discussed in the previous chapter.

The two fabricators are of a similar size; however, Firm 157 has twice the turnover of Firm 28. The protective painter, Firm 196, employs 28 people and is one-third the size of the steel fabricators. All three firms are structured as state and one division. The distributor is designed to distribute nationally and is structured in that manner. However, it is anticipated that this structure and distribution pattern will change in the future as the steel manufacturer divests themselves of this division and then the distributor will then compete against other distributors. This division employs 80 people. The steel manufacturer is a multinational and operates in Europe, Africa, Australia, North and South America, Asia, but not in the Antarctic or Arctic, and employs approximately 48,000 people.

Similar to the glazing manufacture supply chain, there is a large manufacturer who supplies to numerous projects and has the majority of the Australian market share. A difference between the two chains is that there is currently less importing of steel than glass. There is a similar structural organization in that the manufacturer also has a major distribution division which operates as a separate cost accounting division and is located in each state and competes against external merchants or distributors. The similarity between the two chains is that there can be an extra firm in the supply chain.

Firm 157 (steel fabricator on Project 2) sources directly from Firm 458 (the national steel distributor), whereas Firm 28 (steel fabricator on Projects 1 and 4) sources from a smaller steel merchant who in turn has sourced from the national distributor. In the chain there is a relationship between a firm's turnover and the structural organization of the supply chain; the greater the turnover, the greater the likelihood of a higher volume of product purchased and therefore the greater the likelihood of being able to purchase directly from the distributor, thereby reducing commodity owner transferral steps and thus the number of procurement activities.

8.3 Markets, commodities and competitors

The market structure for steel fabricators is stratified – there are three levels of firms related to the size and complexity of the project and both firms confirmed this market stratification. This is perhaps a refinement of the concept of market segmentation or at least allied to the concept. Where market segmentation typically refers to different types of products, these suppliers are often (although not always) supplying a similar product, but because the projects are larger the commercial risk is higher. It is the risk of the project that tends to produce the stratification.

Firms 157 and 28 are in level 1, which typically translates to the larger, more complex steel fabrication projects. The product is somewhat different in that these projects are typically complex architectural designs. This stratification is borne out in the façade market as well. As the demand in the construction industry fluctuates, so too does the size of the steel fabrication market. When there are less projects of this nature, these firms tender for smaller projects. This tends to mean that they are competing with firms in the next level down who have a smaller operation. When this happens, these smaller firms do one of two things: they compete for smaller and less complex projects in the next level down or form consortiums to compete against the 'big five'. The markets are typically related to their clients, which involve the larger contractors; however, they compete for a variety of other smaller and less complex projects. Firm 157 also competes in the heavy civil engineering steel fabrication market, whereas Firm 28 does not.

The steel painter has more than twenty clients; however, the major proportion of their income would be divided between five major clients. These clients are fabricators, two are primarily in the construction building industry, one is heavy engineering (oil rigs, etc.) and is primarily structural and one is in the mechanical process market (refineries/equipment, etc.). The market is uncertain and highly competitive and the nature of this is evident in the relationship with one of the key customers.

The distributor, Firm 458, offers two main services: distribution and processing services. This division has two major competitors, although there are many more smaller steel merchants in the market. At times their customers become their competitors and purchasing volume is a key criterion. That is, the key criterion to being able to purchase direct from the manufacturing plants is to be able to purchase a certain volume, typically a certain amount calculated over a year. Some merchants' customers, that is, fabricators or contractors, can and have achieved that volume, but this is less likely to occur these days. This differs somewhat to the glazing supply chain industrial organization structure where there are national customers purchasing direct from the glass manufacturing plant. The glass industrial organization, which was quite diverse and complex with players at all levels of vertical integration, also differs from the steel industrial organization in that the steel chain is typically less fragmented. Firms tend to provide products and services for a single market.

The majority of steel fabricators purchase the majority of their steel back through the three distributors (in the particular state that Firm 458 is located). Similar to the glass manufacturer state distributing divisions, this division has to act like an autonomous unit and yet is required to purchase their steel from the parent firm. Their competitors have the option of purchasing steel from the steel manufacturing competitors, albeit there being only one in Australia. They have the option to import.

Although this is discussed in more detail in the next section, the importing of steel is becoming more and more significant; as it is for the aluminium sector. Typically, resellers are engaged in this and tend to combine to improve their purchasing power. A broker is often involved.

With this particular steel manufacturer, Firm 135, there are broad groupings of products and services which are mineral, steel and petroleum. The steel products are primarily directed into the white-goods, automotive and the construction industry. There are four main products associated with steel, according to this firm: slab, coil, plate and wire, and they can be described in terms of the stage of production. Firm 135 does not have any competitors from an Australian steel domestic market – the major competition is from imports. There is a small selective range of steel products where Firm 135 competes with Firm 694, the Victorian-based steel producer, but that is in a selected range of products, namely steel reinforcement. This is a very small part of Firm 135's portfolio. Firm 135 prefer to view their competitors on the world stage.

The firm sells within their own group of companies and outside as well. For example, out of the slab (one output along the production chain) that is produced, 50% goes to a separate division and 50% goes to an external customer. Of the remaining 50% of the flat product, another 35% would be processed through another processing division (the tin mill) to produce the cans (tinplate) and then another 10% would go out to the plate market and then the balance would be sold to the export market. So there are numerous customers. Even when the firm sells the product to the next division, there is cash flow and it is based on transfer pricing. Even when they sell it to an alternate division at transfer pricing the real dollars start to flow. They have hundreds of distributors for steel products in the construction industry in Australia.

8.3.1   Competitive advantage

It is difficult to separate the discussion of structural and behavioural characteristics because of the way in which the study participants described themselves. However, the next section on procurement relationships tends to focus on behavioural characteristics that are related to project-specific procurement relationships, whereas the behavioural characteristics in this section are general firm behavioural characteristics as a reaction to their competitors in the market.

Ultimately, all firms tended to describe their firm behaviour within the markets and in particular their competitive advantage in terms of price. They tend to believe that price is the ultimate and final reason why the customer chooses one firm over another, as it is often a major decision criterion for themselves. However, having said that, it is apparent that competitive advantage and thus supplier choice is not as simplistic as this.

The competitive advantage of the firms can vary from project to project and can be attributed to a variety of situations. For example, it may rely upon the nature of current demand in the market; the type of commodity required; the other commodities supplied; the history of the relationship and/or additional design services. Competitive advantage is a complex issue and at times can not really be attributed to one single factor. The reason why a customer makes a decision of one supplier over another is varied and can be as a result of a combination of two factors. It seems that often a small group of suppliers may be chosen to tender within a project market based upon commodity type and then may compete based upon another factor. For example, even if the type of commodity being supplied is of a similar character, the final choice of supplier may rely upon the relationship of the commodity to other commodities that are being supplied – which is project specific. Alternatively, the final choice may rely upon price or lead times or the history of the relationship between customer and supplier.

For example, Project 1 and Project 4 are similar in that the steel structure was unique and complex requiring problem-solving abilities. However, the two projects differ in terms of the nature of the procurement relationship between the upstream contractor and the downstream steel subcontractor: *one is a hard money contract and the other is a construction management fee*. With Project 4, the 'hard money contract' – the steel is on the critical path and the main contractor has organized all packages as lump sum fees. The steel subcontractor's role for five months prior to any steel being erected on-site has tended to be one of pruning back to maintain costs. For Project 1, the design refinement and erection is similar but the risk is not being absorbed by the subcontractor. The following quote from Project 1's steel fabricator general manager clarifies this.

Box 8.2 Steel fabricator design service: business opportunity versus business constraint

In Project 1 it is unique with a unique steel structure – there is a very complicated façade. There are other items on the critical path though and steel is not as critical. We got involved in the design as well.

There was one particular package where the architects reckoned that they had carte blanche to do whatever they wanted and there was one package that came out which was basically a roof structure and it came in about 35% over budget. We were one of three customers and we went to the contractor, Firm 003, and we said 'give us two weeks; I will employ some shop detailers and put those guys in the architect's office and I will come back with a revised scheme that will take 35% off and meet your budget and with no obligation'. If we could not get to that then they could go and retender. We actually did that and we got 30% out. So what happened then is that it saved the project money and an enormous amount of time. If they had to go and get the architects to redesign it and then tender and test again it would have been months and months down the track.

The tender went out and we came back equal lowest and we went back with the proposal and we sat in the architect's office and we co-ordinated all the services, everything. Convinced the architect that the structure wasn't as pristine as he thought it was needed to be. It was pretty good. This is another one of our competitive advantages because we were prepared to do this to win the job. The other guys did not offer this. The competitors were local firms and not one of the usually big ones we compete with.

(Steel fabricator, General Manager)

Both firms have experienced a change in the level of design service that they are now required to do compared to the recent past. One firm takes it as a business opportunity for competitive advantage and the other takes it as a business constraint. Firm 28 sees the potential to take on risks and the potential rewards to redesign the structure to suit their own detailing and erection methods; however, Firm 157 only sees the risks.

Markets and their structural and behavioural characteristics are dynamic because market boundaries can be ill-defined. Firm 196 supplies the protective coating paint for the steel fabricator, which is considered a specialized product. This firm not only considers other protective coating paint applicators as their competitors but also galvanizers – which is a different product and process by which steel can be protected. Similarly, steel fabricators might consider structural concretors as competitors as this is a commodity substitute; however, both steel fabricators did not consider that these types of firms were their competitors. This seems to suggest that product substitutes need to be close substitutes and then one market competes against the other market in parallel. The competitive advantage of the market as a whole is that the firms that do protective coatings tend to take on more complex structures as galvanizers can only take whatever fits in their bath. 'Painters' can take very complex 3D shapes and can also work on-site as well, which is particularly relevant for oil rigs, ships, refineries, turbines, chimney stacks, etc.

Similar to the previous market for steel fabricators, the steel protective painters subcontractors operate in a segmented market based upon size of project. In the state in which Firm 196 operates there are four such firms operating at the highest level. There are smaller firms operating in the market at times; however, Firm 196 does not typically consider that these firms are competitors. Typically, the size of project involves twenty tonnes of steel or more.

Firm 196 tenders on nearly all of their projects and is generally awarded the project because of capability and performance in service rather than price alone. This is similar to the steel fabrication market in that firms are grouped and segmented within an industrial market for project markets based upon factors related to the commodity. Within that project market, price and lead times often become the choice criteria that upstream customers use to select one supplier over another.

Price is a critical factor as to which firms are selected. There are two main methods for pricing used on projects by the steel distributor, Firm 458. The method for pricing impacts upon competitive advantage. The majority of this firm's clients are steel fabricators and Project 2 is a fairly common-place type of project, although at times they may tender for a slightly different type of project.

The main competitive advantage for the distributor is that their scope is national and they are often able to and do call on their inventory in other states to satisfy customers. The other major competitive advantage is the quality and traceability of the product. Imported steel can often be cheaper than Australian steel, similar to the glass and aluminium supply. The path for imported steel is from a broker in Australia who has sourced from an international supplier who then distributes to a trading house. The trading house would actually sell product from several different mills around the world. Typically, you would not know which country or mill the steel originally was sourced from. For imported steel products in Australia this is quite common. As someone once arrogantly and ignorantly said to me. ...'I don't really care where a brick comes from...they are all the same' – which is incredibly naïve – we actually *do* need to care where the products come from – there are differences. Someone actually cares...and to be honest, so do I. It is not only the product that differs – but the impact upon the economies, the way in which projects become organized – the flow on impacts to pricing – if we understand the impact of our upstream decisions we can make holistic and informed decisions.

Box 8.3 Increasing importation – risks associated with lead times, commodity traceability and quality assurance

The agent/broker purchases from a trading house and the agent deals directly with a fabricator. The trading house could be online with several different mills – it could come from Japan and then the next lot could come from Korea. Given time-lags for purchasing, transportation from overseas, storage, then distribution to a merchant (typically, the smaller ones) and then redistribution to a fabricator's shop – it could be some months before the actual steel product finds its way into a fabricator's shop. By this time it would be difficult to trace the original supplier, although not impossible if appropriate systems were in place.

In the past, larger contractors would purchase direct from the merchants; however, in recent years this has become less prevalent. In the last 12 months this particular distributor has had one project where a contractor purchased the steel and then it was reissued to the fabricator. In this particular instance it was the contractor for Project 4 and the fabricator was the fabricator for Project 2, Firm 157, but it was for a different project.

The major competitive advantage for the steel manufacturer in the global marketplace is price. Within Australia there is a monopoly of steel manufacture. Although there is a monopoly in glass, aluminium and steel manufacturer in the country, there is a higher level of importing in the glass and aluminium supply chain and a much greater threat was perceived throughout the interviews by those involved in those two chains.

8.4   Supplier types

The structural steel subcontractors, the steel distributor, protective painter and the steel manufacturer all require suppliers to enable them to fulfil their contracts with their customers. When the fabricators were questioned about who their suppliers were, they each described them in groups, including:

- common core products (steel merchants and painters)
- common non-core products or secondary products (bolts, connexions, cleats); and
- common services (inspectors and/or shop detailers, smaller fabricators).

Typically, they described and grouped them according to whether they commonly/frequently purchased or sourced from them; the volume of purchase and how critical they were to their success. The painters sourced from two main groups: products (abrasive suppliers) and secondary products (fuel and solvent). Table 8.1 summarizes each of the steel subcontractor suppliers and the suppliers to the subcontractor into the firms' supplier-type categories.

There are numerous suppliers to Firm 135, the steel manufacturer, and in total number approximately 3,000–4,000 for the manufacturer of steel alone. They have a total purchasing power of $12b annually. Firm 135 has a large

Table 8.1 Supplier types as categorized by steel subcontractors and their suppliers

Customer firm	Supplier types
Steel fabricator	There are product, service and product and service suppliers to Firm 157. There are seven different types of suppliers, including: steel merchants, painters, fasteners and bolt suppliers, purlin suppliers, specialized fabrication subcontractors, weld inspectors and paint inspectors. Steel merchants and purlin suppliers are the primary core common product suppliers. Painters were core common product and service suppliers. Weld and paint inspectors were generally considered common services. Specialized fabrication was a unique service. Fasteners and bolt suppliers were core product suppliers. The steel fabricators have a number of different suppliers. Two key products include: steel and steel protective paint. Steel is purchased and the protective paint is usually subcontracted. Firm 28 has six main supplier types, including steel merchants, painters, shop detailers, bolt suppliers, connexions and equipment suppliers.
Steel distributor	For this steel distributor all the steel for the construction industry is sourced from the parent firm. Different sections are purchased from different mills and each transaction is treated as if the firm was an external unit. *If we purchase from Whyalla or Port Kembla it is a straight out business transaction. We do not have any national price agreements upstream for projects. A large project could go for longer than twelve months and we have an agreement in the beginning on price. In the meantime if we have a price rise we will go to them and tell them but I shall sell them the old price. I am subject to that price rise from my clients, but what I do is give my clients a window of opportunity to get in as much order as possible before the price rise.* So the price rise does not tend to be passed onto Firm 157 only if there is a variation to the original order and there was a price rise for that particular product. *The biggest problem actually comes in with variations. If for example the order is for 10 tonnes at an old price and then because of variations to the project they now need 20 tonnes then the new quantity will be subject to the new price. I negotiate a price with my client and BHP has set pricing – in most cases if there is a price rise on Friday it comes in effect on Monday. 70% of the product is sourced from the mill in this state – that is only because that is where their most commonly purchased product is manufactured. There are different types of mills in various locations. If you took a very very broad overview we purchase everything from BHP and we then source it from a particular location depending upon the product required.* Sourcing strategies for this firm may change dramatically in the future as the distribution divisions were being divested. *If you took a very very broad overview we purchase everything from BHP and we then source it from a particular location depending upon the product required.*

(Table 8.1 continued)

Table 8.1 Continued

Customer firm	Supplier types
Protective painter	Firm 196 has four major supplier categories that concern Firm 157, which includes: abrasive suppliers, paint, fuel and solvent, and they are all product commodities. Abrasives, paint and solvent are core commodities and fuel is a non-core commodity. With abrasive suppliers there are three types which are metallic products and include garnet, chilled iron and steel. Chilled iron and steel is sourced from a range of suppliers; however, garnet is sourced from one location in Australia. *Garnet – well there is only one place in Australia where you can get it in volume. It all comes from Western Australia – GMA Garnet and they are worldwide. It is an international company – they are Australian but they sell worldwide and are located worldwide.* *Chilled iron and steel is all from overseas. India, England or America or whatever. We deal with an agent who deals with the materials suppliers overseas.* *We deal with a couple of agents that deal with these types of abrasives.* *Sometimes they run out of supply so you have to deal with a couple of them. You can't be sure of them. We deal with about 2 or 3.* The next major supplier type is the protective coating; that is, the paint suppliers, of which there are five nationally operating suppliers. Two other major types of products used by the painters on a large scale are solvent and fuel. Solvent is used to clean the spray equipment and fuel is used to operate the equipment and for transporting of finished painted steel.

community footprint – that is, it impacts on a large part of the Australian business community because of its operations. The common core products are iron ore, coal and limestone. The firm purchases the coal and the iron ore from another division. Some iron ore is purchased externally. Limestone products are purchased externally from the company and are sourced overseas.

In steel-making it is the properties of the steel that is critical. These purchases are not so much driven by price but driven by chemical composition because different blends of iron ore and coal will provide the optimal ratio of raw material into the furnace to output from that furnace. So, iron ore is purchased internally and externally. Externally it is purchased from a firm in Tasmania and also it is purchased from India, South America and Canada.

Firm 135 typically categorizes its commodity suppliers and develops firm–firm procurement relationships based upon an explicit categorization system. The firm is explicit in how they categorize their suppliers in terms of risk and expenditure. Other firms tend to be implicit and less structured, but nevertheless underpinning their categorization is a consideration of volume of expenditure and a risk assessment. Firm 135 sources from 49 different major commodity groups. Firm 135 *'has been into strategic sourcing for some years and therefore have different strategies for different goods and services'*.

Risk is a combination of the internal significance of the commodity and the external supply communities. The Global Procurement Supply Manager assesses each commodity and then develops strategies to match product to market characteristics. Those strategies will be different according to how Firm 135 perceives that particular commodity.

This largely means that commodities are allocated within a 2 ×2 matrix producing the following four categories:

- Low risk/low expenditure: tactical purchasing
- High risk/low expenditure: strategic security
- High expenditure/low risk: leverage
- High expenditure/high risk: strategic critical.

8.5   Procurement relationships

This section describes the sourcing strategies and interactions used by firms involved in the structural steel supply chain, including: fabricators, painters, distributors and manufacturers.

8.5.1   Steel fabricators sourcing strategies

For details of the supplier markets and the sourcing strategies developed by Firm 28 and Firm 157 (fabricators), refer to Tables 8.2 and 8.3 respectively: Matrix of sourcing strategies by steel fabricator Firm 28 and Matrix of sourcing strategies by steel fabricator Firm 157.

Table 8.2 Matrix of sourcing strategies by steel fabricator Firm 28

Supplier types	Sourcing behaviour
Strategic critical core common products high risk/high expenditure	The steel subcontractor for Project 1 was Firm 28. Firm 28 was also one of the suppliers for Project 4. Firm 28 also tendered on Projects 2 and 3 – they are a direct competitor to the steel fabricators for major building projects, particularly projects that require complex structural problem-solving capabilities. Firm 157 were the subcontractors on Project 2. A matrix for both firms and their suppliers is developed as they are direct competitors. Firm 28 views two suppliers as critical to their ability to be able to impact upon their price and ability to win tenders. These two suppliers are the steel merchant and the painters. Firm 28 views the shop detailing supplier as a critical supplier, but not in terms of direct price.

(Table 8.2 continued)

Table 8.2 Continued

Supplier types	Sourcing behaviour
	Firm 28 approaches the steel merchant market project by project. The choice of supplier is based upon price alone and can vary depending upon the current stock levels of the supplier. There are three suppliers in the market located in The state. Firm 28 is much smaller in size and relative power than the merchants and the relationship is typically transactional. Unlike some of Firm 28's competitors, it does not have the purchasing volume and leverage over this particular supplier. The merchants then purchase from the distributor (Firm 458), which is a division of the Firm 135, the national steel manufacturer. The distributor is treated as a separate firm.
	The next category of supplier is the painters, of which there are two that Firm 28 sources from regularly. This is a small market and for protective coatings there are only three suppliers. Firm 28 generally goes to two of these and balances the workload similar to the welding shops that supply the cleats. The painters are small firms; however, a little larger than the welding shops. Both painters were used for Project 4 as one painter alone could not get the volume through in time. The choice of supplier is typically based upon the firm's ability to supply the service in a timely manner. The prices are based upon a rate that is supplied by the firm. Firm 28 calculates the final lump sum that is used for the tender sum. Again, these firms act similar to the welding shops as an extension to Firm 28's factory. It is noted that Project 4 was such a large project that supply was provided by all three merchants. Also, in this instance another steel fabrication shop was subcontracted to do parts of the simpler fabrication of ties and bases. This firm is usually a competitor to Firm 28, but for the simpler less complex projects. Fabrication for Project 1 was all completed by Firm 28 and steel supply was provided by one of the merchants.
Strategic security common services low exp/ high risk	The small connexions (cleats) are usually fabricated by another firm. These small fabrication shops are located close to Firm 28's fabrication factory and there are generally four to five of these suppliers who employ approximately five people. Firm 28 tends to know the workload of these small shops and transacts with the firms based on this. These small shops act as an extension to Firm 28's factory, in a similar manner that the façade subcontractor views the aluminium welding and façade installation suppliers. They are trusted and are on a preferred list of suppliers. The other supplier that is considered to be high risk but is low expenditure is the shop detailers. This fabrication firm subcontracts the shop drawings and does not do them in-house. The cost of this is not high in comparison to the purchase of the steel or the painting; however, the role they play is critical.

Table 8.2 Continued

Supplier types	Sourcing behaviour
Tactical common services and core common products low risk/low expenditure	For equipment hire to erect the steel on-site this is usually a matter of what size cranes are available at the time of erection. This package therefore is not tendered. This is only necessary if there is specialist equipment required. Firm 28 has their own transport and cranes.

If we have specialized cranage requirements we will subcontract that. Usually every job we do requires a specialist crane supplemented by smaller cranes. It is the smaller ones that we own and then we hire the others. There are three crane hire suppliers and the transaction is usually a specialist service that is negotiated just prior to erection. There is a certain degree of risk and variability involved in this transaction. You can't really tender this package – what we have to do is work out what we need. Then we have to work out what crane will do the job and then we talk to the three people and ask what cranes do you have available and what price for the duration of the project. It is a different type of tender – we usually get it on a weekly rate. But it is a different thing because all the crane suppliers don't have the same crane – so we have to take a punt on some things.

For Project 1 the contractor had their own crane on-site and so Firm 28 used their own equipment and the contractors; whereas for Project 4 they subcontracted to two specialist hire firms. This was done on a weekly basis and therefore the firm on-site often changed from week to week.

The last major supplier to Firm 28 is the bolt supplier. They have an agreement with one firm for the supply of all bolts.

Firm 28 does not have any downstream alliances with its suppliers as they tend to view their position in the relationship as more powerful. However, they have in the past discussed and aligned processes and construction methodology with their suppliers. Their own competitive advantage is typically in two areas: innovative construction methods and a lean management structure. Each project is analysed and treated uniquely to identify efficiencies. Firm 28 does employ more engineers than their competitor Firm 157, and they attempt to get involved in early design work on projects and engineer the construction methodology to suit their own erection processes. The methods will impact the programme of the project rather than actual direct cost, although eventually it will impact upon cost. The management of Firm 28 are all engineers and are involved in projects.

Our other strength is that we employ a lot of engineers so we like to get into the project very early and understand the erection methodology from the members programme. A lot of details can change to make the programme a lot quicker or the fabrication is a lot simpler or we can erect things concurrently.

Table 8.3 Matrix of sourcing strategies by steel fabricator Firm 157

Supplier types	Sourcing behaviour
Strategic critical core common products high risk/high expenditure	Firm 157 has similar supplier types as Firm 28, including structural steel and painters. However, unlike Firm 28, Firm 157 does many other activities in-house and therefore do not require suppliers. For example, Firm 157 does not source shop detailers, crane equipment and small connexions.

For structural steel supply, Firm 157 sources directly from the distributor of the major steel manufacturer in Australia, whereas Firm 28 does not. The distributor acts in a similar fashion to the state sales division/distributor in the glazing supply chain, in that they are vertically integrated and part of the national manufacturing firm. However, the distributor does compete with other steel merchants similar to the glazing distributor. In the state where Firm 157 sources steel there are three such merchants; however, Firm 157 only sources from two. These other merchants purchase from the distributor and also directly from the manufacturer. Sourcing directly from the steel manufacturer can only be done if the customer purchases a certain volume. The choice of merchants is based upon quality of service and price. The quality of service is the decision criterion for sourcing from these two merchants and on an individual project basis the price determines which firm supplies steel for the particular project. The steel fabricator interacts with the merchants outside of the boundaries of the project.

The protective coating subcontractor, or 'painters' as they are referred to, have the largest operation in the state. The supplier is the '*main subcontractor that they (we) have*'. Although there are four firms in the market in the state, the steel fabricator is typically compelled to single source from one firm.

The painters that we use are the biggest and they are really about the only ones that can handle our jobs. There would be about three or four of these guys. We still work with this one firm. We don't go shopping around. These guys are local. We single source. Painters if we could we would go to someone else but there is no one else that size that can handle our jobs. The choice is based upon a monopoly really. If we could open it up a little we would – actually if we could do this in-house we would. But don't tell them that when you go there and talk to them.

The interaction between the painters and the fabricator is high and the relationship extends beyond the project transaction. However, it is not one built upon trust; it is built upon quality of service and product and capability to perform.

| Strategic security common services low exp/ high risk | Purlins are sourced from two 2nd order manufacturers; one is a firm within the steel manufacturer's group and the other is independent. The second supplier is typically cheaper than the steel manufacturer's processor; however, they are not a national supplier. Location is critical for the supply of purlins as transport costs are prohibitive. |

Table 8.3 Continued

Supplier types	Sourcing behaviour
	Now Supplier 2 are cheaper than Supplier 1 but it just happens that they are in Melbourne and we might have a job in NSW and if they are cheaper in NSW than Supplier 1 and if we have a job in NSW we will get them to Supplier 2 purlins for an Adelaide job they will transport from Melbourne to Adelaide. Transport costs become an issue and that is what we look at – we don't have to do anything to the purlins just order and deliver and fix.

The market size for purlins is 3 in the state the fabricator typically operates in and 4 to 5 in the other states. Firm 28 did not discuss suppliers for purlins and therefore it is assumed that this is taken care of within the structural steel contract.

For certain projects Firm 157 might require some subcontracting for fabrication. There are two main reasons for this: first, that it is a specialist fabrication project, or second, that there is too much work on within their own fabrication shop. Typically it is because there is specialist fabrication required; for example, rolling of steel pipes that requires a heating process. There is only one firm in Australia that has the machine and provides this service and they are located interstate (Melbourne). Another type of specialist fabrication is cold bending steel. There are three firms in Australia that provide this service.

There are no subcontracts in these cases – you just have an account with them – they just give you a quote a price and that is that. They bill you and well you send the drawings and they do. We single source for both of these guys. It is quality but there is no competition.

| Tactical common services low risk/low expenditure | Weld and paint inspectors are two other types of supplier to this firm. They each supply a service. Weld inspectors test as part of this steel fabricator's Quality Assurance system. There are between five and six firms in this market; however, the fabricator sources from one firm and the decision of which supplier to choose is based upon trust and a long history between the two firms. The degree of interaction between the two firms is quite low – notification of testing and auditing required, distribution of project specification requirements, testing and then notification of results by supplier. |

Similarly, there is third party inspection to confirm that the paint is correct for their clients. There is a subtle difference between the two services. Painting inspection is inspection on another supplier, whereas the welding inspection is inspection on their own work. The choice of supplier and the market is similar, though.

We have another firm that inspects our subbies. Something that costs $50,000 to paint in the workshop may cost $2m to paint on-site. With all the delays and litigation, etc. This gives the builder a comfort factor. This is our proactive approach. We do the same thing in engineering projects if we do them. There are about five or six possible firms. The choice is based upon trust which is based upon history with these guys. The nature

(Table 8.3 continued)

Table 8.3 Continued

Supplier types	Sourcing behaviour
	and degree of interaction is low. The significant role that the project plays in the supply chain as the initiator of the chain and the impact on formation and creation of chains at this level is best described by the fabricator: *All of the suppliers are based upon projects. There is an order and it is always related to a project. It is an order and invoice issue.*

Leverage
  core common
  products
  low risk/high
  expenditure
	We will send an order for say 2,000 bolts and then they will send that.
	With the steel merchants we give them a list of the steel from our takeoff and then they book it in. We need five pieces of this and two pieces of that for that week and then the next week we order another lot. So on the orders we send an order number with X number of pieces we need and they will deliver that. They will then invoice that to that order and then the next order will be filled and so on. There will be a stack of invoices – we are purchasing a product and that is it – it is not a progress claim. The rate for that steel is the rate for whatever you have negotiated for that job or the discount on that standard price list. So one project will have heaps of invoices.
	There are typically long-term accounts set up as part of the procurement relationship sit in the background. There are established prices for products based upon volume demanded by the upstream customer. However, the individual project can play a role in re-establishing these prices on an individual basis.
	Our standard discount might be 10% and you might negotiate on that job and they will say we can give you special deals on the steel so that works out to be 11% on this job. So when you buy it you put that job name and number and when the steel comes it will be at that price. So it becomes a project price. But we have a running account.
	Painters do the same to us – it is based on tonnage. They will invoice for painting 20 tonnes of steel. There are no progress claims involved; they are accounts.

Firm 28 categorizes their main suppliers according to the following:

- high risk/high expenditure (steel merchants and painters)
- high risk/low expenditure (cleats and shop detailers)
- low risk/low expenditure; comparatively low expenditure (crane hire and bolts supply).

Firm 157 categorizes their main suppliers according to the following:

- high risk/high expenditure (steel merchants and painters)
- high risk/low expenditure (purlins and special fabrication)
- low risk/high expenditure (bolts supply)
- low risk/low expenditure; comparatively (welders and paint inspectors).

Figure 8.1 is a sequence diagram for the procurement of suppliers by Firm 28, the steel fabricator, for Projects 1 and 4.

8.5.2 Painters strategies

For details of the sourcing strategies developed by Firm 196, refer to Table 8.4 Matrix of sourcing strategies by steel painter. Suppliers and their interactions are classed according to risk and expenditure in the following manner:

- high risk/high expenditure (protective coating, solvent and fuel)
- low risk/high expenditure (abrasives iron and steel, protective coating)
- low risk/low expenditure (garnet).

The most significant point to make regarding these particular supplier classifications is that protective coating can change. The change occurs because on a particular project the specification is either restrictive and does not allow any alternative products or alternatively the specification does allow a different type of protective coating system being put forward in the tender stage. If an alternative is allowed, then Firm 196 creates a wider market and has less risk and more purchasing power in the market. The same commodity will then be sourced in a different manner and there will be more interactions in the marketplace with a greater number of suppliers.

8.5.3 Steel distributor sourcing strategies

For details of the sourcing strategies developed by Firm 458, refer to Table 8.5 Matrix of sourcing strategies by steel distributor. Many steel fabricators source from the steel manufacturer's (Firm 135) distributor (458), which is within their group of companies. Sourcing strategies for this firm may change dramatically in the future as the distribution divisions were being divested. Currently, the sourcing strategy is considered to be low risk and high expenditure.

A similar dynamic occurred in the glazing industry approximately 12 years ago. The large glass manufacturer was part of a group that was involved in mineral and sand extraction. Part of the production chain was separated as glass manufacturer became a separate firm. This is about diversification and mergers. The multinationals are seeking different methods and different business opportunities as they react to their competitors and their markets. This has an impact to a certain degree on sourcing strategies and which firms are linked to which firms as the tie between the manufacturing and distribution is severed. Firm sourcing policies don't strictly apply. The glass manufacturer chooses to source from their old mineral/sands parent company for sound business reasons: location to source material and close historical relationships which impact upon production knowledge.

The glass distributor and processor currently are required to source from the glass manufacturer. The steel mills may not have to source from the

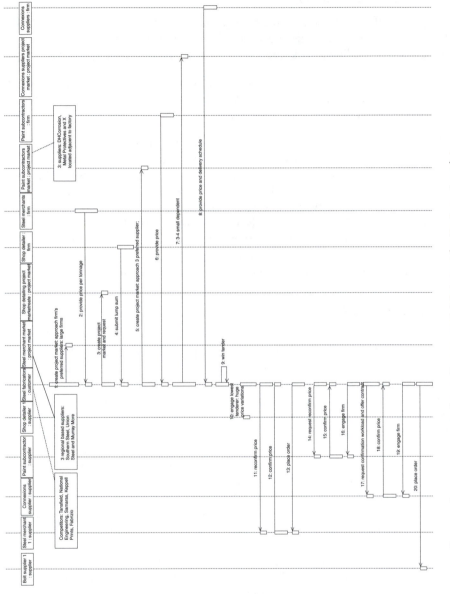

Figure 8.1 Sequence diagram for procurement of suppliers to steel fabricator for Projects 1 and 4.

Table 8.4 Matrix of sourcing strategies by steel painter

Supplier types	Sourcing behaviour
Tactical low risk/low expenditure	The painter for Firm 157 is Firm 196. Firm 196 has four major supplier types: abrasives, paint, solvent and fuel. There are three types of abrasives which are metallic products and they include garnet, chilled iron and steel. Chilled iron and steel are sourced from anyone; however, garnet is a monopoly. *Garnet – well there is only one place in Australia where you can get it in volume. It all comes from Western Australia – GMA Garnet and they are worldwide. It is an international company – they are Australian but they sell worldwide and are located worldwide. Garnet is single sourced. 98% of the product is sourced from one supplier who is the manufacturer. The product is a little more specialized than the other two types of abrasives. Firm 196 approached the supplier and discussed volume, price and quality a number of years ago and little has changed since then. The purchase of this product is not regular and is dependent upon workload as this abrasive is used for different types of work. However, the arrangement still stands.*
Leverage core common products low risk/high expenditure	*Chilled iron and steel are all from overseas. India, England or America or whatever. We deal with an agent who deals with the materials suppliers overseas. We deal with a couple of agents that deal with these types of abrasives. Sometimes they run out of supply so you have to deal with a couple of them. You can't be sure of them. We deal with about 2 or 3. Chilled iron and steel abrasives is largely a transactional relationship and is based upon availability and supply.* The firm will source based upon price and volume and is sourced from agents. It is generally purchased once a month and an order is placed and then it is delivered. There is little discussion and interaction. Firm 196 recycles their own chilled iron and steel but replenishes once or twice a month. The suppliers know that the order will come in.
Strategic critical core common products high risk/high expenditure	The next major supplier type is the protective coating; that is, the paint suppliers, of which there are five. Firm 196 sources from all of the suppliers and *'sometimes (I) shop around'*. In many cases, the source of the paint is dependent upon the project specification paint system which is specified by engineers/architects. The paint suppliers are all national operators and most are international. One supplier manufactures some of its product in Australia. All paint is sourced from the paint manufacturer's agents. The paint supply market can be highly competitive if the paint system is not specified by supplier or specified so exactly that only one supplier provides the product. Each supplier is keen to have their product used and a project can be treated as a highly competitive auctioning and bargaining market. Let us look at this job for Project 2 – I have been told four or five companies I can use and their products. *I went to these people and I said big job and I want a good price. They give me a price and then they come back the next*

(Table 8.4 continued)

Table 8.4 Continued

Supplier types	Sourcing behaviour

| | *day and say how was my price? Then they will come back again and the price gets lower and lower. I say to them well I have got to be honest with you I got a better price and so they will come back to me the next day and say well this is the price now. If you get a legitimate price at the onset and he did not reduce much and it was a good price to start with.* |

Then these other guys meet his price – you know these paint people they are all anxious to get their paint on the job. So we have to make a choice and our assessment and I say well they have all had the opportunity to get me a price and they suddenly give a better price at a later phase than at the beginning when we had to put the price in – whereas another guy gave us a good price at the onset. The only way it will really change is if something happens in the job and there is an alternative and we present it to our client and they are happy. If everyone else tries to come down and meet the other price well I generally stick to my word. I shop around for this but it is more fluid I have more discussion. However, with Project 2 the specifications were exacting, the paint, the product and the colour. Firm 196 had little bargaining power and the paint supplier had a monopoly.

However, for the same customer on a different project any supplier could be used. The chain of supply is typically standardized for the abrasives, but does not have to be for the paint.

Leverage core common products low risk/high expenditure

Two other major types of products procured by Firm 196 on a large scale are solvent and fuel. Solvent is used to clean the spray equipment and fuel is used to operate the equipment and for transporting of finished painted steel. Both products are single sourced from the same supplier. All fuel companies have solvents as well. So it is the same supplier for fuel. Firm 196 transacts with the fuel manufacturers' distributor. There are absolutely no agreements and in actual fact if the prices rise overnight then that new price is passed on immediately. The prices are volatile and fluctuate immensely. There is no negotiation possible for Firm 196.

The same product can change on a particular project from a high-risk product to a low-risk product based upon the specifications of the project. The purchasing power alters as the market is wider and alternative products can be offered up by the painter. In this situation, when the expenditure is high the firm–firm relationship alters and the sourcing strategy and interactions changes.

Now on the convention and the bridge I can use any supplier – they just want this type of paint and this type of topcoat. But I can go anywhere. Now it is Firm X again – same suppliers as for the Project 2 and the prices went way down. Same supplier but different rates. It is unbelievable – same guarantees.

Table 8.4 Continued

Supplier types	Sourcing behaviour
	When we have a big job you need to talk to the suppliers. Now the abrasives are no problem it is just so many tonnes per month and it is transactional. Different paint suppliers operate in a different manner.
	For example, one supplier would respond to a request by Firm 196 for supply of paint by manufacturing it and storing it on their own premises and then transporting the product when requested. Another supplier provides a container on Firm 196's premises and then charge for whatever is used from the container. If it isn't used, then Firm 196 doesn't get charged for it. In terms of logistics and physical distribution, the second supplier's method is preferable to the firm. However, if the project requires non-standard work, that is, *coloured jobs*, it doesn't work as well. The interaction during the job with the painters is reasonably low, but at the beginning it is a bit higher in the negotiation period. During a *coloured job* there is more interaction.

minerals and extraction divisions in the future and the distributors will not have to single source from the steel mills. The sourcing strategies used by the manufacturer, Firm 135, are now considered.

8.5.4 *Steel manufacturer sourcing strategies*

For details of the sourcing strategies developed by Firm 135, refer to Table 8.6 Matrix of sourcing strategies by steel manufacturer. As mentioned previously, Firm 135 sources from forty-nine different major commodity groups. The chart in Figure 8.2 is for the $8b worth of spend and the top forty-nine commodities where they spend their money. This is a very generic grouping and the strategy for procurement depends on Firm 135's own explicit assessment of spend and the risk category of the commodity; that is, each commodity is assessed and categorized by risk and expenditure.

According to Firm 135, risk is a combination of the internal significance of the commodity and the external supply communities. This same risk and expenditure process occurs to some degree with all the firms interviewed; however, in a much less explicit manner than the manufacturer. In some cases the risk versus expenditure assessment process is on a project basis and in some cases it is across the entire firm.

For example, the contractor for Projects 1, 2 and 3 and the mechanical subcontractor for Projects 1, 2, 3 and 4 do not as a general rule strategically source across the entire firm. However, the contractor for Project 5 and the façade subcontractor for Project 1 do strategically source across the entire firm. As you move further down the chain and further away from the project and on-site scenarios, the influence of the project on the procurement

Table 8.5 Matrix of sourcing strategies by steel distributor

Supplier types	Sourcing behaviour
Leverage core common products low risk/high expenditure	Many steel fabricators source from the steel manufacturer's (Firm 135) distributor (458), which is within their group of companies. Sourcing strategies for this firm may change dramatically in the future as the distribution divisions were being divested.
	135 was starting as an offshoot company originally as the company to supply the steel making plants. Over the last whatever years it has now come about that the steel mills and steel manufacturing according to the CEO is holding them back a bit. Although it was what they were all about in the beginning he is saying now that 135 is all about Minerals now. So what was started off as a support business is now what he is saying is the main business. Petroleum, for example. Being Australian, we always think of 135 as steel – the CEO is American and he has come in and looked at the books and the company and with no preconceived ideas of growing up with 135 steel, he has decided that 135 isn't really steel – steel is the part that is holding back the company. He only wants to hang onto the sheet and plate side of the market – which is flat products. He wants to float the Whyalla branch which makes the I beams as well as the 458 Group. Which is not a bad thing. We see it as a good thing. It gives us new opportunities actually. This might change where we source materials and it will change our ideas about how we do things. Just where we go from here who knows – that is, if we implement them.
	A similar dynamic occurred in the glazing industry approximately 12 years ago. The large glass manufacturer was part of a group that was involved in mineral and sand extraction. Part of the production chain was separated as glass manufacturer became a separate firm. This is about diversification and mergers. The multinationals are seeking different methods and different business opportunities as they react to their competitors and their markets. This has an impact to a certain degree on sourcing strategies and which firms are linked to which firms as the tie between the manufacturing and distribution is severed. Firm sourcing policies don't strictly apply. The glass manufacturer chooses to source from their old mineral/sands parent company for sound business reasons: location to source material and close historical relationships which impact upon production knowledge. The glass distributor and processor currently is required to source from the glass manufacturer. The steel mills may not have to source from the minerals and extraction divisions in the future and the distributors will not have to single source from the steel mills.

strategies is reduced. However, it is surprising that the aluminium fabricator thinks about their work in a project and non-project manner as they are very much embedded within the manufacturing sector.

After the Global Procurement Supply Manager for Firm 135 assesses each commodity, he then develops strategies to match commodity to market

Table 8.6 Matrix of sourcing strategies by steel manufacturer

Supplier types	Sourcing behaviour
Strategic critical core common products high risk/high expenditure strategic security common services low exp/ high risk	Firm 135 sources from 49 different major commodity groups. The following chart is for the $8b worth of spend and the top 49 commodities where they spend their money. This is a very generic grouping and the strategy for procurement depends on Firm 135's own spend and the risk category of the commodity; that is, each commodity is assessed and categorized by risk and expenditure. Risk is a combination of the internal significance of the commodity and the external supply communities. The Global Procurement Supply Manager assesses each commodity and then develops strategies to to match product to market characteristics. Those strategies will be different according to how Firm 135 perceives that particular commodity.

This largely means that commodities are allocated within the 2 × 2 matrix:

- low risk/low expenditure: tactical purchasing
- high risk/low expenditure: strategic security
- high expenditure/low risk: leverage
- high expenditure/high risk: strategic critical

If Firm 135 has a high spend for a particular commodity and there is a high risk associated with that commodity, then they attempt to enter into longer-term contracts. In many cases there are strategic alliances developed for the commodities in the top right-hand corner of the diagram; for example, energy, iron ore. For those commodities that are placed in the top left-hand quadrant, Firm 135 doesn't have a high spend and there is not a lot of choice in the market. In this case the commercial advantage is with seller rather than the buyer; that is, the suppliers. Therefore they tend to try to convince suppliers to enter into long-term contracts just to ensure supply. In contrast with the bottom right quadrant, the firm is spending a lot of money and there are a number of firms in the market and therefore competition is high as there are a number of alternative suppliers. Therefore, the buyer – that is Firm 135 – has the commercial advantage rather than the seller; therefore they tend to go for shorter-term contracts to maintain a competitive environment in the market.

There is a high level of automation across the Firm 135's supply. They have identified what they buy and developed detailed descriptions which are specified in the agreements, whether long or short term. Therefore, staff in either plants or offices are able to requisition goods and services by just referring to the Supply Catalogue. The Supply Catalogue specifies a number of items, including the lead times, who the supplier is and the fixed prices. Commodities are then purchased electronically through the firm's systems and the request is sent to the supplier. The bottom left quadrant supply contracts were negotiated for approx 12 months.

For the Strategic Critical suppliers, Firm 135 has developed strategic alliances.

(Table 8.6 continued)

Table 8.6 Continued

Supplier types	Sourcing behaviour
Tactical common services low risk/low expenditure	In the Strategic Critical Quadrant they use exactly the same processes as you would for any of the others, particularly the bottom right, which is termed the Leverage Quadrant – the difference is understanding where the balance of power lies. In the case of the Leverage Quadrant, the firm recognizes that, as the buyer, they have a lot of choice and they can spend a substantial amount of money and clearly the leverage is with the buyer. Therefore, they are more demanding. The process of assessing the market and the commodity is the same – the difference lies with whether or not the power in the relationship is with the buyer or the seller or it is balanced. Firm 135 had just recently put out a global contract on tyres. The process of assessment is similar to that taken with other commodities, for example, stationery, or with any of the chemicals that are of a critical nature strategically. That is, they have representation from the Operations as well as the Supply Community. In this manner, the Global Procurement Manager knows which sites use which commodities and what their requirements are.
	For example, with tyres an assessment was completed to determine which sites spend the most money on what types of tyres. Representatives from both the Supply Community at each of those sites as well as the firm's Operations staff who are benefiting from the use of those tyres, whether it is a mining site or a steel site, form a Cross-functional Team. This team will review where the spend is, what is critical on the particular product and the particular tyre that the firm requires.
	The attributes of each procurement relationship vary, but are typically grouped by how they have been assessed and where they have been positioned. The most important attribute that varies is the time-frame. Leverage contracts tend to be annually reviewed and then they are rolled-over for another 2 years, so they tend to be for 3 years. So, every 3 years the firm approaches the market again.
	However, the Strategic Critical contracts can be for a time-frame of 15 years. This is not the usual case, but it could be possible. The length of the contract is a reflection on how critical it is. For example, in the steel industry there is a particular contract that has a value of approximately $60m. The particular contractor has invested infrastructure and equipment to enable the contract to be executed. Since up-front investment was required for this particular contract, the contract has been established for approximately 15 years. Another long-term contract established is for 6 years. However, many of these contract durations are for 3–5 years.
	There is no established rule that states all contracts will be for a set time period; instead, contracts will vary according to the strategic nature of the contract. A contract which is Strategic Security, positioned in the top left of the quadrant, would tend to go to long term; but in bottom right, where there is a high

Table 8.6 Continued

Supplier types	Sourcing behaviour
	spend, a wide choice in the market and non-critical, would tend to be short term – that is, 1 year with roll-out to 2 or 3. For Tactical Purchasing the relationships are more transaction and the Purchasing card is used. In this manner individual sites determine their suppliers based on the local market characteristics. In the tactical purchasing area, often suppliers are required to tender and the decision criterion tends to be lowest price. 'Price' is typically evaluated in terms of 'cost' and it is the cradle-to-grave concept that is used. This philosophy is applied to all commodities. It is a risk management approach to procurement. As part of the supply procurement relationship there are a number of sourcing strategies that the firm uses and to manage the relationship the strategy of splitting volume between major suppliers is often used to reduce risk. For example, with tyres, initially the firm was sourcing from two major tyre manufacturers. They then decided to introduce a potential third supplier. The purchasing volume and spend was then divided amongst those three suppliers in the ratio of 60:30:10. Another commodity that was being supplied was single-sourced from an Australian supplier and the firm decided to introduce a second player. They searched the market internationally and located another supplier to act as a competitor to what they felt was a monopoly supplier. The spend was then split 50:50. Over time the conduct and performance of firms changes and the firm changes their strategy accordingly. For example, a particular product commodity has four major suppliers in the Australian market and of those the contract was split between 2. The strategy was to share the Australian business 50:50, but in reality it was 40:40 because the other 20% was going to an overseas supplier. 20% of supply was tendered in the international market. At the last review the strategy was changed and 100% of supply was awarded to one of those Australian suppliers, since this *supplier is now acting in a manner that is consistent with how we want them to deal with us – we have the behaviour between the two companies that benefits both organizations.*
Leverage core common products low risk/high expenditure	The attributes of each procurement relationship vary, but are typically grouped by how they have been assessed and where they have been positioned. The most important attribute that varies is the time-frame. Leverage contracts tend to be annually reviewed and then they are rolled-over for another 2 years, so they tend to be for 3 years. So every 3 years the firm approaches the market again. However, the Strategic Critical contracts can be for a time-frame of 15 years. This is not the usual case, but it could be possible. The length of the contract is a reflection on how critical it is. For example, in the steel industry there is a particular contract that has a value of approximately $60m. The particular contractor has invested infrastructure and equipment to enable the contract to be executed. Since up-front investment was

(Table 8.6 continued)

Table 8.6 Continued

Supplier types	Sourcing behaviour

required for this particular contract, the contract has been established for approximately 15 years. Another long-term contract established is for 6 years. However, many of these contract durations are for 3–5 years.

There is no established rule that states all contracts will be for a set time period; instead, contracts will vary according to the strategic nature of the contract. A contract which is Strategic Security, positioned in the top left of the quadrant; would tend to go to long term but in bottom right, where there is a high spend, a wide choice in the market and non-critical, would tend to be short term – that is, 1 year with roll-out to 2 or 3.

For Tactical Purchasing the relationships are more transaction and the Purchasing card is used. In this manner, individual sites determine their suppliers based on the local market characteristics. In the tactical purchasing area often suppliers are required to tender and the decision criterion tends to be lowest price. 'Price' is typically evaluated in terms of 'cost' and it is the cradle-to-grave concept that is used. This philosophy is applied to all commodities. It is a risk management approach to procurement.

As part of the supply procurement relationship, there are a number of sourcing strategies that the firm uses, and to manage the relationship the strategy of splitting volume between major suppliers is often used to reduce risk. For example, with tyres, initially the firm was sourcing from two major tyre manufacturers. They then decided to introduce a potential third supplier. The purchasing volume and spend was then divided amongst those three suppliers in the ratio of 60:30:10.

Another commodity that was being supplied was single-sourced from an Australian supplier and the firm decided to introduce a second player. They searched the market internationally and located another supplier to act as a competitor to what they felt was a monopoly supplier. The spend was then split 50:50.

One of the significant attributes of the procurement relationship that this firm has with its suppliers is that they deal with their supplier's supplier.

Over time, the conduct and performance of firms changes and the firm changes their strategy accordingly. For example, a particular product commodity has four major suppliers in the Australian market and of those the contract was split between 2. The strategy was to share the Australian business 50:50, but in reality it was 40:40 because the other 20% was going to an overseas supplier. 20% of supply was tendered in the international market. At the last review the strategy was changed and 100% of supply was awarded to one of those Australian suppliers, since this *supplier is now acting in a manner that is consistent with how we want them to deal with us – we have the behaviour between the two companies that benefits both organizations.*

Table 8.6 Continued

Supplier types	Sourcing behaviour
	This is the nature of the sourcing strategies; however, there is the management of the particular contract that is considered as well. Senior Managers interact with the Supplier Senior Managers, so it is not just the interaction between the Buyer and the Seller between the two organizations. Part of the interaction at the operational level involves joint Business Plans between the operations people and the suppliers. There is an annual review with key suppliers. It is not just a transactional exchange. This is particularly so for the top right and the bottom right quadrants. Most of those suppliers will have joint Business Plans. The purpose is to get agreement with both organizations, that it would help both if we addressed a particular business opportunity.
	The manner in which the firm formally organizes this interaction is known as CLAN – centre lead action network. CLAN is where employees who are involved in the key operations and buyers at the different sites create a network of contacts across sites. The purpose of that community is to regularly get together and review supply through monthly reviews and develop action plans every 6 months. Involvement and success of such networks is, of course, dependent upon the supplier's commitment, which is largely dependent upon leverage that the customer has. Leverage in the market depends upon amount of purchase and/or duration of contracts. They get a very good sense 6 months into the contract of whether they are going to get an extension and how much extension – because of their performance. We are very driven for performance-based contracts.
	There are normally only two parties to contracts – it is normally between this firm and one supplier. However, there are some arrangements where they will have a number of parties. For example, the hydrochloric disposal contract mentioned previously was an agreement that this firm had with 3 other parties. The general 'rule' is that for product commodities it is one-on-one and for complex scenarios that require integration of products and services where there are 3, 4 or 5 contractual parties a special agreement between all parties will be developed. This is not common and is emerging as a strategy.
	In summary, supplier management of the Strategic Critical and Strategic Security suppliers is typically high, whereas Tactical Purchasing is typically low. Interaction between Tactical Purchasing and the firm is often high, whereas sometimes interaction between Strategic Critical is low. One of the significant attributes of the procurement relationship that this firm has with its suppliers is that they deal with their supplier's supplier.
	You find you can only go so far with your immediate supplier; we find that you need to have your supplier's supplier doing certain things that enable our suppliers to do what we want

(Table 8.6 continued)

Table 8.6 Continued

Supplier types	Sourcing behaviour
	them to do. We take this on – sometimes we may work through a contract it depends on the strategy. *For example, if they work with distributors/agents instead of original equipment manufacturers, then we will go back to the OEM. Also, if the current strategy of working with the distributor is not advantageous because of poor performance, then they will directly with the OEM.* Contracts are generally performance-based contracts. There are separate assessments of the supplier; the firm expects cost savings of X over the contract and they typically expect that the total cost will come down. There is a scorecard built up each week and the supplier can see whether they will get roll-over. There is a high level of visibility to the supplier on how they are performing during the contract. This firm operates as a large multinational with an enormous amount of purchasing power in many markets and uses that power to assess markets, assess their own leverage, assess risk, assess commodities and structure procurement relationships accordingly. Commodities are not categorized by their physical properties, but at the strategic level are categorized according to market-level characteristic and firm-level risk and expenditure attributes. Procurement relationships are then developed according to strategies relying upon attributes related to: supplier sourcing, commodity transacted and transaction exchange management. Ultimately, the supply chain for construction relies upon firms *digging out of the ground*. The sequence of interactions between firms, projects, markets and commodities to form supply chains to link the ground to the site has been described for those chains where a high level of intermediate processing is required.

characteristics. Those strategies will be different according to how Firm 135 perceives that particular commodity. In summary, this largely means that commodities are allocated within the 2 ×2 matrix:

- low risk/low expenditure: tactical purchasing
- high risk/low expenditure: strategic security
- high expenditure/low risk: leverage
- high expenditure/high risk: strategic critical.

8.5.5   *Aggregated project supply chain organization: supply channel*

Figure 8.3 maps the organizational structure for transfer of ownership of steel products in the construction industry from the manufacturer who

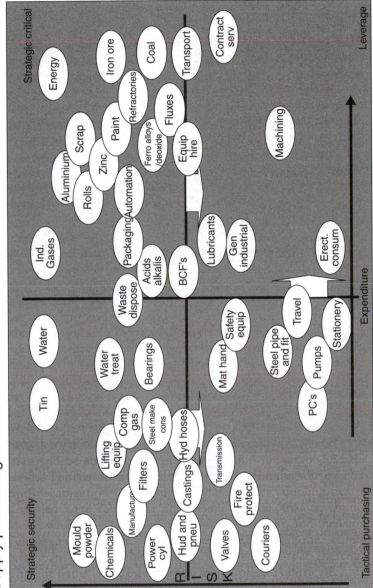

Figure 8.2 Risk versus expenditure supplier categorization by steel manufacturer.

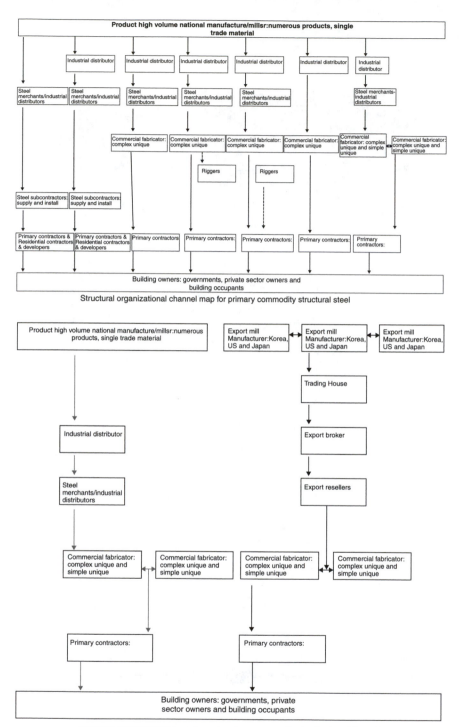

Structural organizational channel map for primary commodity structural steel

Structural organizational channel map for primary commodity structural steel (import)

Figure 8.3 Structural organization channel map for primary commodity structural steel.

extracts raw materials from the ground to the building owners. The channel map indicates the structural organization of primary commodity suppliers and the types of firms that are involved. There are currently seven tiers available for the transfer of ownership of the steel commodity to flow through, disregarding imports; however, there was no chain identified in the study that included all seven tiers. If the steel fabricator imports steel, then the channel does not increase in length, as this type of fabricator usually is importing through a combined purchasing group with another fabricator. They tend to employ their own riggers to install on-site and therefore contract directly to the primary contractors. This class of firm tends to be smaller and can not purchase directly from the major national steel industrial distributor. The leader in the major national steel industrial distribution market has until recently been partially vertically integrated with the manufacturer. However, they still compete with other distributors and now this tier will become even more competitive as they are not partially vertically integrated. Eight chain options were identified, including one standard import channel.

The organization of the channels is relatively simple in that most steel flows through the industrial distributor and then the smaller steel merchants. Fabricators with a larger purchasing volume are able to procure directly from the industrial distributor. Steel merchants who repackage who have a high purchasing volume are able to procure directly from the manufacturer. The map only focuses upon building structural steel elements (I beams, C sections, U beams, etc.); there are a number of other steel products that have not been mapped. Contrary to the previous metal commodity map (Figure 7.8 Structural organization channel map for primary commodity aluminium, in Chapter 7), the primary contractors do not purchase directly from the industrial distributors. Therefore, the steel channel map appears to have a more rigid structure.

8.6   A final word

Chapter summary

1   There is a relationship between a firm's turnover and the structural organization of the supply chain; the greater the turnover, the greater the likelihood of a higher volume of product purchased and therefore the greater the likelihood of being able to purchase directly from the distributor, thereby reducing commodity owner transferral steps and thus the number of procurement activities.

2   The market structures for steel fabricators, merchants, distributors and steel painters are all stratified. Project risk, commodity complexity, purchasing volume and time performance are often the reasons leading to stratification. Once a particular subcontractor has been

selected, then it is typical that a particular path will be taken. At times when there is a downturn in the industry there are movements between these internal market groupings and the boundaries become much more fluid and dynamic – with smaller and less competent firms tendering for projects beyond their capability and larger firms who produce more unique and specialized commodities tendering for smaller projects.

3   Ultimately, all firms tended to describe their firm behaviour within the markets and in particular their competitive advantage in terms of price. They tend to believe that price is the ultimate and final reason why the customer chooses one firm over another, as it is often a major decision criterion for themselves. However, having said that, it is apparent that competitive advantage and thus supplier choice is not as simplistic as this.

4   Competitive advantage is a complex issue and at times can not really be attributed to one single factor. The reason why a customer makes a decision of one supplier over another is varied and can be as a result of a combination of two factors. It seems that often a small group of suppliers may be chosen to tender within a project market based upon commodity type and then may compete based upon another factor. For example, even if the type of commodity being supplied is of a similar character, the final choice of supplier may rely upon the relationship of the commodity to other commodities that are being supplied – which is project-specific. Alternatively, the final choice may rely upon price or lead times or the history of the relationship between customer and supplier.

5   Design complexity in projects – steel fabricator increasing the level of design service on projects is seen as either a business opportunity or a business risk and constraint. Steel fabricators are experiencing a change in the level of design service that they are now required to do compared to the recent past. One firm takes it as a business opportunity for competitive advantage and the other takes it as a business constraint.

6   There is an increase in importation and the issues related to this are the risks associated with lead times, commodity traceability and quality assurance – it increases the length of the supply chain because there are more tiers.

7   Steel fabricators describe their products in groups, including: common core products; common non-core products or secondary products and common services. Typically, they then described and grouped them according to whether they commonly/frequently purchased or sourced from them; the volume of purchase and how critical they were to their

success. Firms source their suppliers according to the following: high risk/high expenditure; high risk/low expenditure; low risk/high expenditure and low risk/low expenditure; which is not an uncommon model – however, not all firms who are direct competitors source the same suppliers on the same basis (i.e. firms competing categorize suppliers differently).

8 Major construction materials manufacturers experience similar dynamics in relation to productive functions along the chain that are either in-house or out-sourced – various divisions of companies are divested and as such distribution can become an out-sourced activity. This does, however, create historical ties that are difficult to explain years after the divestment. It is also a problematic period during which the divestment is occurring, as there is a transition time when internal divisions are pressured to compete in the market and yet do not enjoy the benefits of being internal to the company.

9 The steel manufacturer is perhaps the most advanced of all firms interviewed in this study in relation to supplier management and supply chain management; with many systems in place to co-ordinate, develop and monitor suppliers' suppliers. Specifically, supplier management of the Strategic Critical and Strategic Security suppliers is typically high, whereas Tactical Purchasing is typically low. Interaction between Tactical Purchasing and the firm is often high, whereas sometimes interaction between Strategic Critical is low. One of the significant attributes of the procurement relationship that this firm has with its suppliers is that they deal with their suppliers' supplier. This firm operates as a large multinational with an enormous amount of purchasing power in many markets and uses that power to assess markets, assess their own leverage, assess risk, assess commodities and structure procurement relationships accordingly. Commodities are not categorized by their physical properties but, at the strategic level, are categorized according to market-level characteristic and firm-level risk and expenditure attributes. Procurement relationships are then developed according to strategies relying upon attributes related to: supplier sourcing, commodity transacted and transaction exchange management.

10 Firms do not conduct supply chain management – they co-ordinate and perhaps develop suppliers; but by and large supply chain management is extremely difficult and rare. Even the extremely large firms find it difficult to manage beyond the immediate tier that they contract with. This is not just a construction project problem – it is endemic at all tiers. Different tools are thus required to help deal with this problem. It does not stop firms first mapping their own suppliers in an explicit manner, then understanding their suppliers' suppliers context – particularly the firms which they identify as critical to their core business,

and then from there develop different strategies to 'manage' the supply chain. Different tools are required, for example, for monitoring and diagnostic tools.

11 The metal commodity channel maps differ (i.e. steel and aluminium). The glass industrial organization, which was quite diverse and complex with players at all levels of vertical integration, differs from the steel industrial organization in that the steel chain is typically less fragmented. Firms tend to provide products and services for a single market.

12 A detailed sequence diagram was developed for suppliers to the steel fabricators. Categorizations depended upon an assessment of risk and an assessment of expenditure related to the supplier type, which then impacted upon interactions between customers and suppliers. Risk involved both the market risk and the commodity risk, which is a combination of the internal significance of the commodity and the external supply communities. Upstream firm interactions can be considered in the following five stages: assess what suppliers they need by commodity type and create a description of the commodity (including commodity type, transaction complexity, frequency and financial value); create a project market for the commodity (most typical if firm is supplying directly to site or one tier from site suppliers); assess their own firm attributes (typically, scope and turnover) related to the commodity supplier project market; group by common core/non-core and unique core/non-core; develop sourcing strategies related to market power relationships dependent upon commodity grouping; categorize commodity by risk and expenditure assessments; tender and negotiate with various supplier firms; these negotiations are dependent upon countervailing power, which is based upon perceptions of risk and purchasing power related to the commodity – negotiations differ for each commodity. Increasingly, it is clear that the formation of procurement relationships depends upon commodity, firm and market characteristics and are quite unique and complex.

9 Case studies

Simple and complex core commodity supply chains – mechanical services, formwork, concrete and masonry

What are the structural and behavioural characteristics of simple and core and complex and core supply chains?

9.0 Orientation

> **Box 9.1 Chapter orientation**
>
> **WHAT**: In a similar fashion to Chapters 7 and 8, this chapter provides the results of an investigation into the nature of the procurement relationships that are formed in relation to supply chains; including mechanical services, formwork, concrete and masonry. It involves discussion on firms, projects and market attributes – which underpin the nature of the sourcing strategies and approach and negotiation interactions between customer and supplier firms. Underlying questions which are considered include:
>
> What sourcing strategies are used at various tiers to deliver commodities? Are the chain paths different across the sector? What are the connections between markets, firm types, commodity types and procurement relationships? What is the sequence of events which takes place during procurement?
>
> **WHY**: Again, this chapter provides information about the industrial organization economics of the markets at various tiers in the chain and the way in which this influences sourcing strategies and composition of the chain.
>
> **WHO**: The supply chain clusters described in this chapter are core commodities and then either simple to complex. A complex core commodity supply chain is a commodity which is core to the project contract. A complex commodity chain is one where the nexus of contracts to the project contract is complex in either technology or managerial complexity, that is, requiring unique, customized, specialized

and innovative design and/or construction solutions or a high level of integrative managerial capacity. These types of supply chains can be characterized by innovative design, new materials, juxtapositions of new materials, numerous different types of suppliers, and a requirement to source, integrate suppliers not typically managed previously and often have an element of prefabrication. This is of course an interpretive categorization of supply chains.

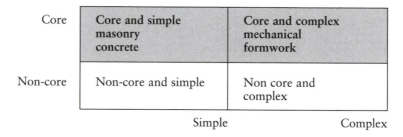

	Simple	Complex
Core	Core and simple masonry concrete	Core and complex mechanical formwork
Non-core	Non-core and simple	Non core and complex

Chapter orientation categorization of chains.

9.1 Introduction

This section presents the results from the interviews with various project managers, firm executives, production and procurement managers involved in the supply chains for commodities that are clustered around four key suppliers to contractors. These form four case studies, which include the following commodities:

- mechanical services
- formwork
- concrete
- masonry.

Fire products and tiles were also investigated and are discussed in Chapter 9 only to indicate the supply channels. After the clients and/or contractors were interviewed, then specific subcontractors were interviewed and various chains were followed in detail related to a commodity product, and they included the following:

Specialist subcontractors supplying complex commodities of products and services: design, supply and install

- Mechanical services for Projects 1, 2 and 3
- Formwork for Projects 1 and 3.

Subcontractors supplying simple or moderately complex commodities of products and services: supply and install

• Concrete for Project 1
• Fire doors and products for Projects 1–5
• Bricks for Project 6.

After the subcontractor interviews, subsequent interviews with fabricators, processors, merchants and manufacturers eventually led to a tracing of more general industry chains for the supply of the following products with the associated projects:

• Concrete for Projects 1 and 3.

9.2 Case study: mechanical services – core commodity

Mechanical services supply is integral to the six projects in varying degrees of complexity and significance. There are three types of markets: residential, commercial and industrial. Projects 1, 2 and 4 were of a commercial nature and Projects 3, 5 and 6 were of a residential nature – Project 3 was a unique situation as it was a high-rise residential project and therefore had elements of both the first market type and the second market type.

9.2.1 Firm details

In Table 9.1 Matrix: mechanical services and equipment firms, describes the details of each of the following firms: three mechanical services subcontractors (Firms 187, 132 and 58); an equipment supplier (Firm 195) and an original equipment manufacturer (OEM) (Firm 197).

In summary, Firm 187 supplies mechanical services to Project 1, Firm 132 supplies mechanical services to Projects 2 and 3 and Firm 58 supplies mechanical services to Project 4. Firm 58 was not interviewed for the project, but they completed a questionnaire. Firm 58 is a competitor of Firms 187 and 132 and sourced from both Firm 195 and Firm 197. During the fieldwork Firm 132 experienced financial difficulties and was unable to complete their contracts and Firm 187 took over these two projects. These firms are located at tier 2 and supply directly to the contractor.

Firm 195 is an equipment product supplier and supplies air handling units and chillers to Projects 2 and 3. Firm 197 supplies air handling units, chillers, coolers and replacement components to Projects 1, 2, 3 and 4 and is located at tier 3 for Project 1 and tier 4 for Projects 2 and 3. Firm 197 is an agent; which means that they order many components on behalf of their upstream clients. Firm 197 also completes a large amount of refurbishment and replacement contracts directly to industrial project clients. The following Figure 9.1 illustrates Project 1 subcontractors' suppliers and Figures 9.2, 9.3 and 9.4 are maps for Projects 2, 4 and 5.

Table 9.1 Matrix: mechanical services and equipment firms

Firm type and project no.	Structure and scope, size – employees and turnover	Commodities and competitors	Competitive advantage
187 Mechanical services subcontractor for Projects 1 then took over for Projects 2 and 3	8 offices, head office is located in Brisbane. Firm is over 100 years old. It is international in scope and has an annual turnover of $90–100m. 600 employees; approximately 100–120 in this division and 30–40 in this office.	Commodities: specialist subcontractor, design supply and install mechanical services for new construction and refurbishment or replacement work for existing buildings. *I guess the market used to be global, non residential construction and new construction where we tender direct to owners. But our new market is refurbishment and we deal direct with owners.* *Competitors sort of vary from one state to the next, there is Tyco that operates in all states that we do but it is a major US company and mechanical services is a small part of what they do. Potentially they are a major threat because they own an electrical company and a fire services company . . . that has resulted from acquisitions. And they don't integrate well and so each company operates quite separately.* *There are probably about 6 competitors of our size, then there is about 8–12 in the next tier down but they compete with us as well. There is a raft of about 6 which are equal in size. They do projects about 20–30M and then next tier do 10–15M and then the rest. I suppose if we went back a few years what we had was the top 6 only being considered for the major projects and because there was such big jobs*	*Basically we are in a commodity market. There is little competitive advantage.* *However it is scope that segments the market but then little competitive advantage amongst the smaller market.*

| 132 Mechanical services subcontractor for Projects 1 and 2 | Scope is state and one division; >20 employees and >50 in firm. Turnover is approx. $6–25m. | *going on then we all chased the big jobs. Then the next tier sort of moved up that is the ones that could cope with the bigger jobs. Trouble is when they moved up they didn't protect the smaller jobs and so when they moved back down there were a lot more competitors and so now there are a lot more competing and there is a blurring of the boundaries between the tiers. Trouble is sometimes on a tender the contractor will then open the tender up for a large project to someone in the third tier and that really stretches the situation. The reason is to bring the price down dramatically. We have just picked up two large projects of 25M which were being done by someone in the third tier they were the top of the tier but nevertheless… well they couldn't cope and have gone in to liquidation and now we are doing all their projects.*

Commodities: specialist subcontractor, design supply and install mechanical services for new construction and refurbishment or replacement work for existing buildings
1 installation of new mechanical air conditioning plant and
2 servicing and refrigeration in existing air conditioning plant | Scope of operations and then price. |
| 195 Equipment supplier for Projects 1, 2, 3 and 4 | International scope; >10 in division. $6–25m turnover of division. 2–5 offices. | Product supplier agent. Nine different air conditioning services products. *We supply everything from a room air conditioner unit to big air handling units. I don't keep a good track of that. We might* | Product and then relations. |

(Table 9.1 continued)

Table 9.1 Continued

Firm type and project no.	Structure and scope, size – employees and turnover	Commodities and competitors	Competitive advantage
		be in the process of supplying at any one time – 20 odd orders for smaller type equipment. But what I would call project type – jobs over $50,000 worth of equipment – we would have 10 or 12 of those at any one time. This is the scale of the Hockey and Netball centre. The contractors who we work for typically group off into the big jobs and the smaller jobs. We have everything from a bedroom in a house to a larger contractor.	
197 Original Equipment Manufacturer [OEM] for Projects 1, 2, 3 and 4	Multinational, 11–20 offices. Division turnover $26–75M. Division number of employees >100 and organization >5000.	We are an OEM. We manufacture for example the compressors for instance. There are similar sorts of components in there – they need replacing – so we have a spare parts operation here. We provide the manufacture, the supply, the after sales service and the spare parts backup. There are 3 markets; residential, commercial and industrial which each have different products. Residential tends to be mass produced. The industrial is specialized and is usually built to custom made specifications and commercial fits somewhere in between. Fixed design and fixed sizes but slight variations incorporated for that design. We have competitors I'd put residential at the bottom there is a lot out there it is mass produced it is easy it is not specialized and	We have a couple of advantages over the other two. We have a more specialized engineering advantage over the other two. We have an extensive backup – we have the largest service operation in Australia. Also by default we are a local manufacturer and so this is important.

it is fairly easy to set up. South East Asia can produce these at a cost effective rate. A 40ft container can be packed to the rafters and can be shipped in fairly cost effectively.

Our commercial – the equipment is a lot bigger and bulkier. It can be produced overseas but transport is prohibitive. So it tends to be built locally and our competitors do this as well. There are five competitors in this group in Australia. We used to be part of one of our competitors up until about five years ago. We left them in 1992–93. Prior to '86 we were owned by Board Warner that is both here and international. They wanted some cash flow so they sold off a few divisions and we were one of them.

Internationally we became a company in our own right – over here the license for our company was bought by our competitor. Anyway our company grew in those years and they bought back the license. Then this competitor sold up about a year ago – the factory was dismantled. A French company bought and all the Australian guys went over there to set up the factor and production so you basically buy an Australian product. In commercial most build locally. In residential everyone sources from overseas and in industrial it tends to be built locally but source some specialist items overseas.

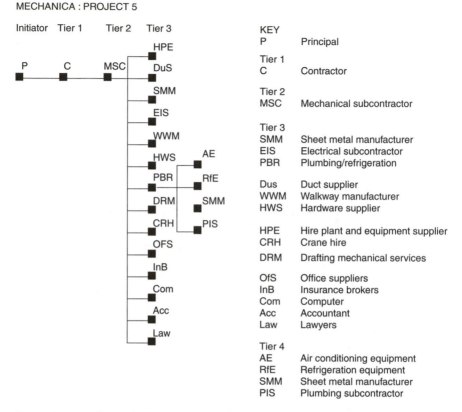

Figure 9.1 Mechanical services structural organization map, Project 3.

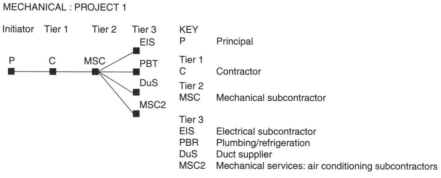

Figure 9.2 Mechanical services structural organization map, Project 1.

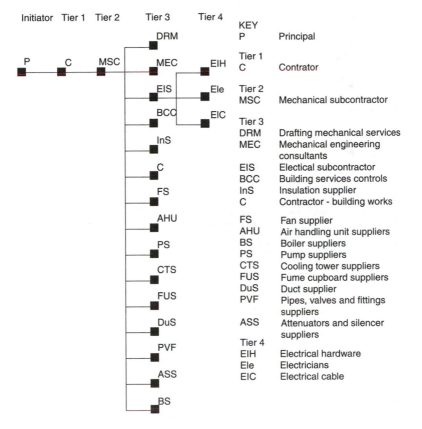

MECHANICA : PROJECT 4

					KEY	
Initiator	Tier 1	Tier 2	Tier 3	Tier 4	P	Principal

Figure 9.3 Mechanical services structural organization map, Project 4.

9.2.2 Firm details

One of the subcontractors (Firm 187) is larger and has an international scope which is wider than the other Firm 132, who only works within the state, but has more than one division. Although Firm 132 competed for larger projects, it became apparent during the fieldwork that this firm was in financial trouble and was operating outside of its capacity. A subsequent interview with the managing director of Firm 187 confirmed that Firm 187 picked up the contracts after Firm 132 determined it couldn't complete them.

Firm 195, the equipment supplier, has an international scope; however, it is quite small in terms of its size and annual turnover. Firm 197, similar to the large steel, aluminium and glazing manufacturer, is a multinational, employs over 5,000 people and has between 11 and 20 offices worldwide.

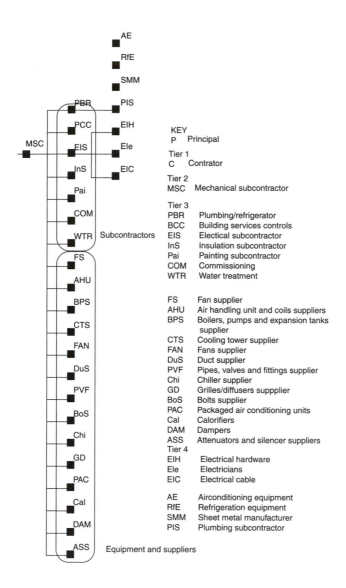

KEY
P   Principal
Tier 1
C   Contrator
Tier 2
MSC   Mechanical subcontractor

Tier 3
PBR   Plumbing/refrigerator
BCC   Building services controls
EIS   Electical subcontractor
InS   Insulation subcontractor
Pai   Painting subcontractor
COM   Commissioning
WTR   Water treatment

FS   Fan supplier
AHU   Air handling unit and coils suppliers
BPS   Boilers, pumps and expansion tanks
        supplier
CTS   Cooling tower supplier
FAN   Fans supplier
DuS   Duct supplier
PVF   Pipes, valves and fittings supplier
Chi   Chiller supplier
GD   Grilles/diffusers suppplier
BoS   Bolts supplier
PAC   Packaged air conditioning units
Cal   Calorifiers
DAM   Dampers
ASS   Attenuators and silencer suppliers
Tier 4
EIH   Electrical hardware
Ele   Electricians
EIC   Electrical cable

AE   Airconditioning equipment
RfE   Refrigeration equipment
SMM   Sheet metal manufacturer
PIS   Plumbing subcontractor

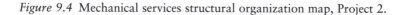

Figure 9.4 Mechanical services structural organization map, Project 2.

9.2.3   *Markets*

The most significant result regarding the mechanical services subcontracting market is the clear understanding of a high level of market segregation in the mechanical services market by the subcontractors who were interviewed (Firm 187 and Firm 58) and that their perceptions of the structure of the market aligned.

The mechanical subcontractor market for Projects 1, 2, 3 and 4 includes six firms. All six firms tendered on each of these projects. Firm 187 won the tender for Project 1, Firm 132 won the tender for Projects 2 and 3, and Firm 58 won the tender for Project 4.

The general industrial market is segregated into three levels based upon *firm scope* and *capability*. Although each of these firms considers that they compete within this market of six firms for the major mechanical services trades packages on major projects, the project tendering market includes more than this number of firms.

The structure of the subcontracting market alters somewhat for each project. Although the contractors stated that there were only five to six subcontractors tendering, both subcontractors stated that they were in competition with approximately fifteen subcontractors for these projects. This is intriguing as the project managers for contractors on these projects stated that there was a small group of tenderers. It seems that at some stage during the tendering process the level two subcontractors were invited to tender on these projects as well as level one.

> **Box 9.2 Dynamic nature of tendering period: negotiation, bargaining and new competitors**
>
> Well, generally I mean, tenders are called, prices go in, and the bargaining starts. I mean the price will come out and you will find out what sort of price you're looking at and generally it's up here [gestures], revising. It's just a matter of how far you're prepared to come down to try and meet the price…and this happens on every project. Some time during the project the word will go out on who else is tendering…we find out through our suppliers as to who they are providing prices to.

Although the markets are defined as residential, non-residential and industrial, the two subcontractors interviewed indicated that they also considered that they operated within the markets: international and maintenance.

Similarly, the manufacturer has three markets which they operate within:

- residential: mass-produced market and one major competitor (retail chain)
- commercial: five local manufacturers

- industrial: one competitor who do not manufacture locally and source internationally.

Not only are the markets categorized by the building type, they are also categorized by the project complexity (and thus firm scope and capability) and then also by whether they are international or domestic and then whether or not they are maintenance or project.

9.2.4   Supplier types

Both subcontractors considered their suppliers in two categories: equipment suppliers and subcontractors. The subcontractors provided a product and a service and equipment suppliers simply provided a product.

Both mechanical services subcontractors did not explicitly state that they categorized their suppliers beyond the above classification and they did not have a rational method for purchasing by volume across their firms. There are three main equipment items and there is one main subcontractor (electrical). Each project manager procured suppliers and subcontractors independently for each project. However, Firm 187 had recently decided to rationalize their procurement across the entire organization and in doing so had implicitly begun to classify their suppliers by volume purchasing.

Box 9.3 Risk transferral for major capital expenditure items

Equipment suppliers, and subcontractors as well. So we supply all mechanical equipment, chillers, boilers, fans, all that sort of stuff... the pipework, ductwork, we generally employ electricians to wire it all up. Laggers to build the lagging, the ductwork, pipework... painters to do the painting. A lot of other minor subbies there as well. We purchase the equipment and all the components, and install it on-site. We source it, get it built, get it delivered.

Occasionally, we'll get the contractor to actually buy the chillers himself because they're a major purchase, a major value, so why should a mechanical services contractor do it and get a margin on it when he can do it himself?

The product agent categorized his suppliers by type of product complexity; that is, it was either standardized or specialized. All their products were imported and in some cases they were the only supplier in Australia with a licence for particular products. Different systems and different relationships were developed for a standardized or specialized product.

Box 9.4 National strategic procurement approach for equipment: leverage purchasing

Well, yeah, I guess when I was looking after Victoria that was one of the things I was changing, because it was just unbelievable. We were going off independently dealing with a supplier, on a range of different jobs getting different prices, we were not even packaging up what we were doing and getting the best deal. And that's one of the things now we've tried to introduce, we've put in a new computer system, one of the things now that I want is that whenever we get a job now all the materials that have been toward it sort of go into a holding account so if I get a job with ductwork, chillers, and pipework, that will go into a holding account and all the other jobs that have got ductwork, pipework will go into this account as well so at any one time we can look at it and say 'What do Firm 187 want as far as chillers around Australia over the next three months? What have we got now, when do we have to order them?' So then we can go to suppliers and say 'Okay, we've got three chillers in Queensland, one in Brisbane, Melbourne, what's the deal?'

9.2.5 *Aggregated project supply chain organization: supply channel*

Figure 9.5 maps the transfer of ownership of mechanical equipment products in the construction industry from the manufacturer who purchases and assembles components to the building owners. The channel map indicates the structural organization of primary commodity suppliers and the types of firms that are involved. There are currently six tiers available for the transfer of ownership and two channels were identified which involved all six tiers. Importation occurs primarily at the original equipment manufacturer (OEM) level. Eleven chain options were identified.

The organization of the channels extends from three main branches: the OEM who imports for the residential market only, the OEM commercial and residential product supplier and the OEM who imports for the industrial market. In this channel the industrial building has highly complex and specialized products, somewhat similar to the specialized products identified in the concrete industrial market.

Similar to previous channels, the purchasing power of higher-order tier firms allows for a firm to leap a tier and purchase directly from the tier above. The unique project market is characterized by a high degree of different types of specialized firms who provide services of installation and design; however, they may not be involved in transfer of ownership.

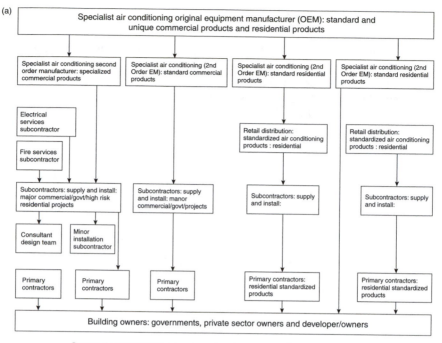

Structural organizational channel map for primary commodity mechanical services

Figure 9.5 Mechanical services aggregated project supply chain organization: supply channel.

9.3   Case study: formwork – core commodity

This section describes the much smaller and more specialized supply chain – the formwork for high-rise buildings.

9.3.1   *Firm details*

The formwork supplier, Firm 560, supplies to Projects 1 and 3 and is a highly specialized firm. They are a subsidiary of a formwork firm in Singapore who in turn are owned by a multinational firm with numerous divisions in various construction commodity markets. This multinational firm is in turn owned by another multinational which has headquarters in Tokyo. There are approximately 300 firms within the group located primarily in Asia. The group is diverse and is involved in land development, construction, construction materials and components and retrofitting of buildings. Production in Australia is undertaken primarily in the Melbourne division. During the interview it felt like I was uncovering a 'sleeping giant'.

(b)

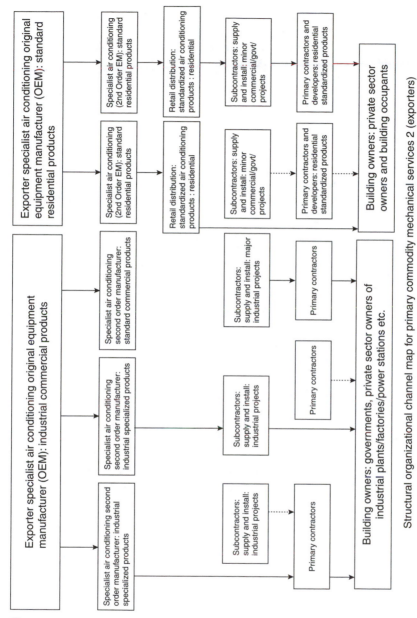

Structural organizational channel map for primary commodity mechanical services 2 (exporters)

Figure 9.5 Continued.

Box 9.5 Formworker

The manufacture of the systems will only take place here, but we also have a manufacturing facility in Singapore. But we do their design work down here. They service primarily the Singapore market. In the Firm 560 Group we have 18 people in the factory and 12 in the office here. In Singapore we have probably a very similar number. Therefore, we have about 60 people. Our turnover for this operation here in Melbourne is about $6m per year.

The firm provides a product and a service for multi-storey construction which involves two products. The first product is the formwork for the concrete wall for the core of the high-rise building and the second is the protective screen system around the perimeter wall (façade) of the building.

9.3.2 Markets and competitive advantage

The competitive advantage of the firm is twofold, in that the core wall formwork is an innovative design developed by this firm, and the perimeter wall screen systems typically involve a highly customized solution to the problem. To achieve greater efficiencies they have developed the product even further in recent years. Projects 1 and 3 both had unusual façades whereby the perimeter wall was highly irregular. In both situations the contractor approached the firm for advice on how to solve the problem and to design a system prior to awarding the contract.

Firm 560 won the tender based upon the savings that the alternative design could provide the contractor (refer to Table 9.2).

This raises the question of the firm's competitors. They have two competitors in Australia in the tower formwork system and in Singapore they have 100% of the market share. For the Singapore market the product is manufactured in Australia and shipped to site, which seems surprising; however, it is cheaper to do so. The factory facilities are better in Australia, primarily because of lack of land in Singapore. This is particularly effective if the components are fairly standardized.

The firm has approximately 95% of the market share in Melbourne and 8% of the Sydney market. The three suppliers in this market all have slightly different variations on formwork systems. One of Firm 560's competitors supplies a jumpform system and another a slipform system. The jumpform system was originally designed by Firm 560 30 years previously. Firm 560 redesigned the system in 1987 with the main aim of achieving a more accurate concrete finish.

Firm 560 has two competitors in the protective screen system. The competitive advantage of Firm 560 is the efficiencies it offers to the contractor on-site through prefabrication and the service.

Table 9.2 Matrix: formwork firm and markets

Supplier types	Firm commodities, competitors and competitive advantage
There are four major suppliers for Firm 560 including the following products: structural steel, timber, plywood and bolts. Structural steel and plywood are the critical commodities, followed by bolts and timber. The steel supplier is a large state-based structural steel merchant/fabricator. Firm560 is not large enough to purchase directly from the major steel distributor Firm 458 discussed in Case study 2. Firm 560 attempts to reuse as much of their product after projects are completed. Firm 560 typically sources from one major supplier, that is, one steel merchant, however at times, they source from '*a couple of smaller suppliers … we only go to them every now and then to keep the big guy honest*'. Their volume of spend is large enough to negotiate an annual purchasing agreement with the steel merchant. Firm 560 typically sources from one major timber merchant for difficult projects, as they are the only suppliers of high quality plywood. If the project does not require high-quality ply they 'shop around'. Firm 560 appears to have developed a particularly strong network of suppliers and customers with innovative solutions to numerous construction problems. Another example of supplier development occurs with the plywood supplier: *As far is plywood we basically buy it off Firm 638 and that is because we*	In the past we have done more of the specialized area and only of late have we decided to bring that into a modularized design so that we can do a reusable hire market, or rather a contracting market. So what we would do is we give the builder a price to do a job, and it is a fixed price contract for the life of the job. We preassemble and prefabricate components and send them out and all they do is pick up the crane and start using it. We used to do a more specialized market and sell it and not bring it back. Projects 1 and 3 had unusual facades, whereby the perimeter wall was highly irregular. For the Philadelphia and Fed Square projects we did the formwork and the screens. *We actually took all the standard components and bolt them together and we made the working platforms, we totally prefabricated them and then bolted them to the frames this was for the perimeter screens. We then transport this into site and they just pick them up and put them in place. They get basically a perimeter screen system section by section around the building normally up to 4.5 metre long segments and that becomes a working platform covering 2 floors and a handrail. They then lift that up the building with a crane while they do their façade. Then above that we have a system called Formwork screens which is basically a protection system that protects people and things falling off the building and it covers 2 floors again. We lift it up by crane we actually bolt all that together here for them and send it in to them as prefabricated platforms. We send it to site ready to use. They pick it up with the crane with the support equipment that is part of our contract and then they just sit it on top of the support system and it is there ready for use. As soon as it gets to site so soon as they lift a unit they have immediate edge protection which is required in this industry.* In both cases the contractor approached the firm for advice on how to solve the problem and to design a system. The firm analyses the drawings and discusses the solution with the contractor. Their alternative is to go back

(*Table 9.2 continued*)

Table 9.2 Continued

Supplier types	Firm commodities, competitors and competitive advantage
developed with them a plywood strong enough to get 76 concrete poured against it. They are the only people with it. In normal jobs when there are only 20 pours we shop around and get prices because it does not have to be the good quality ply only the normal plywood. Timber supply is not as critical to their final product and they source from a number of suppliers and the decision is based upon price. Regardless of this, the volume of spend is particularly high and therefore there is an annual purchasing agreement in place. However, bolt suppliers are sourced from a small group of medium sized Australian firms and the quality is particularly significant. *We tend to use Australian made bolts. Because of the problem with overseas bolts is that they have a high carbon content and more brittle. Particularly in the structural areas where we have to use 8.8 bolts we always buy Australian made because we know we are going to get the quality We source of*	and use conventional scaffolding or cantilevered scaffolding. They weigh that cost up against the firm's cost and the huge fact with ours is that all they have to do is the equipment arrives on site to stand up and it is ready to use, which saves time and ultimately money. This raises the question of the firm's competitors. They have two competitors in Australia in the tower formwork system and, in Singapore they have 100% of the market share. For the Singapore market the product is manufactured in Australia and shipped to site as it is cheaper to do so. The factory facilities are better in Australia, primarily because of lack of land in Singapore. This is particularly effective if the components are fairly standardized. The firm has approximately 95% of the market share in Melbourne and 8% of the Sydney market. The three suppliers in this market all have slightly different variations on formwork systems. One of Firm 560's competitors supplies a jump form system and another a slipform system. The jumpform system was originally designed by Firm 560 30 years previously. Firm 560 redesigned the system in 1987 with the main aim of achieving a more accurate concrete finish. Firm 560 has two competitors in the protective screen system. *Our big advantage is that we provide a full service we send everything fully fabricated ready to use. We prefabricate all the major components they don't have to do it onsite. When it turns up it is ready to bolt together and*

those well we have 4 or 5 little companies around. We used to use Walter Bar but they went out of business last year. AE Baker is a medium sized supplier who we use and another couple about the same size. We try to reuse as many bolts as we can.

Similar to a number of other firms discussed previously, Firm 560 matches the risk of procuring poor quality products to sourcing strategy, supplier market size and supplier choice criteria. Firm 560 also tends to match their firm size to the firms that they source from. Although they are within an extremely large group of multinationals Firm 560 tends to operate as a small to medium-sized firm (60 employees). This is done by choice for one supplier (bolt suppliers) and by market 'rules' (steel distributor) for another supplier. Even though there are annual purchasing agreements in place with the steel, plywood and timber suppliers each project is typically renegotiated on an individual basis.

use and that frees their site up. They don't have to have extra labour onsite and extra supervision. We also send in when it is going to be assembled onsite an extra supervisor onsite who directs their people in how to lift things and how to assemble things how to bolt together. Basically we take on board that part of the structure and once it is up and running there is no work required for us so all we do is supervise every second day or so whatever the foreman wants. We walk around and see if everything is ok. We don't get paid very much extra for it.

Our opposition does not provide the prefabrication to the extent that we do, they don't provide all the timber and the plywood that we do in our contract. They may say 'we have here a bare skeleton of a system and you provide all the necessary consumerables, we provide the labour to bolt together and we will give you a supervisor.' What we are doing is giving them a total package well they still have to provide some labour so not quite a total, but we are taking out all the extra bits that have to be purchased to do the job. We are providing a main service everything in preassembled. The other way they have to do it themselves and they may stuff it up.

Firm 560 typically works directly for the major contractors in the country and have done so for more than 40 years. The director of the firm has worked with the father of some of his clients and therefore the relationship is built on personal trust and relationships coupled with the quality of the product and service. The formwork and the screens are treated as separate packages.

9.3.3   *Supplier types*

There are four major suppliers for Firm 560, including the following products: structural steel, timber, plywood and bolts. Structural steel and plywood are considered to be the critical commodities, followed by bolts and then timber.

The steel supplier is a large state-based structural steel merchant/ fabricator. Firm 560 is not large enough to purchase directly from the major steel distributor, Firm 458, discussed in Chapter 8. Firm 560 attempts to re-use as much of their product after projects are completed. Firm 560 typically sources from one major supplier, that is, one steel merchant. At times they source from *'a couple of smaller suppliers...we only go to them every now and then to keep the big guy honest.'* Their volume of spend is large enough to negotiate an annual purchasing agreement with the steel merchant. The steel merchant sources from Firm 458.

Firm 560 typically sources from one major timber merchant for difficult projects as they are the only suppliers of high quality plywood. If the project does not require high quality ply, they *'shop around'*. Firm 560 appears to have developed a particularly strong network of suppliers and customers that assist them with innovative solutions to numerous construction problems. Timber supply is not as critical to their final product and they source from a number of suppliers and the decision is based upon price. Regardless of this, the volume of spend is particularly high and therefore there is an annual purchasing agreement in place.

Bolt suppliers are sourced from a small group of medium-sized Australian firms and the quality is particularly significant and therefore there is no annual purchasing agreement.

Similar to a number of other firms discussed previously, Firm 560 matches the risk of procuring poor quality products to sourcing strategy, supplier market size and supplier choice criteria. Firm 560 also tends to match their firm size to the firms from which they are sourcing.

Box 9.6   Countervailing power

Although they are within an extremely large group of multinationals, Firm 560 tends to operate as a small- to medium-sized firm (60 employees). This is done by choice for one supplier (bolt supplier) and by market 'rules' (steel merchant) for another supplier. Even though there are annual purchasing agreements in place with the steel, plywood and timber suppliers, each project is typically renegotiated on an individual basis.

9.4   Case study: concrete – core commodity

There are two primary types of concrete supply chains by commodity type, namely precast and in-situ. The in-situ chain is described in detail in this text. In very general terms, firms involved in the chain have similar structural characteristics to the steel, aluminium and glazing supply chains, including:

- a large Australian manufacturer which has a high turnover, employs a large number of staff and typically has a multinational scope
- a distributor/2nd order manufacturer which is their own business unit has an Australian wide network and
- another link in the chain with a fabricator or merchant which may be smaller or of equal size and scope to the previous distributor; and then
- a subcontractor which is small to medium sized and typically has a more Australian focus often state-based divisions that operate as separate business units and a smaller turnover.

However, the configuration differs somewhat again for the concrete chain. To clarify, the glazing and steel supply chains differ from each other in terms of two key structural market characteristics:

- the distributor/2nd order manufacturer market; and
- the relationship between the manufacturing arm of the firm and their distribution business unit.

These differences then give rise to slightly different structural organizations of the supply chain.

Representatives from the concrete subcontractor and the concrete manufacturer were both interviewed. Two staff were interviewed from the manufacturer: one from the contracting and distribution division and one from the manufacturing quarry plant. The relationship between the distribution division and the manufacturing division is much closer than for the steel and glazing manufacturers. The contracting division does not currently have a separate cost accounting and competes within the framework of the larger manufacturer. However, they are required to contract directly to the contractors and this is a strategy by the large manufacturer to increase concrete supply to projects.

The subcontractor and the manufacturer both supply to Projects 1 and 3. The manufacturer's contracting division supplies directly to the contractor and then subcontracts the installation services to the subcontractor. The projects are too large in terms of risk for the subcontractor to manage. The market for the concrete manufacturer is somewhat diverse as the firm is known as a construction materials supplier of concrete, timber and aluminium products. However, they have two competitors for concrete supply and one of these is a supplier of numerous construction materials as well. This other large construction materials supplier also has an in-house project contracting division.

The markets which the concrete subcontractors operate in are residential, commercial and industrial. Similar to a previous discussion on market segmentation for the mechanical services subcontracting market – the concrete subcontracting market is clearly segmented. This subcontractor is actually in the second level in the market and only operates in the first level when they are associated with the manufacturer.

> ### Box 9.7 Chain decoupling
>
> The subcontractors typically purchase their materials from the one concrete manufacturer – so, as soon as a project has a particular subcontractor, then the concrete supplier is known. Sometimes decoupling occurs; however, the subcontractor is then not in the same bargaining position.
>
> The price for concrete supply is typically reliant upon distribution costs which rely upon geographical distance to site. The location of the quarry in relation to the construction site ultimately can play a role in the final coupling of subcontractor and manufacturer in the chain.

However, the three major concrete suppliers purchase concrete products from each other's quarries. The quarries act as independent business units.

9.4.1 Supplier types

The particular subcontractor for Projects 1 and 3 has five types of suppliers in the following categories:

- quarry products (sand/concrete/asphalt, etc)
- waterproofing materials (plywood/patching/polythene)
- steel reinforcement
- specialist products and
- equipment.

The last two supplier categories are sourced less frequently as they don't occur on every project. The subcontractor owns most of their equipment; however, occasionally they are required to hire at short notice various pumps and cranes for special circumstances. Typically, they purchase the quarry products from the manufacturing firm – although not always. As noted previously, price is closely related to location because of transportation costs and delivery times and therefore this plays a critical factor in choosing between suppliers.

The contracting division sources all concrete and quarry supplies from their own manufacturing quarry division. They then source from the subcontractor who supplies all the other items for the project. Their main priority in dealing directly with the contractor is to take a pro-active approach to increasing the volume of their concrete sales and to manage the risk for the subcontractor.

The quarry operations have approximately fourteen groups of suppliers. They likened their operations to that of a building site. In the past they typically owned equipment and employed staff to operate the machinery and equipment; whereas now the majority is outsourced and they manage the operations through contracts which define schedules and payments. The list of suppliers included those that provided a service and those that provided a product and a service. They classed the suppliers as: operations and maintenance, and within both categories there were product suppliers or subcontractors who supplied a combination of a product and a service.

9.4.2 Aggregated project supply chain organization: supply channel

Figure 9.6 maps the transfer of ownership of concrete products in the construction industry from the manufacturer who extracts raw materials from the ground to the building owners. The channel map indicates the structural organization of primary commodity suppliers and the types of firms that are involved. There are currently five tiers available for the transfer of ownership of the concrete commodity to flow through and there was a channel identified in the study that included all five tiers. No import channels were identified. Eight channel options were identified. The organization of the channels extends from three main branches according to final end user; the supply of concrete to home owners, governments or major commercial clients and warehouses.

The critical difference between concrete channels and the previous channels is that the manufacturers transfer ownership amongst themselves quite readily. The majority of concrete flows through subcontractors who supply and install on-site. If the purchasing power of the developer/primary contractor is high – for example, with major national residential developers and civil contractors – then the subcontractor is employed by the contractor to install only. The other major difference with the concrete channel is that the manufacturer also has a contracting division which feeds concrete directly to construction sites. In the previous channels the manufacturer had distribution divisions, but they were quite distant from site supply. This may shorten the length of the chain; for example, one chain has only three tiers: manufacturer to subcontractor to homeowner. The concrete manufacturers have distribution divisions; however, unlike steel distribution, they are highly integrated. The distribution divisions do not compete against other distributors as the steel distributor does. There is a low level of value adding in the channel.

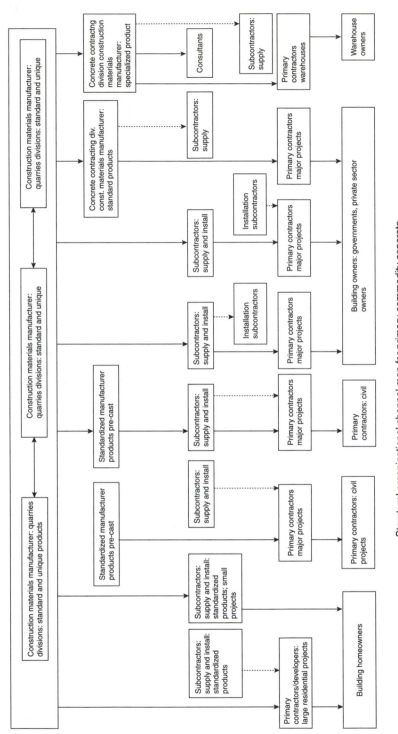

Figure 9.6 Concrete aggregated project supply chain organization: supply channel.

9.5   Case study: brick – core product

The brick supply chain was considered for Project 5, where it involved the contractor directly purchasing from the manufacturer. The relationship between the contractor and the manufacturer is a strategic alliance and there were no other tenderers for this project.

9.5.1   Firm details, commodities and industrial markets

The firm is an Australian manufacturer of clay products, including bricks, brick facing tiles, pavers and terracotta paving tiles, and is located in Melbourne, Victoria. The firm has three quarry sites with two manufacturing plants. They employ approximately 75 people and their annual turnover is $12m. They operate in other states through agents and export approximately 10% of their product. They typically operate in a niche market and their products are considered to be of a high quality.

The strategic alliance was developed because of the quality of the product in the first instance. However, other issues were also important to the contractor; for example, the size of the manufacturer was important. The contractor was not large enough to have any real purchasing power with the larger manufacturers. This factor had little to do with price and more to do with the development of an effective working relationship, which was particularly needed to develop the export market (refer to Table 9.3 Matrix: Brick manufacturer, for further detailed results from the interviews).

The manufacturing process is different to the major manufacturers and their competitors, which results in a higher quality product but a largely inefficient and resource-intensive process that is time consuming. In the residential market the customer is typically the residential homeowner or the developer of a residential subdivision. In the commercial market the customer is typically the subcontractor. At times, if the project is large in the commercial market the contractor will deal direct with the manufacturer and purchase the products. The majority of projects in both the residential and the commercial market have architects involved and it is a selected-tender process.

The firm also supplies to the Japanese market and this is the reason for the strategic alliance between the contractor and the manufacturer. The chain to supply to Japan involves a number of extra links which involves trading houses and distributors. This firm has twelve such Japanese trading houses that they typically deal with as well as the contractor, Firm 590, which is Australian. In recent years it has begun to be less difficult to export to Japan as in the past it was quite controlled. Many of the Japanese construction companies have import arms to their business.

The manufacturer also licences, from a US firm, the technology for an innovative brick facing system and an associated plastic/metal rail which is particularly suited to the high structural constraints placed on brick

Table 9.3 Matrix: brick supply chain firms and markets

Supplier types	Firm commodities, competitors and competitive advantage
There are six supplier categories for this firm, grouped as: 1 core products: white clays, additive products and plastic extrusion and 2 non-core products: fuel 3 services: quarry management, machinery and logistics. Generally the clay is taken from the firm's own quarries; however, they do not have any white clay quarries in Melbourne so this is purchased. The white clays come from quarries in Bendigo (a country town approximately 2 hours west of Melbourne), where farmers have formed small companies and sell to manufacturers. This particular firm sources from one quarry and has done so for the last twenty years. The only additive that the manufacturer uses is manganese dioxide which is sourced from a firm in South Africa, Commercial Minerals. *This is mostly sourced and bought directly from the people who have the quarries and who process it. It is not generally done through an agent. Unless it is from someone who is coming in from overseas then we will buy from an agent.' Large quantities are not purchased and the product is expensive. It is also subject to price fluctuations due to international exchange rates.* There are three major contracts for services. For this firm the most critical supplier is the contractor who	The firm has three quarry sites with two manufacturing plants. They employ approximately 75 people and their annual turnover is $12M. They operate in other states through agents and export approximately 10% of their product. They typically operate in a niche market, and their products are considered to be of a high quality. *Our product is deemed to be a little different to the usual and the major manufacturers. We look at ourselves as being in a niche marketplace really. Our product is different because it really starts with the type of clay that we have – we have three quarries of our own and the clays that we tend to get tend to withstand high temperatures really well.* The manufacturing process is different to the major manufacturers and their competitors, which results in a higher quality product but a largely inefficient and resource intensive process that is time consuming. We have very old style kilns which are not economical by any stretch of the imagination. It is a stop–start process – we put all our material into the kiln, seal it up, fire it up and then cool them down and take them out. Whereas kilns these days with the major manufacturers where you are looking at high volume and high outputs, are called tunnel kilns in which everything consistently goes through them. It is 24-hour continuous production as everything flows through. They are much more economical on fuel. With ours we have to heat it up, then cool it down and then heat it up, etc. There is also a lot more labour involved in our kilns because it varies a lot and we have to bring it out into our yards, cross-mix it and blend it. But that has also opened up a lot of opportunities that the major manufacturers can't offer. That is what sets us apart. It is the manufacturing process. All our product is good quality but it is the aesthetic quality rather than the strength etc. It is the richness in colour and the variation in texture and colour. The fact that it varies naturally from one tone to another and it is a good drifting in colour in a wall.*

manages the quarry. The two other major contracts for services include maintenance contractors for the machinery that works the clays and maintenance contractors for the electronic and computer equipment that controls the kilns. Finally, the last supplier is the logistics company who transports the bricks.

This firm

'... used to run our own trucks and have our own drivers. But we have moved out of that. The responsibility is now completely with another contractor.'

Their competitors are the large firms and include Nubrick, Sellkirk and Boral. There are a lot of other smaller companies like Glenn Thompson and interstate manufacturers coming down from Sydney. It is a competitive market. The firm primarily concentrates on the local state market and the export market is emerging. The bricks are provided to the residential market, the brick facing tiles to the commercial market and the pavers are in the both markets. The facing tiles are exported to Japan.

In the residential market the customer is typically the residential homeowner or the developer of a residential subdivision. In the commercial market the customer is typically the subcontractor. This does not happen in residential market – the developer is usually the customer. At times if the project is large in the commercial market, the contractor will deal directly with the manufacturer and purchase the products. The majority of projects in both the residential and the commercial market have architects involved and it is a selected tender process. The chain to supply to Japan involves a number of extra links which involves trading houses and distributors. This firm has 12 such Japanese trading houses that they typically deal with and then Stonehenge which is Australian.

Japan is actually changing it was fairly strict lines to get product into Japan. You had to go through trading houses and then the houses would sell to somebody else and then it would sell to the distributors and then it would sell to someone in the retail sector. At various levels that there were in the process you can imagine that by the time it got to the user the price had gone way out of the reach of most people. That is the way it is – you still have to come in through this. With business over there business is often a gentleman's agreement with a handshake deal. Loyalty is very highly thought of in Japan and once they have an agreement with somebody that is it – that is the channel that the business is done through.

It was until quite recently much more difficult to export to Japan; however in recent years this has begun to change although still is fairly controlled. Many of the Japanese construction companies have import arms to their business.

(Table 9.3 continued)

Table 9.3 Continued

Supplier types	Firm commodities, competitors and competitive advantage
	Like business people anywhere they are looking for cheaper sources of material so what they will do is import themselves directly and thus cutting out the middlemen.
	So the building companies have their own import houses trading houses – some of the building companies are huge, very very large. So a lot of them have been doing that. But with the electronic age we are in now – I get emails either directly from the people themselves or from Australia or somewhere – someone is interested in importing. We have had quite a few where we have dealt directly with the customer, either the company doing the project – the builder developer. They are looking for something and they will send out emails all over the place. Normally what they have done in the past is just go to someone that they buy through and they would make the inquiries.
	The brick facing system and the plastic/metal rail is particularly suited to the high structural constraints placed on brick construction with Japanese earthquake codes. The system was not developed by this firm however they have just purchased the license from the US firm that did develop it. The metal and brick facing tiles are particularly well suited to tilt up precast concrete panel construction. The precasting companies are providing a new customer for this firm in Australia. The Japanese have a similar technology. We didn't develop that system we took that from someone else. One of our companies that are making the tiles for them who developed the whole system and we now have the rights to market it. It is not too different from existing Japanese systems.
	It was until quite recently much more difficult to export to Japan however in recent years this has begun to change although still is fairly controlled. Many of the Japanese construction companies have import arms to their business. Like business people anywhere they are looking for cheaper sources of material so what they will do is import themselves directly and thus cutting out the middlemen.

> **Box 9.8 Reduced chain tiers and reduced cost**
>
> Like business people anywhere, they are looking for cheaper sources of material, so what they will do is import themselves directly and thus cutting out the middlemen.

construction with Japanese earthquake codes. The metal and brick facing tiles system is particularly well suited to tilt up precast concrete panel construction. The precasting companies are providing a new customer for this firm in Australia. The approach to innovative solutions and specialist quality brick is one of the competitive advantages of this manufacturer.

9.5.2 Supplier types

There are six suppliers for this firm, grouped as such:

- core products: white clays, additive products and plastic extrusion
- non-core products: fuel
- core services: quarry management, machinery and logistics.

Generally, the clay is taken from the firm's own quarries; however, they do not have any white clay quarries in Melbourne, so this is purchased. The white clays come from quarries in Bendigo (a country town approximately 2 hours west of Melbourne), where farmers have formed small companies and sell to manufacturers. This particular firm sources from one quarry and has done so for the last 20 years. The only additive that the manufacturer uses is manganese dioxide, which is sourced from a firm in South Africa.

There are three major contracts for services. For this brick manufacturing firm the most critical supplier is the contractor who manages the quarry. The two other major contracts for services are: maintenance contractors for the machinery that works the clays and maintenance contractors for the electronic and computer equipment that controls the kilns. Finally, the last supplier is the logistics company who transports the bricks.

9.5.3 Brick suppliers procurement relationships

As noted in the previous section, the brick supplier to the contractor for Project 5 has seven supplier categories, grouped as such:

- core products: white clays, additive products and plastic extrusion
- non-core products: fuel
- services: quarry management, machinery and logistics.

This particular brick manufacturer sources from one quarry and has done so for the last 20 years. The relationship is well and truly based upon trust and product quality. The prices are set annually and an agreement is drawn up after some negotiation.

The only additive that this manufacturer uses is manganese dioxide, which is sourced from a large mining firm in South Africa.

Box 9.9 One firm and one commodity – alternate chain configurations

This is 'mostly sourced and bought directly from the people who have the quarries and who process it. It is not generally done through an agent. Unless it is from someone who is coming in from overseas; then we will buy from an agent.'

Large quantities are not purchased and the product is expensive. It is also subject to price fluctuations due to international exchange rates.

For the brick manufacturer the most critical supplier is the contractor who manages the quarry. There are only three to four contractors who are capable of providing a quality service. However, this contract is not tendered and each year a negotiation takes place with the current contractor. This has been a long-term contract and has been in place for 20 years.

There are two other major contracts for services and these are: maintenance contractors for the machinery that works the clays and maintenance contractors for the electronic and computer equipment which controls the kilns.

Finally, the last supplier is the logistics company who transports the bricks.

Box 9.10 Divesting internal services and increasing supply chains

This firm 'used to run our own trucks and have our own drivers. But we have moved out of that. The responsibility is now completely with another contractor.'

This relationship is once again built on trust based upon the quality of the service and it seems that the market is not competitive.

Box 9.11 Security of supply and trust

We simply have gone to him for a number of years. Pricing comes up for discussion every now and then; he doesn't really push that a lot. We give him a huge amount of business. He is conscious of that as well.

He just started in the business by buying bricks and delivering for other companies; he has a second-hand brick yard. And then he bought his own yard. He has a huge delivery business. He delivers for other companies as well as his own.

The man who runs our business is one of the most conscientious people I have ever come across. He runs his business extremely well – his own approach runs throughout the business – everyone works hard for him. He constantly tells his people to look after the customer.

The brick manufacturer exports. One of the main methods for their export business is through the contractor. One of the critical suppliers for export is the logistics supplier. The logistics supplier for the domestic business does not do their export distribution. Logistics is competitive for their export market and they usually try to look for the best possible price.

Box 9.12 Price criteria for specific markets

'Keen pricing is more important in the export business – I shop around for that.'

The suppliers were categorized as such:

- high risk/high expenditure (clay products and quarry management)
- high risk/low expenditure (additives)
- low risk/low expenditure (logistics)
- low risk/high expenditure (fuel and machinery).

9.5.4 Aggregated project supply chain organization: supply channel

Figure 9.7 maps the transfer of ownership of masonry products in the construction industry from the manufacturer who extracts raw materials from the ground to the building owners. The channel map indicates the structural organization of primary commodity suppliers and the types of firms that are involved. There are currently seven tiers available for the transfer of ownership of the masonry commodity to flow through, including the exporting of masonry products. There was one channel identified in the study that included all seven tiers; however, this was the export channel. Six chain options were identified.

Direct procurement from the manufacturer is encouraged and more common for the masonry channel. The commodity is a niche market of

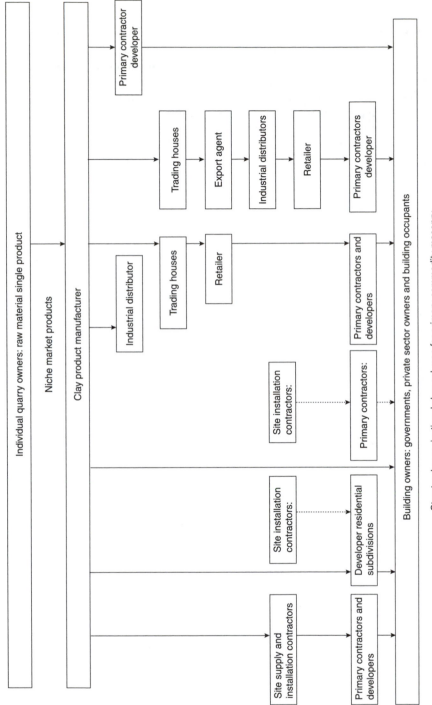

Figure 9.7 Masonry aggregated project supply chain organization: supply channel.

masonry products; that is, high quality and a specialized product (due to manufacturing processes).

9.6 A final word

Chapter summary

1 Firm diversity: Firm scope varied from multinational, international to regional. Perhaps what is most surprising is that firms operate on a national and international level much more than first anticipated – perhaps challenging the small to medium enterprise 'dog in the back of the ute' stereotype for unsophisticated, low-key, small-scale 'one-man-band' show. (Australian expression – a 'ute' is a utility car similar to the US SVU.)

2 The steel and concrete supply chain firm's scope and sizes are similar at each tier of the chain; however, the concrete materials manufacturers also have an internal subcontracting division which supplies directly to contractors – making the chains much more cohesive and the decision point of choice of subcontractor and the eventual path a much more transparent environment (for specific major projects or for selected markets that the major concrete suppliers have chosen to penetrate and be active in). I have termed this chain coupling and when this doesn't occur for some reason – chain decoupling. However, what is much more interesting is the diffuse purchasing which is occurring at the quarry level – they all may purchase from each other's quarries at various times.

3 Mechanical services categorized their suppliers into equipment suppliers and subcontractors, but had limited levels of understanding about value of strategic procurement in relation to leverage purchasing – however, this appeared to be changing. The formworker identified their suppliers in terms of critical commodities (four commodities: structural steel, timber, plywood and bolts – structural steel and plywood were critical, followed by bolts and then timber). They are aware of their volume of spend and use this to negotiate an annual purchasing agreement with a steel merchant – another firm of equal market power (countervailing power is equal). Plywood is a product which is important for quality and this is strategic critical. Timber purchasing volume is also high, but not as critical in terms of quality, and so it is considered leverage purchasing.

4 Segregated markets: The most significant result regarding the mechanical services subcontracting market is the clear understanding of a high level of market segregation in the mechanical services market by the subcontractors and that their perceptions of the structure of the market aligned. The general industrial market is segregated into three levels based upon *firm scope* and *capability*. Project markets, however, can change, and this is perhaps a diagnostic signifier for an upstream client/customer for the status of the chain – when this occurs it is a sign

that the contractor is wishing to achieve a certain level of quality, but applying pressure to the top level group of suppliers to drive the price down during the tendering stage. By knowing a competitive price from a second level supplier, it provides information to the contractor to renegotiate prices with top level suppliers.

5 Suppliers: Aggregated supply channels mechanical services: There are currently six tiers available for the transfer of ownership of mechanical services and five tiers for concrete. There are currently seven tiers available for the transfer of ownership of the masonry commodity to flow through, including the exporting of masonry products, and six chain options were identified. Importation occurs primarily at the original equipment manufacturer (OEM) level, but was not identified for concrete. Eleven chain options were identified for mechanical services, but only eight for concrete. The organization of the channels extends from three main branches for both concrete and mechanical services.

6 Highly complex and specialized products were identified in the industrial market for both concrete and mechanical services chains.

7 Similar to previous channels, the purchasing power of higher-order tier firms in the mechanical services supply chains allows for a firm to leap a tier and purchase directly from the tier above. The unique project market is characterized by a high degree of different types of specialized firms who provide services of installation and design; however, they may not be involved in transfer of ownership. Direct procurement from the manufacturer is encouraged and more common for the masonry channel. The commodity is a niche market of masonry products; that is, high quality and a specialized product (due to manufacturing processes).

8 The critical difference between concrete channels and the previous channels is that the manufacturers transfer ownership amongst themselves quite readily. The majority of concrete flows through subcontractors who supply and install on-site. If the purchasing power of the developer/primary contractor is high – for example, with major national residential developers and civil contractors – then the subcontractor is employed by the contractor to install only.

9 The other major difference with the concrete channel is that the manufacturer also has a contracting division which feeds concrete directly to construction sites. In the previous channels the manufacturer had distribution divisions, but they were quite distant from site supply. This may shorten the length of the chain; for example, one chain has only three tiers: manufacturer to subcontractor to homeowner.

10 The concrete manufacturers have distribution divisions; however, unlike steel distribution, they are highly integrated. The distribution divisions do not compete against other distributors as the steel distributor does. There is a low level of value adding in the channel.

10 Conclusions and future directions
Supply chain specialization and integration blueprint

How do we move forward given the case study observations?
Drawing together a framework based upon the observations in relation to projects project portfolios and supply chain procurement modelling

10.0 Orientation

> **Box 10.1 Chapter orientation**
>
> **WHAT:** Chapter 10 draws together conclusions from the case studies in an holistic manner and discusses issues at the project-chain and the sector-chain level. It provides classifications of supplier firms, procurement relationships and supply chains and the descriptions of the key characteristics which differentiate the types. It then develops an overall 'blueprint' for a process for an organization to develop an economic model of their approach to the supply chains they are located within. The interactions that occur on a project for the formation of customer–supplier firm–firm procurement relationships are described in relation to the sequences of events and various negotiations between firms, markets and projects is described. The chapter concludes with a discussion about the interdisciplinary nature of the study and its contributions to various knowledge domains and related to that future research.

10.1 Introduction

The intention of this text was to describe what firms actually do in practice in relation to procurement across a range of supply chains within an economic context. Procurement plays such a large part of what we do in all industries – not just construction. Perhaps those involved in constructed systems are more fixated on procurement as we are in a time of a high degree of specialization – we don't tend to take on more productive functions along the chain unless we are absolutely certain that not only can we create value but, more importantly, that we can be profitable.

For this reason, that is, the focus on firm profitability and notions of the value of 'supply chain management' needs to be intrinsically embedded within the context of the characteristics of the market, the firms involved in that market, the type of commodities which are to be exchanged, the project and thus the firm–firm relationships that evolve. Therefore, the approach has been to explore the economic aspects to the supply chain – thus, construction supply chain economics rather than construction supply chain management has been developed as a new area for us.

Perhaps what is important to those who are involved in the industry is the project. The project primary procurement contract between client and the various key project team members is a specialized and unique knowledge domain and an important part of our industry – the different procurement strategies which we have developed form part of our culture – our shared language and understanding. Less is known about the relationship of the project and the procurement strategies which individual firms deploy in relation to servicing the contract which leads to the project contract. Ultimately, it is the project which is of central importance to many in the property and construction industry. However, firms in the industry do not operate on individual projects – they are embedded within a multi-project environment and for this reason supply chain management is relevant to our industry. In many cases it is actually supplier management that is of more central concern; however, it is proposed that larger organizations (at whatever tier in the chain) will increase their competitive advantage with an explicit supply chain procurement strategy which is project and project-portfolio focused.

The chapter is organized as follows:

- Classifying supplier firms, procurement relationships and supply chains
- Chain Specialization and Integration (CSI) Blueprint
- Firm–firm procurement relationship processes
- Future research interdisciplinary intersections.

The examples reported in this text have enabled us to now develop much more detailed views of the supplier firms, customer–supplier procurement relationships and overall supply chains. It is now clear that there is not one aluminium window supply chain, nor one glazing supply chain, nor one concrete supply chain, and similarly not one way of procuring suppliers. However, even given this multiplicity between firms in the supply chains, we do have a greater clarity on some patterns on how the industry classifies and conceptualizes their suppliers and the way in which they make decisions in relation to procurement of suppliers. We can then also step back and see that there are some common themes upon which we can begin to group various supply chains.

The project industrial organization economic model of supply chain procurement which was proposed in Chapter 5 Section 5.3 Final Word, and presented in Figure 5.15 Project industrial organization economic model of supply chain procurement can now be developed further.

10.2 Supplier firm and procurement relationship classes

The interaction of the market structure and firm behaviour produces ways in which suppliers can be classed: firms can be categorized as either a customer firm or a supplier firm. Each firm acts as a customer or supplier in nearly every procurement scenario in some form or another and ultimately the attributes are the same. The attributes of the firm do not change. However, the behaviour of firms will change when they are in each of these roles. These classifications are only the beginning, but it serves as a useful starting point as it represents the behaviours of approximately forty firms acting in this manner.

Market structure and firm conduct interact to provide supplier firm classes. Firm conduct is governed by attributes of the firm and attributes of the particular market. In the supply chain model the key firm attributes are commodity type, scope and turnover. Associated with commodity type is the project commodity object, which is a further definition of the commodity related to the project and primary attributes include transaction frequency and complexity. The type of commodity is important, that is, product, service or product and service, as is the particular industrial market. However, these supplier attributes are considered when that first classification is undertaken; for example, two different products or a product and a service can both be similarly classed in terms of commodity significance and customer–supplier power relationship in the market. The way a customer and a supplier can behave in the relationship is based upon their individual relative power to each other and also the relative power of one market to the other market; that is, a group of firms supplying from one tier to another.

The market structure is a broad term which becomes more meaningful if we realize that there is a larger industrial market for each commodity and within that market for each project a unique project market may be formed which is smaller and more specialized. The project markets are typically more prevalent closer to the client tier of the chain; although even at tier 4 project markets can be found. In some cases the tracing of a specific project commodity is much more difficult – for example, the actual aluminium billet that is sourced from the aluminium manufacturer really only becomes part of the project when it reaches the aluminium fabricator firm. However, this is a rather significant discovery, that suppliers down at this level in the supply chain, that is, fabricators and at times industrial distributors, consider the project environment. For major projects the national steel industrial distributor considered individual projects. The specialist glazing manufacturer, although a supplier to long-run production markets, still considered the construction project environment as a particular market.

The classification of supplier firms by customer firms is based upon the following:

- commodity significance
- countervailing power

Within commodity significance the following four types of supplier firms have been identified: common core, common non-core, unique core or unique non-core. Further to this, the customer firm then considers the power in the relationship and the supplier firms are categorized by who has the power in the relationship. The countervailing power is related to demand and can be attributed to firm size, turnovers, level in tier, or closeness to client, size of individual contract, that is, purchasing volume. An individual firm may find its location alters, depending upon the firms involved in the relationship. This typology of supplier firm is equally applicable at all levels of the construction industry supply chain. As you move further away from the project environment, it does not affect the classification of the supplier firm. Upstream in the supply chain this classification is still equally applicable; it may simply mean that the procurement of unique type suppliers is more prevalent.

This then brings us to procurement relationship classes which are formed. The procurement relationship process involved is one of formation, transaction and management. The rhetoric for the last decade or two has been that the industry is highly adversarial and fragmented and that strategic industrial sourcing methods which are more collaborative and less competitive would improve the performance of the industry. There has been a tendency to assume that associated with this is that tendering is an activity where large groups of suppliers are competing to drive the price down. Contrary to this, the majority of suppliers are located within small groups of specialized suppliers.

There is also the assumption that, because of the fragmented supply chain, the approach to procurement is non-strategic and that the answer to low productivity problems in the industry lies in integrating supply chain participants. The focus has been particularly on integrating at tier 1, that is, the project team. The case studies indicate that the approach to procurement in the supply chain is strategic at each tier level; however, that integrating the project team participants will have little real impact upon the procurement relationships for the majority of the industry as there was clear indication that there was little supply chain management. There is some evidence that there are supplier strategic procurement decisions being made and high levels of supplier co-ordination at each tier and at times supplier management. The strategic approach can be seen in the manner in which procurement relationships are individually considered and formed, which can be grouped by the following three categories:

- risk and expenditure
- transaction significance
- negotiation.

The details of each of these categories are summarized in Figure 10.1. The categorization of procurement relationships by risk and expenditure dimensions is not new. The four categories are fairly well documented in the literature:

- high risk and high spend (strategic critical)
- high risk and low spend (strategic security)
- low risk and high spend (leverage purchasing)
- low risk and low spend (tactical purchasing).

Perhaps what is not as well documented is the extent of this approach in relation to the construction industry. We can now see that there are cases where this is explicit and contractors and subcontractors and component suppliers are putting into practice this quite simple yet powerful thinking. There are situations where this is implicit and it is not well articulated by the CEOs or project managers which were interviewed but nevertheless they were categorizing their suppliers in this manner. Finally, there are cases where this is just not happening at all and this is somewhat surprising – the mechanical subcontractors were a surprise to me – but perhaps this is changing. The philosophy in the past and still to this day is that individual project managers within firms manage their own accounts and thus project performance may be enhanced on an individual basis, but with little regard for multiple project performance and the 'power' of supply chain management across many projects within an organization has had little regard. On the other hand, there were some large manufacturers who were not as sophisticated as one would have thought – and again in the interviews there were claims that this was changing.

Perhaps where the research reported on in this text has taken a step forward in this respect is developing ideas about the significance of the transaction and negotiation tactics. The transaction significance brings together the level of risk and expenditure dimensions as well as the characteristics of the power relationship between customer and supplier and the characteristics of commodity significance. The significance of the transaction is taken from the customer perspective and involves the following categories:

- Strategic critical and equal control, that is, the power is equal between the customer and the supplier because although there is high spend from the customer there is also high risk due to supplier dependence (for various reasons, e.g. small, highly specialized market with specialized and unique products).

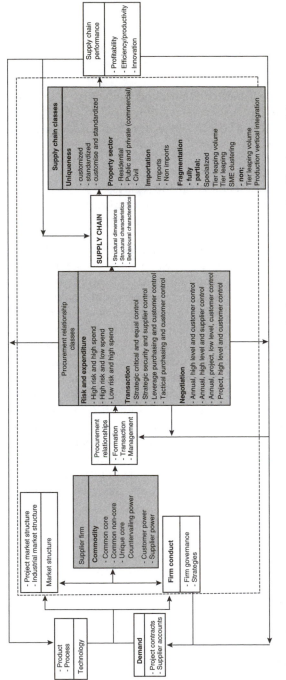

Figure 10.1 Project industrial organization economic construction supply chain procurement classification model.

- Strategic security and supplier control, that is, the power is with the supplier as the customer endeavours to develop security in supply because there is high risk but low spend
- Leverage purchasing and customer control, that is, the customer endeavours to gain control as there is high spend and relatively low risk
- Tactical purchasing and equal control, that is, the power is with the customer or the supplier as the risk is low but also the spend is typically low.

There are, of course, unique situations within each of these scenarios where the 'control' is relative or the risk is relative. Because of the significance of the transaction to the customer, there is then a response by the customer to how and when negotiations take place to establish the procurement relationship, including:

- Strategic critical procurement relationships were typically annual, approached at a corporate level in the organization, multi-project orientated and driven by the customer
- Strategic critical relationships were annual, approached at a corporate level, multi-project orientated and driven by suppliers
- Leverage purchasing relationships were negotiated and approached both annually and on a project level basis and approached both at a corporate level and on an individual project-orientated basis for strategic and significant projects and driven by customers
- Tactical purchasing was primarily negotiated at a project level supported with corporate level policies and customer driven.

With reference to the Figure 10.1, supply chain performance problems are probably the next step in this area of research – now that we have laid some groundwork on the structural and behavioural characteristics of key supply chains, it is possible to begin to see where improvements can be made. This is more complex and each supply chain should be considered on its own merits (that is, each commodity grouping of chains). More credit should be given to the self-organizing capacity of the individual markets rather than some outdated notion of integration; which ultimately alludes to a mythical vertical integration or just as bad a 'happy families' normative collaborative partnership concept. This is not to say that normative integration does not achieve improved performance, but the economic context must support such relationships and this is a very narrow view of the reality of the construction supply chain and ways to improve performance. One can not consider supply chain management and improved performance without considering supply chain economics.

10.3   Supply chain classes

Performance measures could be considered in relation to other project-based industries and this represents another study all together. The work conducted in this study has allowed us to now have a more detailed view of the whole supply chain and supply chain performance along various chains can be considered. Market intervention by policy makers and/or large organizations or organizations empowered to make change and to improve supply chain performance can be considered and comparisons can be made across similar supply chains with similar characteristics. This gives rise to a consideration then of how we consider overall properties of a supply chain and whether or not we can find structural and behavioural characteristics associated with different types of supply chains. The structural channel organization maps assisted in developing some themes which we can consider and thus characteristics for types of supply chains which can now be described.

Chain structures map the transfer of ownership of commodities on individual projects and various project-supply chains were aggregated into supply channel structural maps and presented in Chapters 6, 7 and 8 in relation to various case studies. Eight channel structure maps for the construction industry were developed for the study, including: aluminium, steel, concrete, glass, fire products, mechanical services, tiles and masonry. The maps are of the primary commodity – for example, glass, steel, aluminium – and do not include all the subsidiary product and raw material chains required at the manufacturing tier, nor the gathering together that occurs at the subcontractor level for site installation. For example, aluminium windows require numerous suppliers, including rubber gaskets, silicon, fastenings, framing, fixings, as well as the aluminium window component. The previous maps have served to highlight the complexities of chains for individual firms and the immediate competitors. This has assisted in being able to develop less abstract channel structures than previously used when describing the construction industry. The chain structure that is commonly used is contractor to subcontractor to materials supplier/manufacturer. We now have a much richer description of industrial market structure. Of course, the challenge then is 'so what?' Quite an appropriate question, really. A prominent academic researcher in our discipline said to me quite recently: 'I don't care where the bricks come from. Who does?' Of course, as I have spent a number of years thinking about where various materials, products and firms do come from and the various paths that are taken to arrive at a site and thus are integrated to produce beautiful buildings – perhaps was aghast and could not reply any more intelligently much later in the dead of the hours to no one in particular – 'but I do'. Upon reflection – does it matter to anyone else, though?

The structural organization maps for primary commodity channels have developed an overall perspective of the industrial markets. These were derived by building up a picture from the individual chains identified in the empirical study. The maps serve a number of purposes and these are now summarized.

There is a tendency in construction management and economics research to focus upon subcontractor and contractor relationships and contractor and client relationships. The maps provide a picture of the wide variety of procurement relationships and allow a more realistic view of the impact of the construction sector and also the opportunity to understand the impact of construction industry policy development.

The claims that the industry is highly fragmented which were discussed in Chapter 2 can be seen in perspective. High fragmentation is typically viewed as a negative attribute of the industry. Fragmentation has often been an ill-defined descriptor of the industry and it has been discussed as fragmentation of process and fragmentation of market. Aims to improve process fragmentation pursue design and construction integration strategies; for example, integrated project procurement strategies such as strategic alliances, joint ventures and design and construct contracts. Aims to improve market fragmentation pursue two strategies: first, vertical integration and, second, horizontal integration.

Short-term solutions have pursued the process fragmentation strategies. The premise in this study was that a deeper understanding of the supply chain and its structural and behavioural characteristics would enable more informed policies and strategies related to market fragmentation. Fragmentation in industrial organization economics and supply chain explores the market fragmentation concept through the degree of firm integration along the supply chain in terms of productive functions, and structural fragmentation in terms of smaller and more numerous firms in a highly competitive environment. The results of this study indicate that it is not such a simple situation as stating that the entire industry is highly fragmented and really of what value is that to informing the way we might like to improve profitability, efficiency, productivity or innovation? The structural organization maps indicate that there is a continuum of vertical fragmentation that occurs in all the channels. All channels exhibit a high degree of fragmentation and they also exhibit a high degree of vertical integration. There are a variety of supply chain options available to customers.

The concept of multiplicity allows us to organize our thoughts in relation to the variety of supply chains and the temporary project organization whereby it seems that there are so many firms continuously and constantly associating and disassociating for short-term relationships – which, incidentally, this study served to dispel this very popular description of the industry. A firm can have multiple interactions between suppliers and customers, however; the commodity is typically located within a specific market which has its own set of rules and therefore multiplicity is not as chaotic as one may first imagine. The other major factor that impacts upon the path of supply is the volume which a firm is purchasing.

Integration is not static and it is not limited to one tier only. For example, fragmentation is not simply a phenomenon related to the subcontracting tier. The large manufacturers, such as glass and steel, are fragmenting their services as they shed distribution from their core business. Distributors which had strong dependencies on manufacturers and close ties as either an internal division and vertically integrated or as an external division but with monopoly markets are now being asked to compete. This will impact upon that market, allowing smaller distributor/ merchants to gain access to this market. In future, this will impact upon the relationship that these merchants have with fabricators who supply directly to site.

The policies of governments towards supply chain management have been of a direct and indirect type and have focused upon improvement of performance for small- to medium-sized enterprises (SMEs). The direct type tended to focus upon reducing contract sizes so that SMEs could engage in contracts. There was little evidence of this in the study conducted. On the contrary, smaller contracts may provide greater access upstream for more firms in the market but may not impact upon the chain organization at other levels. The results of the study clearly show that purchasing volume is one of the key structural factors that affects fragmentation: the greater the purchasing power the more likely a firm can directly access larger lower-tier manufacturing and distribution firms. The increase in volume for more firms in the market only serves to redistribute the volume across more firms and it is speculation as to whether or not this will cause a change in those firms and their channel structures. However, regardless of this, the structural organization maps of the industry now provide a first step to benchmark and monitor future changes.

The indirect construction industry policies relate to supporting a clustering of SMEs in a co-ordinated manner around larger contractors. To some extent this already occurs as the results indicate that in many chains there are a core group of tenderers for each project. The policies suggest rewards and incentives for such strategic procurement behaviour. The maps developed can serve to monitor the claims of the extent of supply chain management. To date, the organization of the industry has been largely a guarded secret.

Rewards and incentives are mechanisms which can be used to improve supply chain management performance and behaviour by any larger firms in the chain. At this stage it is clear that contractors do not know what occurs beyond the tier that they are contracting with. The study indicated that there was little real supply chain co-ordination, development or management beyond the tier that they were directly engaging and that there was an extremely low level of contractor supply chain management at this tier. Supply chain management was evident at the manufacturing end of the chain, although in varying degrees. Supplier development and co-ordination (Hines, 1994) is evident, although supply chain management is less so.

Another major indirect policy discussed in Chapter 2 was electronic procurement (e-procurement). E-procurement is primarily intended to manage the tendering phase more efficiently. The first stages of e-procurement have focused on clients and contractor contracts and contractor and subcontractor contracts, which are focal firm to tier 1 suppliers and tier 1 to tier 2 suppliers. The other more long-term view of e-procurement is that all levels of the chain can in the future be accessed directly with auctioneering websites and portals, which is theoretically an electronic marketplace. There is potential for e-procurement in the construction industry. The structural organization maps clearly indicate this as well. Direct procurement of commodities is possible when the purchasing volume reaches a certain level; thus, the nature of demand is critical. Once the threshold of a customer's demand is reached, they then can place themselves in a different location in the chain. However, a firm may not automatically decide to purchase directly from a distributor and may still maintain conventional structural relationships.

However, I suspect this may be a naive view of procurement in the construction industry as other decision events related to the commodity need to be considered. Commodities are not simply a homogenous product; they are embedded within historical and social relationships as well as economic ties – there is intellectual and social capital shared within the procurement relationships.

This study has served to reinforce the specialization of commodity supply. Commodities possess a range of attributes. The diversity between commodities is significant. Even with a simple commodity such as sand for glazing manufacture, there are certain properties that are required for the manufacturing process. The physical proximity to the manufacturer is also deemed to be an important characteristic of the procurement relationship.

Each commodity has a level of uniqueness which gives rise to a supply chain which is 'geared' towards that commodity; for example, when the project chains were aggregated, it was typical that there were standardized products within the broad commodity group of chains alongside customized products and each formed a chain type which may have different firms and markets and thus different conditions than the other chain. The links in the chain are the firm to firm relationships and there are two main dimensions to the relationship which need to be considered in relation to strategic procurement which are summarized in Table 10.1. The first major dimension is the supplier firms and their attributes attached to the relationship which includes the significance of the commodity and countervailing power. The second dimension to be considered when developing a strategic procurement plan is the procurement relationship and the three attributes of risk/expenditure, transaction siginificance and negotiation policy and practice.

It was possible to draw out at least four properties of chains which can serve to describe and perhaps enable us to begin to benchmark various chains, refer to Table 10.2. Supply chain classification can be developed upon the following basis:

- uniqueness
- sector
- importation
- differentiation/fragmentation.

First, the level of uniqueness of the chains is perhaps to some extent similar to past work in other sectors where subcontractors are categorized; for example, in the automobile manufacturing sector. However, in the property and construction sector there are groupings of firms that consistently work together along the entire chain and the chain is mobilized towards supplying either a customized, standard or customized and standard commodity. The firms who tend towards supplying both project special unique commodities and standardized commodities can be identified as a distinct grouping.

It follows then that the chains are orientated towards supplying to a particular property market sector; residential, commercial, civil or industrial. At times the firms are supplying to more than one market sector.

The supply chains are then distinguished by firms that are importing or non-importing. This is perhaps an area rich for further exploration. The internationalization of the supply chain is becoming increasingly important in numerous diverse ways; information management, data security, commodity traceability, disaster management, etc.

Finally, the channel is either fully fragmented, partially fragmented or non-fragmented. Within the partially fragmented they are either fragmented because firms are leaping tiers because of an ability to purchase directly from the next tier through a high volume purchase. They may be leaping a tier because smaller firms choose to cluster around the larger firm and provide the services required or there may be fragmentation because there are particularly specialized services involved in the chain. Non-fragmentation occurs because firms can purchase directly from the larger manufacturers or distributors or because they choose to perform all productive functions along the chain. Tables 10.1 and 10.2 summarize key attributes arising from this study which organizations need to understand about their supply chains if they are serious about supply chain management.

Table 10.1 Blueprint for project strategic procurement plan: supplier and procurement relationships

Supplier firm		Procurement relationship		
Commodity significance	Countervailing power	Risk/ expenditure	Transaction significance	Negotiation policy and practice
Firm 1_Firm 2				
Firm 2_Firm 3				

Table 10.2 Blueprint for multiple project organizational chain scan

Sector * chain	Importation	Fragmentation	Uniqueness
Structural steel * steel chain 1			
Structural steel * steel chain 2			
Non-structural steel * steel chain 3			
Aluminium * aluminium chain 1			

For an individual organization construction supply chain economics can be viewed along a spectrum of multiplicity of procurement relationships; that is, to what extent are the relationships repeated within the organization on multiple projects and therefore at what level can we govern these relationships – at the project or corporate level? What activities tend to be those that are primarily considered on a multi-project portfolio basis or focused on the project? The following section draws together thoughts on a framework for how an organization can respond to supply chains.

10.3.1   Chain specialization and integration blueprint

It is worthwhile to consider even further the practical relevance of construction supply chain economics in two ways. First, how can an organization think about the supply chain they are located within in terms of management and economics? Second, how can an organization develop a position in relation to the supply chains they are located within?

Figure 10.2 organizes key activities into either portfolio multiple-project focused activities and then those activities that are typically project focused.

There are four main activity types:

- Supplier group strategy maps
- Strategic sourcing
- Supplier co-ordination
- Supplier development.

Of particular importance to this study was the multiple-project environment; that is, the portfolio-based activities. The chain analysis and supply strategy is key to the development of the Group Supplier Strategy Map:

1. **Economic Client Demand analysis**: Industry partner's purchasing history, future portfolio expenditure (regional/state) economic policy impacts upon the supply chains.

2. **Supplier Market Chain analysis**: market characteristics, underlying structure and behaviour.

3. **Demand/Supply Groups' Strategic alignment**: development of Key Performance Indicators for the supply chain.

4. **Client Organizational SCM Capacity audit:** internal business processes and policy, including corporate supply chain management and project supply chain management policy and processes.

5. **Development Supply strategy:** Risk versus expenditure categorization of suppliers mapped against specific sourcing strategies.

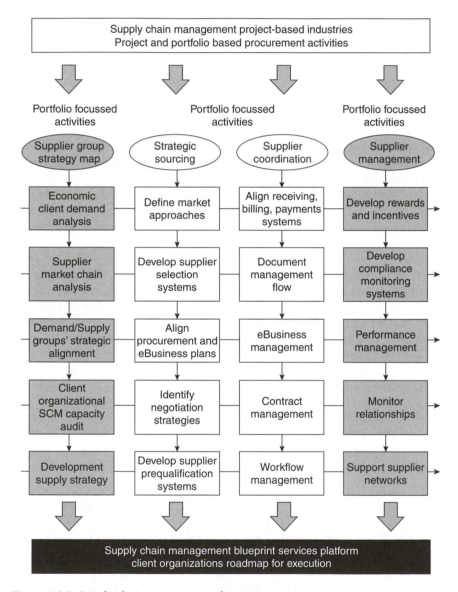

Figure 10.2 Supply chain management for project environments.

10.4 Supply Chain Information Procurement Model (SCIP Model)

The conceptualization of the supply chain has always relied upon the idea of commodity, cash and information flow. We now have a framework to understand commodity flow and also empirical data which provide rich and detailed pictures of the how, what and why of commodity flow. This provides a way forward for modelling of the other two dimensions of the supply chain; cash and information flow.

The study did also describe a theoretical model for supply chain procurement in relation to an information sciences methodology; namely, object-oriented modelling. This, as stated previously, has not been reported on in this text; however, it is important to briefly summarize it now.

An object-oriented model relies upon describing a system using a duality of nomothetic and ideographic approaches. A system of objects and their relationship to each other is described, whereby each object represents a group or class of objects. There are many instances of objects which fall within that class; for example, there is a project class which has attributes and then each individual project object has unique attributes; there is a firm class which has attributes and then each individual firm object has unique attributes and so on.

Each object possesses both an underlying structure (the data about its attributes) and a behavioural aspect (how it interacts with other objects and then how the data changes after it interacts with other objects, i.e. class operations). The findings of this study have allowed the class operations for the creation of the firm–firm procurement relationship objects to be described. The case studies and the statistical analysis have contributed to the following object-oriented class model (Figure 10.4). This is discussed further in Section 10.4.1 Firm–Firm Procurement Relationships.

With further development, one of the important points to remember is that now using this model we can relate the project supply chain to the overall supply chain. This is essentially a data management model and thus areas for further exploration include data capturing, data management and simulation for more informed decision making.

In particular, the interaction of this model with building information models and product and supplier databases will become more important. The integrated industrial organization economic project-based object-oriented model of procurement is a potential link between future BIMs and product and supplier databases, refer to Figure 10.4.

In the construction sector this level of sophisticated modelling of business processes – that is, the sourcing of suppliers and the interaction between internal organizational decision making and business models of behaviour

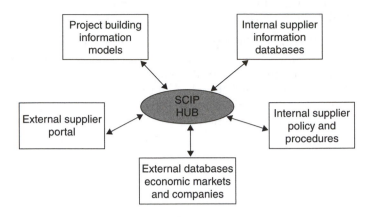

Figure 10.3 Supply Chain Information Procurement Model (SCIP).

and external market and project conditions – has not arrived. However, it has arrived in other sectors and with the advent of BIMs the SCIP model is an important aspect of our future (Figure 10.3).

The approach taken in this text is to describe supply chains within the context of individual projects and I began with the project and then 'moved out'. The following section reorientates ourselves back towards what is happening on individual contracts. Organizations studied have indicated to varying degrees that there are corporate level supply chain management policies and processes and then project supply chain management policy and processes. The following section explores at a project level the sequence of activities that typically takes place in the formation of firm–firm procurement relationships.

10.4.1   *Firm–firm procurement relationships*

In Chapter 5, a project industrial organization economic model of supply chain procurement was proposed which outlined the key elements in terms of structural and behavioural characteristics. Structural characteristics – supply chain entities:

*   project attributes
*   firms, their commodities and their market structure; and
*   attributes of firm–firm relationships.

Behavioural characteristics – mapping relationships between the supply chain entities:

*   organization of firms; and
*   firm procurement events.

There are three main ideas in relation to respond to in relation to this model, multiplicity, interaction and types:

- *multiplicity* of associations between entities
- *interaction* between structural and behavioural characteristics of entities and
- *types* of entities which have common characteristics.

First, multiplicity refers to the manner in which there are multiple associations between firms, between firms and projects, between firms and commodities; arising due to the constant forming and re-forming of firm–firm procurement relationships for each unique project and the sheer volume of transactions on each project. The sheer volume of associations between entities is immense. Although there are numerous associations that arise in the real world, the firm's characteristics do not change; nor do, for example, the project's characteristics. Therefore, there are numerous associations between entities that can be described once. Entities couple and decouple as projects arise, firms work on these projects and then projects are completed.

Second, further to the industrial organization economic concept the interaction between structure and behaviour as each object possesses a duality of structural characteristics and behavioural characteristics at the same time. The industrial organization methodology attempts to overcome this problem by simply separating the two concepts of market structure and firm conduct and market performance. Although a useful abstraction, this problem becomes more acute when we begin to think about the objects that have been described in the model and how they interact. Objects, such as firms, commodities and firm–firm procurement relationships have attributes that describe underlying structural characteristics; however, when they interact with other objects to varying degrees the underlying structural characteristics are affected and may ultimately change. Each entity in itself has structural attributes that describe the object in its static state. However, each object also has a way of behaving or operating in the real world. The discussion on channel organization assists in understanding the interaction between structure and behaviour. Further to this Figure 10.3 Procurement Relationship Class Model: operations, incorporates results from the case studies summarizing the interactions between key objects associated with each other during procurement and thus provides a procurement relationship class model.

Finally, types of entities refers to the way that the system has a group of similar objects with similar characteristics. For example, an object known as a firm is unique; however, there are many firms that have similar characteristics and therefore may behave in a similar manner. Even if the industry is project-based and each project provides for unique circumstances, there are patterns to this seemingly highly diverse world. The critical point is that both uniqueness and similarity needs to be accommodated at the one time. The industrial organization methodology has been taken as the underlying theory for the development of the ideas in this text and then pursued further to accommodate a more dynamic view of the way in which objects behave in the project environment economic system as opposed to many long-term, long-run, stable economic systems. The Procurement Relationship Class Model: operation rules, attempts to develop the procurement relationship class model further by arriving at decision rules based upon the results of the case studies and a statistical analysis which was also undertaken as part of this study. The statistical analysis has not been reported in this text. The following section synthesizes typical procurement relationship events and interactions which occur during the creation of the contractual relationship.

10.5   Procurement relationship events

The sequence diagrams developed in previous Chapters 6, 7 and 8 for various procurement relationships assist in defining the operations for objects. The sequences were developed for the formation of the firm–firm procurement relationship object.

The following discussion details the operations developed from the results of the empirical study. As noted previously, the model is a development of the proposed model in Chapter 5; however, with the addition of the operations based upon the empirical evidence. This section describes the interactions between various entities and assists in describing the behavioural characteristics in supply chain formation. The model is described using object objected Unified Modelling Language (UML) and is a very abstract interpretation of events.

Figure 10.4, Class Model with Firm–Firm Procurement Relationships Operations based upon Object Interactions, summarizes the operations which are now briefly explained. The first operation would be for the firm–firm object to retrieve the upstream ID from the firm object and create the upstream ID, as this is an integral part of the firm–firm link. Then the firm–firm object would be able to retrieve and create the project ID from the project object.

The customer upstream firm assesses the 'project' commodity. Of course, as we progress down the chain the 'project' influence diminishes and the commodity becomes more and more part of a suite of commodities transacted, but nevertheless it is always traceable to a project. The firm–firm object will retrieve and create the commodity ID from the project commodity object. The firm–firm object will then retrieve information about the commodity from the project commodity, it will retrieve the commodity type; that is, whether or not the exchange is of a product, a service or a product and a service.

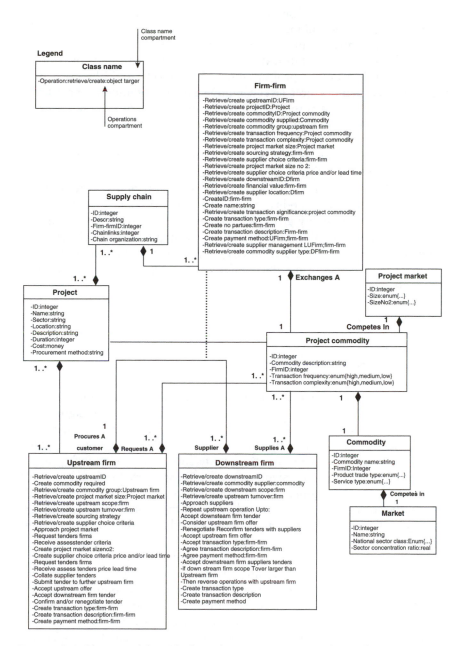

Figure 10.4 Class model with firm–firm procurement relationships operations based upon object interactions.

The empirical study indicated that a unique project commodity is created in many cases. For unique commercial projects, similar to the major projects investigated, a project commodity would be created for many relationships. Even at fabricator/component supplier and distributor/processor levels a unique project commodity object would be created. In most cases at manufacturer level a unique project commodity may not be created; however, even at this level it is possible, depending upon the size of the project, the complexity or significance of the project and/or the commodity.

Further to the refinement of the model presented in Chapter 5, an additional attribute in the firm–firm object is required as well and is somewhat related to the significance attribute of the firm–firm object. The attribute has been termed the commodity group. The firm–firm object would retrieve and create the commodity group by querying the customer firm commodity group attribute that describes the various suppliers and their grouping in the following terms. The attribute is a refinement of the commodity type and describes whether or not the commodity required by the upstream firm is considered:

- common core
- common non-core
- unique core
- unique non-core.

The attribute gives an indication of the significance of the commodity to the customer. The commodity group attribute differs to the significance attribute of the project commodity as one is concerned with the significance of the relationship to the customer firm seeking out suppliers, whereas the other attribute was the significance of the relationship to the supplier. The attribute of significance in the project commodity is an indication of how significant the commodity is to the supplier only. Significance to supplier was indicated by either the value of the contract, the prestige of the project or the opportunity that the transaction represented; for example, ability to enter a new market.

The project commodity differs to the commodity object in that it provides unique attributes that are related to the specifications from the project. These attributes are transaction frequency and complexity for the commodity for the project. They were included in the firm–firm object in the original model; however, it was clear that the association between the project and the commodity objects was not direct enough. The association between the supplier and customer firm and the project commodity is now more direct as well.

The firm–firm object would retrieve and create transaction frequency and complexity information. Following the customer firm's assessment of the project commodity, then the customer object creates project markets for the various commodities required. The firm–firm object retrieves and creates the commodity project market size for each commodity. To do so the customer firm identifies which commodity group the commodity is located in.

If the commodity is a common core commodity then historically the customer firm will have developed a method for procurement; that is, a way to define the market, approach the market and choose a supplier. Typically, the customer firm has already assessed the market and a smaller group of suppliers is selected and approached to provide tenders. The customer firm has dealt with the majority of the tenderers for a number of years. However, one of the suppliers may be one that the customer has included to maintain a check on the tenders that are submitted by long-term suppliers. This decision is related to either the business environment of the customer firm at the particular time of the project or the market environment of the suppliers at the time. For example, there may be numerous other projects available to the suppliers at the time and this impacts upon the market environment and the behaviour of the suppliers.

The common core group is often a standardized product; however, not always. For example, the specialized glazing and aluminium extrusions fabricator for Project 1 façade subcontractor were not standardized products. The commodity is required for every transaction. The important criteria are that the expenditure is typically large and demand by the customer is ongoing; that is, high volume or repetitive and steady purchasing. The firm–firm object then retrieves the scope attribute of the firm. If the customer firm is a multinational, international or national in scope, they typically have a supply agreement across the entire firm and the agreements are typically for 12 months. However, for large multinationals and raw materials commodities the supply agreements can be up to 5–6 years.

The firm–firm object will retrieve the attributes of the customer firm object including scope and turnover, which will assist in the creation of future attributes. For example, if the firm's scope is multinational or national or the purchasing volume is high or the firm's turnover is high, then the upstream firm has more power in the supplier market during the procurement process. The greater the purchasing power then the more strategic the sourcing method; for example, alliances, single or dual sourcing and the greater the likelihood that downstream firms are more likely to be dependent upon the upstream customer firm.

The supplier choice criteria are then developed by the upstream customer firm and the firm–firm object creates this attribute. Supplier choice criteria are used to accept and/or reject tenders received or renegotiate with tenderers.

10.6 Negotiation chain of events

The customer firm object approaches the project market and requests tenders from potential downstream firms. They in turn assess the requirements of the project commodity, assess their own firm, assess their potential suppliers and create commodity groups. Tenders are received and assessed against the criteria.

At this particular time, customer firms who are at tier 2–4 may go through a second tendering process. The downstream firm at the tier supplying to the primary contractor is typically asked to reconsider their price and/or lead time on supply. The primary contractor selects key supplier groups and creates a second project market and undergoes negotiation with this smaller group of suppliers. Typically, in the first project market are a smaller group of firms who are considered to produce quality commodities. In this first project market there is often at least one firm that does not produce at the same quality level. The price and lead time component of the tender offered by the firm that is not in the same segmented market as the previous group are used as the criteria to begin a new round of negotiations. The renegotiation process is primarily dependent upon the upstream contractors' perception that the tenders offered by the firms in the segmented market are too high and they then use the lowest price tender as a benchmark criterion. Of course questions arise on whether this is legal or ethical – regardless of this it is a common practice at certain times with certain suppliers.

A smaller second project market is then created with a select group of suppliers. The smaller group may or may not include the lowest priced tenderer. The new criterion is price and/or lead time. This then typically produces a chain of events in particular supplier markets to the subcontractor. The downstream firm reassesses the tenders received from their suppliers and activates a similar renegotiating process. At tier 2 (typically subcontractors) level the renegotiation process may take place with large suppliers, for example, distributors or agents. New tenders are assessed at each level of the supply chain and progressively passed up. If the project is large, an upstream firm that has a large purchasing volume will be able to engage in a renegotiation with a downstream distributor; however, if the project or the purchasing volume is not large, then the upstream firm has no power to renegotiate. The distributor typically does not renegotiate with the manufacturer or second order processor. This process only takes place in selective markets where there is a possibility to achieve a renegotiation process.

The upstream firm accepts a tender from a downstream firm. The initial financial value of the transaction is then established. The firm–firm object can then retrieve and create the supplier location. The firm–firm object is created and the object retrieves and creates a downstream ID and a firm–firm object ID is finally created.

The name of the transaction can be created. The significance of the transaction can be retrieved from the project commodity associated with the downstream firm object and created in the firm–firm object. The type of transaction can then be confirmed and this primarily relies upon the degree of power in the relationship between upstream customer and the downstream supplier. Once the type of transaction has been agreed upon, the number of parties to the contract can be created. The description of the transaction can then be created as well as the payment method, supplier management (if any) and the commodity supplier type.

Clearly, the operations for the objects during procurement and the structural organizational maps developed in chapters for the case studies describe much of the structural and behavoural characteristics for construction supply chains. The structural maps defined types of firms, types of commodities and chain structural position. From the structural position we can also infer other general behavioural characteristics; for example, tier leaping typically occurs due to consistent high demand and therefore longer-term contractual agreements or when subcontractors are sometimes not large enough to purchase primary products in sufficient quantities and so only supply services. In this situation they act in a typically clustering manner around larger fabricators or contractors.

10.6.1 *Future research and practice intersections:*
not the final word

This study was twofold in its aim; not only was it important to complete a comprehensive study which gave us some grounding in procurement practices, but also to explore contributions to our own disciplines from other disciplines. This section explores what the future might hold for construction supply chain economics and since this study was interdisciplinary then the discussion is focused on the interdisciplinary research context. Increasingly, research is conducted in collaborative arrangements with industry and government and this is not particularly new, but it often is done in an ad hoc manner where we stumble along trying to figure out how to work with/against or for other disciplines. The study in this text reports on an interdisciplinary synthesis. It was conducted by one person developing the skills and knowledge. It is one dimensional in that respect – the real challenge for the future is how to integrate a team of interdisciplinary researchers and create a synthesized response to a research problem.

10.7 Interdisciplinary context and discourse

Three topics are discussed briefly and related to this study in this section to provide an overview on the current discourse on interdisciplinary research, including: defining interdisciplinarity, aims and problems.

Interdisciplinary research is not new, having been discussed probably for the last fifteen years in earnest; however, in the last 5 years it has become quite important in the international community and well supported by national research grant funding agencies. But, as many of us who have been involved in interdisciplinary research have found – it isn't easy. Some questions to consider:

> Why is interdisciplinarity important?
> What position do we take in construction management and economics?
> How did this study contribute to the interdisciplinary debate?

> Disciplines are kinds of collectivities that include a large proportion of persons holding degrees with the same differentiating specialization name, which are organized in part into degree-granting units that in part give degree-granting positions and powers to persons holding these degrees; persons holding degrees of this particular specialized kind are employed in positions that give degree-granting powers to them, such that there is an actual exchange of students between different degree-granting institutions offering degrees in what is understood to be the same specialization.
>
> (Turner, 2000, p. 47)

This definition brings with it a number of associated themes, including the ideas of professional bodies, markets, identity, boundaries and a body of knowledge. If we consider two very strong ideas of boundaries and body of knowledge it is here that we can provide some form of a simplifed model of interdisciplinarity. A number of writers in this field have suggested that it is too difficult to define interdisciplinarity. However, a useful conceptualization is that provided by Maasen (p. 174, 2000), '...interdisciplinarity presupposes.

- a realization that certain topics cannot adequately be approached by a single discipline and
- an identification of various disciplinary activities that converge on topics that – at first sight – might be capable of being conceptualized as a joint problem.

This type of interdisciplinary research can then be described as the act of transferring insights from different disciplines into a set of problems and a set of methods for approaching them. In the course of conceptually relating problems and methods, a certain something we call 'inter' may emerge with respect to the overall topic in question.

Disciplinary boundaries are rooted in academic undergraduate disciplines and interdisciplinary research is rooted in those undertaking research with a disciplinary background(s) transgressing other disciplinary boundaries when they approach a particular industrial, societal or research problem.

Discipline boundaries and the interdisciplinary research is now considered in relation to this research study. In relation to the study reported in this text, the core discipline was considered to be construction management and economics. However, having stated this, it already becomes problematic as the industry and academia often divide this into two separate disciplines.

Therefore, the topic under study, procurement in the construction supply chain, can been located in construction management. However, as the literature review in Chapter 3 uncovered, this is probably a key problem with the understanding and further development of the field. It is the economic context of the construction supply chain that has had little consideration. Therein lies the first merging of a discipline; construction economics to construction management for the supply chain field.

Further to this, within the supply chain field it was considered that it was a particular field of the discipline of economics that could contribute, namely industrial organization economics to construction supply chain research. A couple of significant studies had already borrowed from the industrial organization economics field previously. The most notable being empirical studies investigating the automotive and electronics industries. These industries are within the manufacturing sectors and are typically characterized by long-run production. The construction industry is known as a project-based industry. Industrial organization economics studies had typically focused on manufacturing or retail sectors. The rigid and static nature of these industrial sectors differ somewhat to the character of the construction industry; or those industries which are characterized by the project life cycle; that is, design, construction and operation.

This theoretical and methodological problem was solved in two ways. First, by uncovering through empirical evidence what happens on project supply chains by beginning with the project and mapping the structural and behavioural characteristics of the various chains. This action identified the different chains that have developed in response to the different types of projects and the different sectors that operate within the construction industry (residential, public and private commercial and civil). I developed further the industrial organization economic concept in particular by categorizing supplier firms types, procurement relationship types and supply chain types on selected attributes.

Second, by borrowing from an information sciences methodology known as object-oriented modelling which is explicitly aimed at exploring real world problems where structural characteristics are integrated with behavioural characteristics simultaneously using objects and classes; instances versus patterns. This allows a conceptualization of a research problem in a dynamic and static manner simultaneously; both a nomothetic and ideographic approach. This material was not included in this text but was a part of the overall study and has highlighted that borrowing from other disciplines can contribute to an investigation of a research problem in two key ways; there is both content or knowledge contributions which disciplines can make as well as methodology and technique contributions.

Porter and Rossini (1984) noted that there is no satisfactory definition of what a discipline is and so suggested an idea of intellectual skills needed for problem-focused research. They distinguished between two types of skills: substantive knowledge and technique. It is assumed that this is similar to theoretical concepts/principles and research methods/techniques. They developed a knowledge-technique skills chart whereby each point in the chart represents a knowledge-technique pair – a substantive knowledge area and a technique for processing knowledge of the area. An established research area comprises a fixed set of knowledge and technique skills, defined as a discipline.

This previous discussion has focused on the boundaries between disciplines and how to locate the research in order to understand and identify what is the knowledge-technique in one discipline that will contribute to the research problem at the heart of the study. This highlights the next point of discussion in relation to interdisciplinary research, which is that the underlying aim of interdisciplinary research is for innovation in knowledge production.

The discourse on interdisciplinary research can be understood through various metaphors that are used to describe disciplinary versus interdisciplinary research. First, knowledge is often viewed in a territorial manner; we describe research areas or fields, which is partitioned into disciplines that are separated from one another by boundaries (Weingart and Stehr, 1995; Klein and Thompson, 1996). Interdisciplinary work is considered to be venturing to the 'borderlands' and to the 'frontiers' of knowledge.

It is also considered through organic metaphors, growth areas and with images for knowledge diffusion. According to Wiengart and Stehr (2000), the crucial feature of the discourse is the polarity of value. Disciplines carry images of being static, controlled, rigorous and conservative, therefore well grounded. Interdiscplinarity carries connotations of and is valued as being dynamic, flexible, liberal and innovative. However, the positive valuations also are the source of negative views. For example, interdisciplinary research antagonists claim that it is research which is vague, lacks 'discipline' and rigour. Likewise, discipline research is considered too specialized and rigid and lacking in innovation – that the structure of disciplines is too simple to deal with representations of the real world in all its complexity.

Of course, there is merit in all arguments and interdisciplinary research can be both innovative and lacking in rigour and disciplinary research can be both rigid and innovative – this is considered to be the paradoxical discourse in interdisciplinary research (Wiengart and Stehr, 2000). Perhaps what is more fruitful is to consider that the process of knowledge production (which takes place in disciplinary and interdisciplinary studies) is a process of specialization and differentiation. Each interdisciplinary study can recombine bits of knowledge from other fields, where it has been determined that something is lacking and needs innovation to explore the particular research problem.

The following section explores how the interdisciplinary approach contributed to the understanding of the research problem in this study by highlighting what was lacking and the innovation inherent in tackling the problem from an interdisciplinary perspective.

10.8 Interdisciplinary patterns of study: borrowing, hybridization and common ground

In considering how interdisciplinary studies are undertaken, Bechtel (1986) identified five patterns of disciplinary relations:

- developing conceptual links using a perspective in one discipline to modify a perspective in another discipline
- recognizing a new level of organization with its own processes in order to solve unsolved problems in existing fields
- using research techniques developed in one discipline to elaborate a theoretical model in another
- modifying and extending a theoretical framework from one domain to apply in another
- developing a new theoretical framework that may reconceptualize research in separate domains as it attempts to integrate them.

This study operates on numerous levels of interdisciplinarity. Clearly, in an interdisciplinary study each discipline and/or sub-field can contribute to the research problem in a certain manner, as indicated by Bechtel. The contribution can take place in terms of theoretical/substantive knowledge and research methods/techniques and it can arise because one discipline/sub-field is deficient in relation to the research problem.

Borrowing concepts from other disciplines or fields creates a hybrid character to the research. 'Hybridization is a biological metaphor connoting formation of new animals, plants or individuals or groups' (Klein, 2000, p. 9). This creates an intersection between fields and many such intersections occur involving techniques, specialized skills and instruments. However, intersections can also occur in interpretive acts, such as borrowing language and ideas. The borrowing of concepts and theories are generally much more influential than the simple borrowing of tools, data, results and methods (Klein, 2000).

Although the claim is made that disciplinary research can represent specialization and interdisciplinary research can represent generalization through over-simplification – neither discipline is well represented. The counter to this argument is that rigour and a depth of understanding can create an extremely specialized interdisciplinary study. The choice of disciplines becomes critical and it is the common ground and careful balance between each discipline's ontology, epistemology and methodology. Typically, many researchers working across boundaries are not seeking to change their ontology; rather, researchers seek to merge disciplines that align ontologies and create compatible methodologies. The interdisciplinary research requires a careful sifting and selection process to create a new common ground that converges disciplines and integrates partial spheres.

The need for interdisciplinary research arises typically because a particular field is lacking or deficient. A search into other disciplines creates a move across traditional disciplinary boundaries. The view of a discipline as a static, well-defined entity is problematic as disciplines can not be so simplistically defined, as they vary in the way they operate. A discipline/sub-field can be a 'shifting and fragile homeostatic system' that evolves and adapts to changing environments (Heckhausen, 1972; Easton, 1991).

The following Table 10.3 Interdisciplinary patterns of study, describes the interdisciplinary patterns in this particular study and summarizes each discipline/sub-field in respect to its theoretical knowledge, research methods, contributions, deficiencies and common ground in relation to the research problem of modelling of procurement in the construction supply chain. The Information Sciences contribution to the model and the ensuing discussion was completed for this study but has not been reported on in this text.

The table highlights new common ground and suggests that further research work on the role of government policy in market intervention and market/industry supplier co-ordination and development can be empirically grounded. Further to this, Box 10.2 outlines a recent nationally funded study which I have initiated and which is attempting to achieve this.

Box 10.2 Further reading

Government agencies change direction – perhaps a cynic might say in direct response to new leaders. However, the following is a study which was being conducted at the time of this text going to print and is an extension to the blueprint that was developed as a result of the work reported in this text. All Group Supplier Strategy Map activities were conducted in an action research study.

I was the Project Leader for this study and it was funded by the Collaborative Research Centre for Construction Innovation in Australia to explore the development of a Group Supplier Strategy Map for two government departments in Queensland. The two agencies were Brisbane City Council and Queensland Department of Main Roads and the two sectors which we explored were construction and demolition waste and precast concrete.

The general research question was:

How do public sector clients develop sustainable supplier group strategy maps?

The objectives are to:

1   Investigate the productivity and performance problems and the associated actions or changes of two supply chains (precast concrete and construction and demolition waste) to indicate to industry and government what can be achieved

2   Develop, trial and evaluate a Supplier Group Strategy Map for the two chains
3   Document the development, trial and evaluation process to develop a Supplier Strategy Map
4   Develop a benchmarking guide to monitor market performance post implement (monitor policy and practice) to inform decision making to monitor business environment changes triggered by federal, state and local government policy
5   Develop a best practice guideline for government supply chain management

The study was entitled Supply Chain Sustainability. The study began in mid 2005 and is anticipated for completion in mid 2007. The following are papers which have been completed to date:

London, K. and Chen, J. (2006) 'Construction supply chain economic policy implementation for sectoral change: moving beyond the rhetoric', COBRA RICS annual conference, Sept 06, London.

Invitation to reprint in book 'Collaborative Relationships in Construction', Blackwell Publishing, in print (Brown, S., Pryke, S and Smyth, H. eds).

The fifth part of the Supplier Group Strategy Map involves the organizational audit and I invited Professor Steve Rowlinson on to this grant to provide a cultural organizational perspective to critique the ability of the organizations to be able to interact effectively with their sectors and implement ongoing structural and behavioural change to supply chains for performance improvements. Steve Rowlinson and Fiona Cheung had completed a cultural study on one of the organizations previously and so this was to build upon that capacity.

By the time this text goes to print the study will be completed and further publications will be forthcoming. The CRC_Construction Innovation will host material related to this study but the CRC_CI has a sunset clause – therefore it may 'wind up' in mid 2008. The University of Newcastle, Australia through the Centre for Interdisciplinary Built Environment Research (**CIBER**), of which I am currently the Director, will provide material related to this study. http:newcastle.edu.au

There are limitations to the research presented in this text and one of them is the lack of analysis on the relationship of the project market to the industrial market. At this stage only the project market was captured. The project market has an association with the industrial market and this could form an area of future research, thus allowing a connection to national databases.

Table 10.3 Interdisciplinary patterns of study

Discipline, sub-field	Theoretical and substantive knowledge	Research methods/ techniques	Deficiencies	Contributions to construction economics; supply chain concept	New common ground
Construction management Supply chain management	Drawing from supply chain management: Lean construction/ lean production; materials logistics; channel maps; industrial sourcing	An array of research methods/techniques which are empirical and interpretive relying upon typically qualitative techniques for data collection and analysis	Industry-wide studies	Developing conceptual links using a perspective in one discipline to modify a perspective in another discipline	supply chain concept from mainstream management discipline modifies the perspective of the construction industry
Construction economics Supply chain economics	Construction economics borrows from industrial organization economics No defined concept/sub-field known as supply chain economics	No defined techniques using economic theory in relation to SCP, TCE or firm theory	No supply chain study based upon economic theory supply chain concept is the flow of product, cash and information and yet little work explores the transfer of ownership based upon real world situations Lack of methodology	Developing conceptual links using a perspective in one discipline to modify a perspective in another discipline Modifying and extending a theoretical framework from one domain to apply in another Developing a new theoretical framework that may reconceptualize research in separate domains as it	An emergence of a concept/sub-field supply chain economics which uses the structure conduct performance concept An ability to describe supply chains in terms of markets, firms, projects and links. An ability to reconceptualize the multiplicity and complexity inherent in the 'fragmented' or 'specialized' construction industry as a series of

				attempts to integrate them	objects that are associated. The objects and associations can be defined
Industrial organization economics methodology	Market structure – conduct – performance model – view of the inter-relationships of SCP	Econometric analysis Industry studies using panel data Various static measures of market structure	Non-project-based industry perspective of industrial organization economics	Developing conceptual links using a perspective in one discipline to modify a perspective in another discipline developing a new theoretical framework that may reconceptualize research in separate domains as it attempts to integrate them	Further research work on role of government policy in market intervention and market/industry supplier co-ordination and development can be empirically grounded
Information Sciences/object-oriented modelling	Object and class; types and categories; structural and behavioural concepts; multiplicity sequences and interactions	Unified Modelling Language	Little application of modelling language (UML) to describing real-world problems in construction	Using research techniques developed in one discipline to elaborate a theoretical model in another	UML technique, developed in IS, elaborates the theoretical model of construction supply chain procurement An integrated IO/OO language to simultaneously comprehend overall patterns of behaviour and individual scenarios; that is, a nomothetic and ideographic ontology

There is always the limitation that the study is not comprehensive enough – in this case the study is fairly exhaustive and for this reason it may serve as a limitation. The extensive rigour and data collection may negate further replication studies. However, this can be overcome as well by taking smaller portions of the model to examine in detail and refine.

The rules for procurement behaviour in terms of patterns of behaviour and various interactions between firms, markets and projects to create the firm–firm procurement relationship entities have been developed based upon the fieldwork in a particular region, in the southern hemisphere with particular market characteristics and within a particular sector of the construction industry (major building projects). The supplier firm classes, procurement relationship classes and supply chain classes are open to debate and further empirical work would be required to improve the credibility of these typologies.

The structural and behavioural model views were developed further and recast in an object-oriented modelling language which supports dual-structural and behavioural conceptualization of real-world problems. The original dissertation which much of the material in this text has drawn from documents these views. These are still limited in that there was no computer modelling. However, an important and necessary step has been taken in developing these descriptions and re-presenting the supply chain.

10.9   Further studies

There are opportunities to take up other research techniques afforded by the industrial organization econometrics researchers – the limitation of this study is that such quantitative measures were not embraced. The research reported in this dissertation is timely. It raises the possibilities of further research. The international construction economics research community has recently made moves to establish an identity through a CIB research agenda through the W55 Building Economics Working Group.

* The role of construction in economic growth
* Construction industries and markets in different countries
* Government policy and institutional arrangements affecting construction
* Competition, competitiveness and business strategy.

(CIB W55 Research Forum, 2004)

Clearly, these broad topics align with an industrial organization economic methodology as they are all topics that find resonance within the field.

The Australian Procurement Construction Council has released a National Framework for Procurement which outlines that the strategic importance of procurement is on the national agenda (APCC, 2003).

Skills of government employees is perhaps one of the most pressing issues; which is similar to what Kelman (2003) found when he did his national reform in the United States. Coupled with this, the Australian Department of Industry, Science and Resources has also released its report on the Evaluation of the Action Agenda for the construction industry (DITR, 2004) and, supported by consultation with the National Australian Building and Construction Council (NatBACC), the report clearly outlines supply chain management as a major initiative that was not satisfactorily addressed through the Supply Chain Partnerships programme. Further to this – the Australian construction industry is currently undergoing a major skills crisis in a variety of areas.

The model proposed that various supply chain classes performance should be measured in terms of profitability, efficiency/productivity and innovation. We now have a framework to deal with conceptualizing supplier firms, customer and supplier firm interactions and categories of different types of supply chains which can provide some order to the seemingly disordered fragmented picture of the industry. It is perhaps more fruitful to begin to ask in what ways can strategic management of the supply chain enhance profitability, efficiency, productivity and innovation? What do we already know about improving performance related to these concepts and does the supply chain concept have a role to play?

The development, adoption and diffusion of various technological, organizational, product and process innovations across the industry as a whole (Langford and Male, 2001) can be considered through the supply chain. Perhaps one of the most important developments in the next decade will be the global introduction of building information models (BIMs). We can not even contemplate this happening at the moment on a sector wider scale.

Developing building information models on projects would need an electronic literate sector – where much more sophisticated electronic data management underpinned everyday business practices than is evident currently. This would have to begin from the client – which has begun in the United States.

Why isn't e-business being adopted and diffused in the industry at what tier in the chain is it impeded and can the behaviour of firms be changed through supply chain management strategies? Is there too little customer power in a particular tier to demand change? In what way will electronic procurement decouple previous customer–supplier links and how will this impact upon construction performance?

Conducting business electronically on a widespread scale was the subject of a research project which was underpinned by considerations of the influence of the supply chain and upstream and downstream pushes and pulls (refer to Box 10.2). Many people think that diffusion of e-business is a cultural and social change – of course, this may be true; but fundamentally it is also about the business and economic factors influencing firm behaviour and industry structural change. This research project explored social, cultural and economic contexts in relation to adoption of e-business innovations – both process and product innovations. The study resulted in a series of case studies focusing on suppliers to key large organizations who were interested in improvement of e-business adoption practices of these suppliers in relation to their web-based portals. The final outcome of the study was an Adoption Profile.

Box 10.3 Supply chain influence on e-business adoption profiles

Another study which I was invited to lead for the CRC_Construction Innovation in early 2005 explores the influence that the supply chain has on construction business' adoption profiles of electronic business.

The results of this study can be found in the following publications:

E-Business Adoption Case Study Series Research Reports:

Brisbane City Council, Queensland Department of Public Works and John Holland Group

A literature review was also produced in the early stages of the project. The final outcome was a research report entitled: **E-Business Adoption Profile**

I led the project whilst at the University of Newcastle and RMIT was also a research partner on the project. The common thread through the two teams was the identification of impediments and drivers to e-business adoption. RMIT's team was led by Professor Ron Wakefield and the reports they produced were on the following case studies:

Melbourne City Council, Queensland Department of Main Roads and Victorian Building Commission

The reports can be obtained from the CRC_CI or from our **ciber** website.

Supply chain performance raises questions of comparisons between supply chains delivering the same product whereby customers and suppliers alike wish to benchmark their performance in the interest of competition. International benchmarking is constantly finding favour with researchers and comparisons across countries of various features of construction supply chains would be of interest to practitioners and academics alike.

Transparency of the industry through supply chain descriptions and procurement modelling is important as more and more players enter markets. Reduction of international trade barriers impacts upon the competitiveness of industries in a country and increases importation. How does importation impact upon the structural organization of the supply chain and the traceability of products in the chain? What role does international procurement and traceability play in on-site testing of structural materials?

The increasing internationalization of firms in the construction industry provides fertile ground for future research as supply chains take a completely different form.

There are also implications for policy research and construction supply chain economics. Perhaps one of the greatest implications of this work is for governments, particularly for policy development and policy monitoring. This study simply begins the development of a language to investigate the further development of using technology to develop decision support tools, e-business process techniques and simulation modelling techniques. Such tools and techniques would focus upon strategic analysis of commodity markets in relation to construction supply and could involve chain supply analysis, purchasing history, spend management and market environmental factors to assist governments provide policy makers and large procurers of infrastructure with information about the supply chains they initiate. It would provide them with something more tangible than a vague notion of a large industry with numerous small- to medium-sized enterprises 'out there' over the contractor hurdle.

Construction industry policy development in this manner would assist construction industry policy makers to interact with economic policy makers who typically focus on macroeconomic indicators with greater clarity. Regional economic development and growth in the construction sector can be underpinned by market intervention through supplier co-ordination and development. The ramifications of such interventions can be evaluated through supply chain performance indicators specific to the construction industry rather than be left to the vagaries of macroeconomic indicators.

Information technology/information sciences is one of the most marked areas of growth in all facets of daily life. In particular for this study, IT has emerged in both industry and academia as an area of growth. Complex systems are modelled using IT with increasing frequency. The translation of the complex system of procurement modelling of the construction supply chain into a computer language will enable decision support or simulation tools to now be developed. The object-oriented model discussed elsewhere is of course only the beginning. Connecting a building information model to a supply chain information model is the next step in this research – that is, bringing together an integrated industrial organization economic object-oriented methodology for supply chain procurement modelling with a project building information model which includes a supplier procurement plan.

This study did not model cashflow in the supply chain and this is an important step as well. Modelling of cashflow in supply chains will enable an understanding of economic growth in a region – a particularly useful approach when trying to stimulate growth for support of SMEs.

Visualization has been a hallmark of this study when representing the entities and connections between the entities further visualization is an important next step in research for this particular model.

10.10   A final word

Chapter summary

1   This study addressed the industry problem of:

The widely held belief that improving the performance of the construction industry is achieved through overcoming fragmentation problems by supply chain integration.

2   This study addressed the research problem of:

The lack of a supply chain positive economic model and methodology to understand the structural and behavioural characteristics of the supply chains that underpin the construction industry to enable more credible normative integration/specialization supply chain management models.

3   An interdisciplinary approach using and applying concepts from the field of industrial organization economics and concepts and research techniques from the field of object-oriented modelling to construction management and economics – the results of the application of industrial organization economic concepts was reported in this text. An inductive empirical epistemology was used to develop descriptions of real-world scenarios. Using a combined qualitative and quantitative method for data collection and analysis structured interviews were conducted and questionnaire interviews were distributed. Structural and behavioural model views of real-world procurement in construction supply chains were developed based upon data collected related to six major building projects in an Australian capital city involving seven hundred and twenty-four firms. One thousand two hundred and fifty-three procurement relationships were mapped using data collected from forty-seven structured interviews (thirty-nine different participants) and forty-four questionnaires.

4   The supply chain structural and behavioural characteristics which were identified can be summarized as follows: supply chains can be classed according to a number of attributes including: uniqueness, property sector, importation and fragmentation

- uniqueness: supply chains tend to be grouped by the degree of uniqueness of the commodity into highly customized, standardized and customized and standardized

- property sector: supply chains tend to be grouped by residential, commercial and civil; however
- importation: supply chains are either impacted by importation or not and import supply chains are longer and have a lesser degree of traceability
- fragmentation: there is a high level of diversity in channel structure maps for transfer of ownership of products within commodity types and across commodity types and therefore supply chain structural and behavioural characteristics which were identified are dependent upon interactions between customer firm, supplier firm, project commodity, project market and industrial market which ultimately rely upon supplier firms being classed as follows:
- commodity significance: commodity is either common core, common non-core, unique core or unique non-core and then
- countervailing power: the power in the transaction either lies with the customer or the supplier; and power could be dependent upon volume of purchasing of uniqueness of commodity.

5 The classification of supplier firms into these classes then relates to the type of firm–firm procurement relationships that can be identified. It was found that these typically were reliant upon an if/then scenario and could be summarized and classed according to three characteristics, as follows:

- formation: risk and expenditure: high risk and high spend; high risk and low spend; low risk and high spend and low risk and low spend
- transaction: significance: strategic critical and equal control; strategic security and supplier control; leverage purchasing and customer control and tactical purchasing and customer control
- management: negotiation: annual, high level and customer control; annual, high level and supplier control; annual and project, low level and customer control and project, high level and customer control.

6 The perception that the construction industry is fragmented, unstructured, unpredictable and high risk clearly is a simplistic view of a complex set of varied and numerous markets. There are sections of the supply chain that are quite predictable, structured and low risk and sections that are unpredictable, unstructured and high risk. The project nature of the industry and the short-term contracts certainly seems to increase the unpredictability of firm–firm relationships in comparison to the very stable nature of quite long-term contracts in the manufacturing sectors. This is apparent even as you move down the supply chain and move into manufacturing sectors.

7 However, even given this and the seemingly arm's length relationships in comparison to other sectors, there is a structure and order to the way in which supply chains are organized and there is predictability in how they are formed. The results of this empirical study provide a detailed understanding of firms, their commodities, the project market, the project and the contractual relationship between the firms that make up the chain. Clearly, the industrial organization economics model has a high degree of relevance to procurement modelling of the construction supply chain.

8 The second central theme of this research was to establish the intellectual basis and practical framework for guiding the description and analysis of procurement related to a project-based industry across many firms, markets and chains. This brings us to the conclusions of the study in terms of interdisciplinary contributions.

9 The study has contributed to the construction management discipline and in particular the field of supply chain management and those that use industrial organization economic concepts as it has provided an industry-wide study and borrowed concepts from the generalist management discipline and in particular the supply chain management field. It has primarily contributed to the construction economics discipline and in particular in formalizing the field of construction supply chain economics in terms of theory and methodology.

10 The extent to which an interdisciplinary study borrowing from the fields of industrial organization economics can contribute to procurement modelling in the construction supply chain can be summarized as follows:

- the project-based industrial organization object-oriented methodology for construction supply chain procurement modelling provides for a conceptual framework of structure-conduct-supplier-procurement relationships-supply chain-performance rather than a simple structure-conduct-performance
- It has also contributed back towards industrial organization economics in that there has been a reliance towards non-project-based industry studies in relation to the supply chain concept.
- industrial organization economics does not typically embrace a project-based industry with numerous interactions and the conceptual links from object-oriented modelling contribute to a new language for industrial organization economics in terms of firm objects with classes and market objects within market classes; multiplicity can be accommodated through the class model with firm–firm procurement relationships.

11 There is strong alignment conceptually between industrial organization and object-oriented modelling (information sciences methodology) in that both fields support a structure-behaviour perspective; however, the industrial organization economic field has struggled with encapsulating the duality of objects; markets have both structure and behaviour and firms have both structure and behaviour; the object-oriented model provides a mechanism to accommodate this duality

- the object-oriented methodology provides a set of representational techniques for capturing, specifiying, visualizing, understanding and documenting objects associated with procurement in the construction supply chain from an industrial organization economic perspective
- the language and framework which has been developed can be extended to investigate, develop and refine further relationships and classes; supply chain performance, project market structure, demand classes and technology classes.

12 The wider study did recast the findings in this light and, although not reported in this text, this is where fruitful further research has been identified in the following:

Decision support modelling
Information modelling and
Resources modelling (time and cash) for chains at a sectoral and project level

13 Although there has been a step forward in our understanding of the supply chain, there are limitations to this research and it is important to acknowledge these.

14 This text is only a very small step in the development of a more detailed understanding of the role that procurement modelling of the construction supply chain research can play in thinking and acting strategically in relation to improving the performance of the construction economy.

Bibliography

Agapiou, A., Clausen, L. E., Flanagan, R., Norma, G. and Notman, D. (1998) The role of logistics in the materials flow control process. *Construction Management and Economics*, **16**, 131–137.

Alarcon, L. (ed.) (1997) Lean Construction: Compilation of 1993–1995 IGLC Proceedings, Balkema, Rotterdam.

Alter, C. and Hage, J. (1993) *Organisations Working Together*, Vol. 1. Sage Publications, London, UK.

Andersson, N. (2001) *A meso-economic analysis on the Swedish construction industry*. CIB W055-W065 Construction Sector System or Construction Industry Cluster Analysis for Construction Industry Comparative Analysis Project Group.

Aouad, G. (1999) *The development of a process for the construction sector*. Joint Triennial Symposium CIB W65 and W55. Cape Town, South Africa.

Atkins, B. (1993) *A programme for change*. Europe Construction Office of Official Publications.

Australian Bureau of Statistics (ABS) (2001) Private Sector Construction Industry, Australia, Catalogue No. 8772.0.

Australian Bureau of Statistics (ABS) (2003) ABS Statistical Concepts Library ANZSIC Ch1: About the Classification, Classification Principles.

Australian Bureau of Statistics (ABS) (2003) Private Sector Construction Industry, Australia, Catalogue No. 1292.0.

Australian Construction Procurement Council (APCC) (1997) *Construct Australia – Building a better construction industry in Australia*, Australian Procurement and Construction Council, Perth, Australia, www.apcc.gov.au (accessed June 2001).

Australian Construction Procurement Council (APCC) (1999) *National management framework: Delivering opportunities for small and medium enterprise in government procurement and contracting*. Australian Government Publishing, Canberra, Australia, www.apcc.gov.au (accessed June 2001).

Australian Constructors' Association (ACA) (1999) Australian Construction Productivity International Comparison. Prepared by Access Economics and World Competitive Practices Pty Ltd, August, www.constructors.com.au/main/index.htm (accessed May 2007).

Australian Expert Group for Industry Studies (AEGIS) (1999) *Mapping the Building and Construction Product System in Australia*. Department of Industry, Science and Technology, Canberra, Australia.

Australian Construction Procurement Council (APCC) (2001a) *Construct Australia*. Australian Government Publishing, Canberra, Australia.

Australian Construction Procurement Council (APCC) (2001b) *Government Framework for National Cooperation on Electronic Procurement.* Australian Government Publishing, Canberra, Australia, www.apcc.gov.au (accessed June 2001).

Australian Procurement Construction Council (APCC) (2002a) *Government framework for National Cooperation on Electronic Procurement.* Australian Government Publishing, Canberra, Australia, www.apcc.gov.au (accessed June 2002).

Australian Procurement Construction Council (APCC) (2002b) *About APCC,* www.apcc.gov.au (accessed June 2002).

Australian Procurement Construction Council (APCC) (2003) *National procurement reform principles,* www.apcc.gov.au (accessed June 2004).

Australian Pacific Projects Corporation (APPC) (1998) *Procurement and Project Delivery,* National Building and Construction Committee, Australia.

Baker, M. (1990) *Dictionary of Marketing and Advertising.* 2nd edn Nichols Publishing, New York, NY.

Ballard, G. and Howell, G. (1999) *Bringing light to the dark side of lean construction: a response to Stuart Green.* Proceedings 6th Annual International Group for Lean Construction Conference. Berkeley, CA.

Banwell, H. (1964) The placing and management of building and civil engineering contracts report. HMSO, UK.

Barlow, J. (1998) *From craft production to mass customisation: customer focussed approaches to housebuilding.* Proceedings 6th Annual International Group for Lean Construction Conference. Berkeley, CA.

Barrett, P. and Aouad, G. (1998) *Supply chain analysis for effective hybrid concrete construction.* Proceedings of Joint Triennial Symposium CIB Commissions W65 and W55, eds Bowen, P. and Hindle, R. Cape Town, South Africa.

Berger, S. and Piore, M. (1980) *Dualism and discontinuity in industrial societies.* Cambridge University Press, Cambridge, UK.

Betts, M. (1999) *Strategic Management of IT in Construction,* Blackwell Publishing, UK.

Bon, R. (1986) Comparative stability analysis of demand-side and supply-side input–output models. *International Journal of Forecasting,* 2 (2), 231–235.

Bon, R. (1988) Supply-side multi-regional input–output models. *Journal of Regional Science,* 28 (1), 41–50.

Bon, R. (1990) Historical comparison of construction sectors in the United States, Japan, Italy and Finland using input–output tables. *Journal of Construction Economics and Management,* 8, 233–247.

Boulding, K. (1985) *The World as a Total System.* Sage Publications, Beverly Hills, USA.

Bowersox, D. and Closs, D. (1996) *Logistics Management: The Integrated Supply Chain Process.* McGraw-Hill, New York, NY.

Bowley, M. (1966) *The British Building Industry.* Cambridge University Press, UK.

Briscoe, G. H., Dainty, A. R. J., Millett, S. J. and Neale, R. H. (2004) Client-led strategies for construction supply chain improvement. *Journal of Construction Management & Economics,* 22, 193–201.

Burt, D. and Doyle, M. (1993) *The American Keiretsu: Strategic Weapon for Global Competitiveness.* BusinessOne Unwin, Homewood, IL.

Callaghan, M. (1998) *Supply Networks.* www.labs.bit.com/people/callagjg/ion/supply.htm (accessed June 2003).

Carassus, J. (2001) *Macroeconomic data on the construction industry: Construction Industry A sector system approach applied to the French Construction industry*. CIB W055-W065 Construction Sector System or Construction Industry Cluster Analysis for Construction Industry Comparative Analysis Project Group.

Cardoso, F. (1999) *Entrepreneurial Strategies and new forms of rationalisation of production in the building construction sector of Brazil and France*. Proceedings of Annual IGLC Conference. Berkeley, CA.

Chamberlin, E. (1933) *The Theory of Monopolistic Competition*. Harvard University Press, Cambridge, MA.

Christopher, M. (1998) *Logistics and Supply Chain Management: Strategy for Reducing Costs and Improving Services*. 2nd edn Pitman, London.

CIB W55 Research Forum (2004) *Summary of Key Initiatives and Directions: Reporting of Commission Meeting*, de Valence, G., International Working Commission Co-ordinator, CNBR Discussion List, accessed June 2004.

Clark, J. (1899) *The Distribution of Wealth*. New York, NY.

Clausen, L. (1995) *Report 256: Building Logistics*. Danish Building Research Institute, Copenhagen, 4, Denmark.

Coase, R. (1937) The Nature of the Firm. *Economica*, **4** (16), 386–405.

Construction Industry Development Agency (CIDA) (1995) In Principle, Construction Industry Reform, 1991–1995 – A celebration of the Work of the Construction Industry Development Agency, a report by Barda, E., Chair, Australian Government Publishing, Canberra, Australia.

Construction Industry Development Board (CIDB) (1999) *Towards Transformation and Development of the South African Construction Industry*. Report by Inter-Ministerial Task Team on Construction Industry Development. Pretoria, South Africa.

Construction Client Forum 1998.

Co-operative Research Centre – Construction Innovation (CRC-CI) (2001) *CRC Strategic Plan 2001–2003*. Australian nationally funded Research Centre, www:construction-innovation.info (accessed 2001).

Co-operative Research Centre – Construction Innovation (CRC-CI) (2003) *CRC Strategic Plan 2003–2008*. Australian nationally funded Research Centre, www:construction-innovation.info (accessed 2004).

Copacino, W. (1997) *Supply Chain Management: The Basics and Beyond*, Vol. 1, 1st edn St Lucie Press, Boca Raton.

Cox, A. and Lamming, R. (1995) Procurement Management in the 1990s. Ch 1 in *Strategic Procurement Management in the 1990s: Concepts and Cases* (eds) Lamming, R. and Cox, A. Earlsgate Press, London.

Cox, A. and Townsend, M. (1998) *Strategic Procurement in Construction: Towards Better Practice in the Management of Construction Supply Chains*, Vol. 1, 1st edn Thomas Telford Publishing, London.

Coyle, J., Bardi, E. and Langley, C. (1996) *The Management of Business Logistics*. 6th edn West Publishing, MN.

Dainty, A., Briscoe, G. and Millett, S. (2001) Subcontractor perspectives on supply chain alliances. *Journal of Construction Management and Economics*, **19**, 841–848.

Day, M. (1998) Supply chain management: business process effects in the ceramics industry. *Logistics Research Network 1998*. Cranfield School of Management, Cranbrook, UK.

Department of Industry Trade Resources (DITR) (1997a) *Supply Chain Partnerships Program – Building More Profitable Inter-organisational relationships*. GovernmentPrinters, Commonwealth of Australia, ACT, Australia.

Department of Industry Trade Resources (DITR) (1997b) *Investing for Growth*. GovernmentPrinters, Commonwealth of Australia, ACT, Australia.

Department of Industry Trade Resources (DITR) (1998) *Building for Growth: A Draft Strategy*. GovernmentPrinters, Commonwealth of Australia, ACT, Australia.

Department of Industry Trade Resources (DITR) (1999) *Building for Growth: An Analysis of the Australian Building and Construction Industries*. GovernmentPrinters, Commonwealth of Australia, ACT, Australia.

Department of Industry Trade Resources (DITR) (2004) *Building and Construction Agenda Evaluation*. GovernmentPrinters, Commonwealth of Australia, ACT, Australia.

Department of State and Regional Development (DSRD) (2000) *Strategic Industry Audits of Victorian Industry*. Victorian Government Publishing, Melbourne, Victoria, Australia, www.business.gov.au (accessed 2002).

de Valence, G. (2001) *Comparison of traditional and cluster models of construction industry structure*. CIB W055-W065 Construction Sector System or Construction Industry Cluster Analysis for Construction Industry Comparative Analysis Project Group.

Dore, R. (1976) Goodwill and the spirit of market capitalism. *British Journal of Sociology*, **34** (4), 459–482.

Dore, R. (1987) *Taking Japan Seriously*. Stanford University Press. Stanford, CA.

Drucker, P. (1992) *Managing for the Future: The 1990s and Beyond*. Truman Talley Books/Plum Inc., New York, NY.

Easton, D. (1991) The division, integration and transfer of knowledge, in *Divided Knowledge*. (eds) Easton, D. and Schelling, C. Sage Publications, Newbury Park, CA.

Eccles, R. (1981) The quasi-firm in the construction industry. *Economic Behaviour and Organisation*, **2**, 335–357.

Economic Development Committee (EDC) (1994) *The First Report to Parliament: The Corruption of the Tendering Process, Inquiry into the Victorian Building and Construction Industry*, L. V. North, Government Printer, Melbourne, Australia.

Eden, C. and Radford, J. (1990) *Tackling Strategic Problems – The Role of Group Decision Support*. Sage Publications, London.

Edgeworth, R. (1881) *Mathematical Psychics*, London.

Egan, J. (1998) *Rethinking Construction: The Report of the Construction Task Force*. Department of the Environment, Transport and the Regions, London.

Ellram, G. (1972) The organisation of industry. *Economic Journal*, **82**, 883–896.

Ellram, L. (1991) Supply chain management: the industrial organisation perspective. *International Journal of Physical Distribution and Logistics Management*, **21** (1), 13–22.

Emmerson, H. (1962) *Survey of Problems Before the Construction Industries*. Ministry of Works, HMSO, UK.

Finch, E. (2000) *Net Gain in Construction*. Elsevier, UK.

Forrester, J. (1961) *Industrial Dynamics*. MIT Press, Cambridge, MA.

Fujita, K. (1965) *Japanese Industrial Structure and Small Business*. Iwanami Shoten, Tokyo.

Gailbraith, K. (1952) *American Capitalism: The Concept of Countervailing Power*. Houghton-Mifflin, Boston, MA.

Gale, T. and Eldred, J. (1996) *Getting Results with the Object-oriented Enterprise Model*. SIGS Books, NY.

Gann and Slater (1998) *Learning and innovation management in project based firms*, Conference Proceedings 2nd Annual Conference Technology Policy and Innovation, 3rd–5th August, Lisbon.

Gold (1991) as cited by Bowersox, D. and Closs, D. (1996) *Logistics Management: the Integrated Supply Chain Process*. McGraw-Hill, New York, NY.

Gomes-Casseres, B. (1996) *The Alliance Revolution: The New Shape of Business Rivalry*. Harvard University Press, Cambridge, MA.

Graves, A. (2000) *Agile Construction Initiative*. University of Bath, Bath, UK.

Green, S. (1999) The dark side of lean constriction: exploitation and ideology. *IGLC 7th Annual Conference*. Berkeley, CA.

Green, S. and Lenard, D. (1999) Organising the project procurement process. Ch 3 in *Procurement Systems: A Guide to Best Practice in Construction*, (eds) Rowlinson, S. and McDermott, P. E&FN Spon, London.

Groak, S. (1994) Is construction an industry? Notes towards a greater analytic emphasis on external linkages. *Journal of Construction Management and Economics*, **12**, 287–293.

Groat, L. and Wang, D. (2002) *Architectural Research Methods*. John Wiley & Sons, Inc, New York, NY.

Hall, R. (1983) *Zero Inventories*. Dow Jones Irwin, Homewood, IL.

Harland, C. M. (1994) *Supply chain management: perceptions of requirements and performance in European automotive aftermarket supply chains*. PhD thesis, Warwick Business School, UK.

Harland, C. M. (1996) Supply chain management: relationships, chains and networks. *British Journal of Management*, 7 (Special), 63–80.

Hart, O. (1989) An Economist's Perspective on the Theory of the Firm. *Columbia Law Review 1757*.

Hasegawa, H. and Ueki, K. (1978) Case explanation of the Law on the Prevention of Delay in the Payment of Subcontracting Charges and Related Matters. Tokyo, Japan.

Hay, D. A. and Morris, D. J. (1979) *Industrial Economics Theory and Evidence*. Oxford University Press, Oxford, UK.

Hay, D. A. and Morris, D. J. (1980) *Industrial Economics Theory and Evidence*. Oxford University Press, Oxford, UK.

Heckhausen, H. (1972) *Discipline and Interdisciplinarity Interdisciplinarity Interdisciplinarity in Problems of Teaching and Research in Universities*. OECD, Paris, France.

Hillebrandt, P. (1982) *Economic Theory and the Construction Industry*, Vol. 1, 2nd edn Macmillan Press, London.

Hines, P. (1992) Supplier Associations: Creating World Class Manufacturing Performance in the Supplier Network. Boosting the Competitiveness of the Motor Industry Conference. London.

Hines, P. (1994) *Creating World Class Suppliers: Unlocking Mutual Competitive Advantage*. Pitman Publishing, London.

Hines, P. (1996) Network sourcing: a discussion of causality within the buyer–supplier relationship. *European Journal of Purchasing and Supply Management*, **2** (1), 7–20.

Hines, P., Rich, N., Bicheno, J., Brunt, D., Taylor, D., Butterworth, C. and Sullivan, J. (1998) Value stream management. *International Journal of Logistics Management*, **9** (1), 25–41.

Hobbs, J. (1996) A transaction cost approach to supply chain management. *Supply Chain Management*, **1** (2), 15–27.

Hoek, R. (1998) Reconfiguring the supply chain to implement postponed manufacturing. *International Journal of Logistics Management*, **9** (1), 95–103.

Hong-Minh, S., Barker, R. and Naim, M. (1999) *Construction supply chain trend analysis*. Proceedings of Annual IGLC Conference. Berkeley, CA.

Horman, M., Kenley, R. and Jennings, V. (1997) *A lean approach to construction: an historical case study*. Proceedings of Annual IGLC 5 conference. Gold Coast, Australia.

Howell, G., Laufer, A. and Ballard, G. (1993) Interaction between cycles: one key to improved methods. *ASCE Journal of Construction Engineering and Management*, **119** (4), 714–728.

Huston, C. (1996) *Management of Project Procurement*. McGraw Hill Companies, Inc. New York, USA.

IBIS Business Information Service (2000) *Data contained in IBIS Industry Sets*.

Introna, L. (1991) The impact of information technology on logistics. *International Journal of Physical Distribution and Logistics Management*, **21** (5), 32–37.

ISCP (Integrated Supply Chain Program) (1998) *Department of industry science resources industry action agenda programs*, Commonwealth of Australia, Australia.

Ito, T. (1957) *Chusho kigyo ron (On small and medium enterprises)*. Tokyo, Japan.

Jennings, I. (1997) *Systemic integration in construction project organisations*, Unpublished Phd Thesis, University of Melbourne, Australia.

Jevons, S. (1880) *Theory of Political Economy*. London.

Kaklauskas, A. and Zavadskas, E. K. (2001) *Working out a rational institutional model of Lithuanian construction industry development*. CIB W055-W065 Construction Sector System or Construction Industry Cluster Analysis for Construction Industry Comparative Analysis Project Group.

Kamata, S. (1982) *Japan in the Passing Lane: An Insider's Account of Life in a Japanese Auto Factory*. Unwin Paperbacks, UK.

Kelman, S. (2003) *Implementing Federal Procurement Reform*. University of Harvard Innovations in American Government Occasional Papers, Kennedy School of Government, www.ksg.harvard.edu/research/working_papers/index.htm (accessed 2007).

Kenley, R., London, K. and Watson, J. (1999) Strategic procurement in the construction industry: mechanisms for public sector clients to improve performance in the Australian public sector. *Journal of Construction Procurement*, **6** (1), 4–19.

Kenney, M. and Florida, R. (1993) Beyond mass production: production and the labor process in Japan. *Politics and Society*, **16** (1), 121–158.

Klein, J. (2000) A conceptual vocabulary of interdisciplinary science. Ch 1 in *Practising Interdisciplinarity*. University of Toronto Press, Toronto, Canada.

Klein Thompson, J. (1996) The dialectic and rhetoric of disciplinarity and interdisciplinarity in *Interdisciplinary Analysis and Research: Theory and Practice of Problem-Focused Research and Development*. (eds) Chubin, D., Porter, A., Rossini, F. and Connolly, T. Lomond, Mt Airy, MD.

Knight, F. (1921) *Risk, Uncertainty and Profit*. New York, NY.

Koskela, L. (1992) *Application of the New Production Philosophy to Construction*. Technical Report 72, Center for Integrated Facility Engineering, Department of Civil Engineering, Stanford University, CA.

Krafcik, J. (1988) Triumph of the lean production system. *Sloan Management Review*, **30** (1), 41–52.

Lai, L. (2000) The coasian market-firm dichotomy and subcontracting in the construction industry. *Journal of Construction Management and Economics*, **18**, 355–362.

Lambert, D. M., Pugh, M. and Cooper, J. (1998) Supply chain management. *The International Journal of Logistics Management*, **9** (2), 1–19.

Lamming, R. (1992) *Supplier strategies in the automotive components industry: development towards lean production*. Unpublished PhD thesis, University of Sussex, UK.

Lamming, R. (1993) *Beyond Partnership: Strategic for Innovation and Lean Supply*. Prentice-Hall, Hemel Hempstead, UK.

Lamming, R. and Cox, A. (1995) *Strategic Procurement Management in the 1990s: Concepts and Cases*. Earlsgate Press, London.

Langford, D. and Male, S. (2001) *Strategic Management in Construction*. 2nd edn Gower, Hants, UK.

Latham, M. (1994) *Constructing the Team*. HMSO, London.

Lawrence, P. and Lorsch, J. (1986) *Organisation and Environment: Managing Differentiation and Integration*. Harvard Business School, Boston, MA.

Le Compte, M. and Preissle, J. (1993) *Ethnography and Qualitative Design in Education Research*. 2nd edn Academic Press, New York, NY.

Leontief, W. W. (1936) Quantitative input and output relations in the economic systems of the United States. *The Review of Economics and Statistics*, **18** (3) (August, 1936), 105–125.

Lincoln, Y. and Guba, E. (1985) *Naturalistic Inquiry*. Sage Publications, Beverly Hills, CA.

London, K. (2002) Supply chain management. Ch 7 in *Construction Management: New Directions*. 2nd edn (Eds) McGeorge, W. D. and Palmer, A. Blackwell Publishing, Oxford, UK.

London, K. (2004) *Construction Supply Chain Procurement Modelling*, PhD dissertation thesis, University of Melbourne, Australia.

London, K. (2005) *Construction Supply Chain Procurement Modelling*. PhD dissertation thesis, University of Melbourne, Australia.

London, K. and Chen, J. (2006) *Construction Supply Chain Economic Policy Implementation for Sectoral Change: Moving Beyond the Rhetoric*. The Construction Research Conference of the Royal Institute for Chartered Surveyors. London, 6–8 September.

London, K. and Kenley, R. (1999) Client's role in construction supply chains – a theoretical discussion. *CIB Triennial World Symposium W92*, Cape Town, South Africa.

London, K. and Kenley, R. (2000a) The development of a neo-industrial organisation methodology for describing and comparing structural and behavioural characteristics of construction supply chains. *6th Annual Lean Construction Conference*. Brighton, UK.

London, K. and Kenley, R. (2000b) Mapping construction supply chains: widening the traditional perspective of the industry. *7th Annual European Association of Research in Industrial Economic EARIE Conference*. Lausanne, Switzerland.

London, K. and Kenley, R. (2001a) The evolution of an alliance network to develop an innovative construction product: an instrumental case study. *CIB World Congress 2001*, Wellington, NZ.

London, K. and Kenley, R. (2001b) An industrial organisation economic supply chain approach for the construction industry: a review. *Journal of Construction Management and Economics*, **19** (8), 777–788.

London, K., Kenley, R. and Agapiou, A. (1998a) Theoretical supply chain network modelling in the building industry. *Association of Researchers in Construction Management Conference* (ARCOM). Reading, UK.

London, K., Kenley, R. and Agapiou, A. (1998b) The impact of construction industry structure on supply chain network modelling. *Institute of Logistics, Logistics Research Network Annual Conference*, Cranfield School of Management. Cranfield, UK.

Lopes, J. (2001) *A mesoeconomic analysis of the construction industry*. CIB W055-W065 Construction Sector System or Construction Industry Cluster Analysis for Construction Industry Comparative Analysis Project Group.

Maasen, S. (2000) Inducing interdisciplinarity: irresistible infliction? The example of a research group at the Center for Interdisciplinarity Research (ZiF), Bielfeld, Germany. Ch 9 in *Practising Interdisciplinarity* (eds) Weingart, P. and Stehr, N. University of Toronto, Buffalo, Toronto, Canada.

Martin, S. (1993) *Industrial Economics: Economic Analysis and Public Policy*. 2nd edn Prentice Hall, NJ.

Masten, S., Meehan, J. and Snyder, E. (1991) The costs of organisation. *Journal of Law, Economics and Organisation*.

McGeorge, W. D. and Palmer, A. (2002) *Construction Management: New Directions*. 2nd edn Blackwell Publishing, Oxford, UK.

Miles, R. and Snow, C. (1992) Organisations: New concepts for new forms. *California Management Review*, **28** (3).

Milgrom, P. and Roberts, J. (1992) *Economics, Organisation and Management*. Prentice Hall, NJ.

Miller, J. (1999) Applying multiple project procurement methods to a portfolio of infrastructure projects. Ch 9 in *Procurement Systems: A Guide to Best Practice in Construction*. (eds) Rowlinson, S. and McDermott, P. E&FN Spon, London.

Min, H. and Zhou, G. (2002) Supply chain modeling: past, present and future. *Journal of Computers and Industrial Engineering*, **43**, 231–249.

Minato, T. (1991) *The development of Japanese parts supply relationships: past, present and future*. Customer–supplier Relationship Study Tour, EC-Japan Centre for Industrial Co-operation.

Mitusi, I. (1991) *A unique Japanese Subcontracting system from a global point of view*. Customer–supplier Relationship Study Tour, EC-Japan Centre for Industrial Co-operation.

Murray, M. and Langford, D. (2002) *Construction reports 1944–98*. Blackwell Science, Oxford, UK.

Murray, M., Langford, D., Hardcastle, C. and Tookey, J. (1999) Organisational design. Ch 4 in *Procurement Systems: A Guide to Best Practice in Construction*. (eds) Rowlinson, S. and McDermott, P. E&FN Spon, London.

National Building and Construction Committee (NatBACC) (1997) *A Report for Government by the National Building and Construction Committee*. Department of Industry, Science and Resources Building for Growth: Key Initiatives of the Building and Construction Industries Action Agenda. Australian Government Publishing, Canberra, Australia, www.isr.gov.au/industry/building (accessed 2002).

National Building and Construction Committee (NatBACC) (1998) *Initial Report for Department of Industry Science Tourism (DIST) on Procurement & Project Delivery in the Building and Construction Industries.* Prepared by Australian Pacific Projects. Australian Government Publishing, Canberra, Australia, www.isr.gov.au/industry/building (accessed 2002).

National Economic Development Office (NEDO) (1988) *Achieving Quality on Building Sites.* HMSO, London, UK.

New, S. (1997) The scope of supply chain management research. *Supply Chain Management,* 2 (1) 15–22.

Newman, I. and Benz, C. R. (1998) *Qualitative Quantitative Research Methodology: Exploring the Interactive Continuum.* Southern Illinois University Press, Carbondale, IL.

New South Wales State Government (NSW) (1993) *NSW Codes of Practice.* NSW State Publisher, Australia.

Nicolini, D., Holti, R. and Smalley, M. (2001) Integrating project activities: the theory and practice of managing the supply chain through clusters. *Journal of Construction Management and Economics,* 19, 37–47.

Nischiguchi, T. (1987) Competing Systems of Automotive Components Supply: An Examination of the Japanese Clustered Control Model and the Alps Structure. First Policy Forum, Tokyo, Japan.

Nischiguchi, T. (1991) Beyond the honeymoon effect, INSEAD Information, Summer, 8–9.

Nischiguchi, T. (1994) *Strategic Industrial Sourcing: The Japanese advantage,* Vol. 1, 1st edn Oxford University Press, New York, NY.

O'Brien, W. (1997) Construction supply chains: case study, integrated cost and performance analysis. In *Lean Construction* (ed.) Alarcon, L. Balkema, Rotterdam.

O'Brien, W. (1998) *Capacity costing approaches for construction supply chain management.* PhD thesis. Department of Civil and Environmental Engineering, University of Stanford, USA.

O'Brien, W., Fischer, M. and Jucker, J. (1995) A project view of economic coordination, *Journal of Construction Management and Economics,* 13 (5), 393–400.

O'Brien, W., London, K. and Vrijhoelf, R. (2004) Construction supply chain modelling: an interdisciplinary research agenda. *ICFAI Journal of Supply Chain Management.*

Office of Federal Procurement Policy (OFPP) (1993) Federal Acquisition Streamlining Act (1994) and Federal Acquisition Reform Act (1995). USA Congress, USA.

Olsson, F. (2000) *Supply chain management in the construction industry: opportunity or utopia?* Unpublished licentiate thesis. Department of Engineering Logistics, Lund, Sweden.

Ouchi, W. (1986) Markets, bureaucracies and clans. *Administrative Science Quarterly,* 25, (1).

Philips, A. (1974) Commentary. In *Industrial Concentration: The New Learning,* (eds) Goldschmid, H. J., Mann, H. M. and Weston, J. F. Little Brown, Boston, MA.

Pietroforte, R. (1997) Building International Construction Alliances: Successful partnering for construction firms. E&FN Spon, London, UK.

Piore, M. and Sabel, C. (1984) *The Second Industrial Divide: Possibilities for Prosperity.* Basic Books, New York, NY.

Poirier, C. C. and Reiter, S. E. (1996) *Supply Chain Optimisation: Building the Strongest Total Business Network*, Vol. 1, 1st edn Berret-Koehler, San Francisco, CA.

Porter, A. and Rossini, F. (1984) *Interdisciplinary Analysis and Research. Theory and Practice of Problem Focused Research and Development.* Lomond, Mt Airy, MD.

Porter, M. (1985) Competitive Strategy: Creating and Sustaining Competitive Advantage, Vol. 1. Free Press, New York, NY.

Posner, R. A. (1974) The Chicago school of antitrust analysis. *University of Pennsylvania Law Review*, **127** (4), 925–948. Reprinted in Sullivan, T. E. (1991) *The Political Economy of the Sherman Act: The First One Hundred Years.* Oxford University Press, Oxford, UK.

Reder, M. W. (1982) Chicago economics: permanence and change. *Journal of Economic Literature*, **20** (1), March, 1–38.

Reve, T. (1990) The firm as a Nexus of internal and external contracts. Ch 7 in *The Firm as a Nexus of Treaties.* (eds) Aoki, M., Gustafsson, B. and Williamson, O. Sage Publications, London.

Roenfeldt, R. (1999) Personal communication, Director Victorian State Government Office of Major Projects.

Rogan, A. (1999) *Inter-organisational relationships in the construction process.* Phd thesis, University of the West of England, UK.

Rosefielde, S. and Mills, D. (1979) Is construction technologically stagnant?, in *The Construction Industry: Balance Wheel of the Economy.* (eds) Lange, J. and Mills, D., Lexington Books, Lexington, Mass.

Ross, D. R. (1997) *Competing through supply chain management*, Vol. 1, 1st edn Chapman & Hall, New York, NY.

Ruddock, L. (2001) *An analysis of construction activity in the UK.* CIB W055-W065 Construction Sector System or Construction Industry Cluster Analysis for Construction Industry Comparative Analysis Project Group.

Ruddock, L. (2002) Measuring the global construction industry: improving the quality of data. *Journal of Construction Management and Economics*, **20**, 553–556.

Runeson, G. and Raftery, J. (1997) Neo-classical microeconomics as an analytical tool for construction price determination. *Journal of Construction Procurement*, 118–130.

Saad, M. and Jones, M. (1998) Improving the performance of specialist contractors in construction through a more effective management of their supply chains. *Proceedings of 7th International IPSERA Conference.* London.

Strategic Audit of Victorian Industry (SAVI) (2001) *Strategic Audit of Victorian Industry: The Metals Fabrications Industry.* Victorian Government Publishing, Melbourne, Victoria, Australia. www.business.gov.au (accessed 2002).

Scollary, R. and John, S. (2000) *Macroeconomics and the Contemporary New Zealand Economy.* 2nd edn Pearson Education NZ Ltd, New Zealand.

Scott, A. and Storper, M. (1986) *Production, Work, Territory: The Geographical Anatomy of Industrial Capitalism.* Allen & Unwin, Boston, MA.

Seymour, D. (1996) Developing theory in lean construction. *Proceedings of 4th Annual IGLC Conference.* Birmingham, UK.

Simon Report (1944) *Placing and Management of Building Contracts.* HMSO, UK.

Slack, N. (1991) *The Manufacturing Advantage.* Mercury Business Books, London.

Smith, A. (1776) *An Inquiry into the Nature and Causes of the Wealth of Nations.* The Modern Library, New York, NY.

South African Department of Public Works (SA DWP) (1999) White Paper Creating an Enabling Environment for Reconstruction, Growth and Development in the Construction Industry in collaboration with the Departments of Transport, Water Affairs & Forestry, Housing and Constitutional Development. South Africa.

Stilwell, F., Rodley, G. and Rees, S. (1993) *Beyond the Market: Alternatives to Economic Rationalism.* Pluto Press Australia, Leichardt, NSW, Australia.

Stock, J. (1990) Managing computer, communication and information technology strategically: opportunities and challenges for warehousing. *The Logistics and Transportation Review,* **26** (2), June, 132–148.

Strategic Audit of Victorian Industry (SAVI) (2001) *Strategic Audit of Victorian Industry: The Metals Fabrications Industry,* www.business.gov.au (accessed: 2002). Victorian Government Publishing, Melbourne, Victoria, Australia.

Sugimoto, Y. (1997) *An Introduction to Japanese Society.* Cambridge University Press, Melbourne, Australia.

Tan, M. (1999) *Construction 21 Study.* Singaporean Ministry for National Development Study, www. mnd.sg.gov (accessed October 1999).

Taylor, J. and Bjornsson, H. (1999) Construction supply chain improvements through internet pooled procurement. *IGLC 7th Annual Conference.* Berkeley, CA.

Terry, C. and Forde, K. (1992) *Microeconomics: an introduction for Australian students.* 3rd edn Prentice Hall, Sydney, Australia.

Tombesi, P. (1997) *Travels from Flatland: The Walt Disney Concert Hall and the Specialisation of Design Knowledge in the building sector.* Unpublished thesis dissertation, University of California, Los Angeles, CA.

Tommelein, I. and Yi Li, E. (1999) Just-in-time concrete delivery: mapping alternatives for vertical supply chain integration, *IGLC 7th Annual Conference.* Berkeley, CA.

Turner, S. (2000) What are disciplines? And how is interdisciplinarity different? Ch 3 in *Practising Interdisciplinarity.* (eds) Weingart, P. and Stehr, N. University of Toronto, Buffalo, Toronto, Canada.

US FARA, United States Federal Acquisition Regulation Act 1994. USA Congress, United States.

Victorian Government Purchasing Board (VGPB) (2003) *VGPB Policies and Strategic Procurement Planning,* www.vgpb.vic.gov.au.

Victorian State Government (VSG) (2003) *Victorian Metal Fabrication Industry Strategic Plan,* www.business.vic.gov.au (accessed 2003).

Vrihjhoef, R. and Koskela, L. (1999) Roles of supply chain management in construction. *Proceedings of IGLC 7th Annual Conference.* Berkeley, CA.

Walker, A. (1996) *Project Management in Construction,* Blackwell Science Pty Ltd., Oxford, UK.

Walker, A. and Lim, C. (1999) The relationship between project management theory and transaction cost economics. *Journal of Engineering, Construction and Architectural Management,* **6,** 166–176.

Walker, D. H., Hampson, K. and Peters, R. (2002) Project alliancing vs project partnering: a case study of the Australian National Museum Project. *Supply Chain Management: An International Journal,* **5,** 7.

Wantuck, K. (1989) *Just-in-time for America: A common sense production strategy.* The Forum. Milwaukee, Wisconsin, USA.

Warren, M. (1993) *Economics for the Built Environment.* Butterworth Heinnemann, Oxford, UK.

Weingart, P. and Stehr, N. (2000) *Practising Interdisciplinary.* University of Toronto Press, Toronto, Canada.

Williamson, O. (1975) *Markets and Hierarchies: Analysis and Antitrust Implications.* Free Press, New York, NY.

Williamson, O. (1985) *The Economic Institutions of Capitalism: Firms, Markets and Relational Contracting.* Free Press, New York, NY.

Williamson, O. (1996) *The Mechanisms of Governance.* Oxford University Press, Oxford, UK.

Winch, G. (1989) The construction firm and the construction project: a transaction cost approach. *Construction Management and Economics,* **13**, 3–14.

Winch, G. (1995) *Project Management in Construction: Towards a Transaction Cost Approach.* Le Groupe Bagnolet Working Paper No. 1, University College, London.

Winch, G. (2001) Governing the project process: a conceptual framework. *Journal of Construction Management and Economics,* **19**, 799–808.

Womack, J. P., Jones, D. T. and Roos, D. (1990) *The Machine That Changed the World,* vol. 1, 1st edn Rawson Associates/Macmillan, New York, NY.

Wood Report (1975) *The public client and the construction industries.* NEDO, UK.

Index

Notes: References such as '178–179' indicate (not necessarily continuous) discussion of a topic across a range of pages, whilst '123f5.1' indicates a reference to figure 5.1 on page 123, '134t5.1' a reference to table 5.1 on page 134 and '145b6.1' a reference to box 6.1 on page 145. Wherever possible in the case of topics with many references, these have either been divided into sub-topics or the most significant discussions of the topic are indicated by page numbers in bold.